FRACTALS in
BIOLOGY and
MEDICINE

T. F. Nonnenmacher
G. A. Losa
E. R. Weibel
Editors

Springer Basel AG

Editors

T. F. Nonnenmacher
Mathematische Physik
Universität Ulm
Albert-Einstein-Allee 11
D-89069 Ulm
Germany

G. A. Losa
Laboratorio di Patologia Cellulare
Istituto Cantonale di Patologia
Via in Selva 24
CH-6604 Locarno
Switzerland

E. R. Weibel
Anatomisches Institut
der Universität Bern
Bühlstrosse 26
CH-3012 Bern
Switzerland

A CIP catalogue record for this book is available from the Library of Congress, Washington D.C., USA

Fractals in biology and medicine / edited by T. F. Nonnenmacher, G. A. Losa, E. R. Weibel
Includes bibliographical references and index.
ISBN 978-3-0348-9652-8   ISBN 978-3-0348-8501-0 (eBook)
DOI 10.1007/978-3-0348-8501-0
1. Medicine-Mathematics-Congress. 2. Biology-Mathematics-Congress. 3. Fractals-Congress.
I. Nonnenmacher, T. F. (Theo F.), 1933- . II. Losa, G. A. (Gabriele A.), 1941- .
III. Weibel, Ewald R.
R853.M3F68 1993
574'.01'51474-dc20

Deutsche Bibliothek Cataloging-in-Publication Data

Fractals in biology and medicine / ed. by T. F. Nonnenmacher
... - Basel ; Boston ; Berlin : Birkhäuser, 1993

ISBN 978-3-0348-9652-8

NE: Nonnenmacher, Theo F. [Hrsg.]

© 1994 Springer Basel AG
Originally published by Birkhäuser Verlag, PO Box 133, CH-4010 Basel, Switzerland in 1994
Softcover reprint of the hardcover 1st edition 1994
Camera-ready copy prepared by the editors
Revision and layout by *mathScreen online*, CH-4123 Allschwil
Printed on acid-free paper produced from chlorine-free pulp

ISBN 978-3-0348-9652-8

9 8 7 6 5 4 3 2 1

Contents

Foreword

This volume contains the scientific presentations given at the International Symposium on *fractals in biology and medicine* held in Ascona, Switzerland, from February 1–4, 1993. Scientists from the United States, Austria, Canada, Japan, France, Germany, Great Britain, Italy and Switzerland came together to present and discuss their research papers as well as to exchange information on their experimental findings and theoretical interpretations.

When we started to prepare for this conference, we had the feeling that the time had come to gather prominent scientists from around the world to discuss this topical subject. To our knowledge, this symposium has been the first comprehensive meeting of scientists on fractals in biology and medicine. Since the publication of Benoit Mandelbrot's book *The Fractal Geometry of Nature*, the fractal concept has been rapidly pushed forward essentially by mathematicians and physicists alike. However, apart from a few exceptions, it needed still more time to enter the fields of biology and medicine. One of the reasons for this delay might have been the *language barrier* between the different fields: fractals, and mainly their graphical representations, are appealing and even accessible to intuitive understanding, but to translate the theoretical concepts at their basis into terms that can be interpreted in a biological context requires a level of mathematical understanding that is not typical for the group of biologists, for example. In order to bridge this language barrier, we invited an interdisciplinary group of scientists to Monte Verità in Ascona. The members of this group have been working in different areas of science but most have focussed some or all of their activities on biomedical research problems.

Almost all 90 participants could be accommodated in the conference building at Monte Verità and thus it was easy to make personal contacts. The exchange of information between different groups of researchers took place in small circles and within an exciting atmosphere, bringing the family of fractal researchers closer together.

We are particularly grateful to all the renowned institutions who accepted to confer their scientific patronage and also to the sponsors who made the achievment of this symposium possible.

Our thanks are also due to Mrs. Antonella Camponovo and Graziana Pelloni, Dr. Danilo Merlini, member of the scientific and organizing Committee, the collaborators Riccardo Graber, Lorenzo Leoni, and Cristoforo Dürig and to Mr. Luca Albertini, managing director of the Centro Seminariale Monte Verità, who made the conference run smoothly, and to Dr. Gerd Baumann for his editorial assistance while preparing this proceedings volume.

In presenting the different contributions in this volume, we did not follow the chronological sequence of sessions. Rather, we arranged the proceedings as to grouping similar topics together.

Ascona 1993   The editors

The symposium took place at the Centro Seminariale Monte Verità, Ascona, Switzerland, from February 1–4, 1993.

Under the Auspices of

- INTERNATIONAL SOCIETY FOR STEREOLOGY
- SWISS ACADEMY OF SCIENCES
- SWISS NATIONAL SCIENCE FOUNDATION
- SWISS SOCIETY FOR CELL AND MOLECULAR BIOLOGY
- COMMITTEE FOR DIAGNOSTIC QUANTITATIVE PATHOLOGY
- WORKING GROUP OF THE EUROPEAN SOCIETY OF QUANTITATIVE PATHOLOGY
- DIPARTIMENTO DELL'ISTRUZIONE E DELLA CULTURA AND DIPARTIMENTO DELLE OPERE SOCIALI, REPUBBLICA DEL CANTONE TICINO

Financial Support was Received from

- DIPARTIMENTO DELL'ISTRUZIONE E DELLA CULTURA DEL CANTONE TICINO
- SWISS NATIONAL SCIENCE FOUNDATION
- SWISS ACADEMY OF SCIENCES
- FRATELLI MONZEGLIO, CARROZZERIA SA, LOCARNO
- CITTÀ DI LOCARNO
- BANCA DEL GOTTARDO, LOCARNO
- SOCIETÀ ELETTRICA SOPRACENERINA, LOCARNO
- RADIO DELLA SVIZZERA ITALIANA, RETE 2, LUGANO
- PHILIPS AG, ZÜRICH
- LEICA AG, ZÜRICH

Preface
Summary of the Symposium

The Significance of Fractals for Biology and Medicine An Introduction and Summary

Ewald R. Weibel
Department of Anatomy
University of Bern
Bühlstrasse 26
CH-3000 Bern 9, Switzerland

In his seminal book *The Fractal Geometry of Nature* Benoit Mandelbrot made the point that the mathematical construct of a geometry which allows for fractional dimensions will prove most useful in the characterization of natural phenomena, structures and processes alike. This should be particularly true in biology for a number of reasons. First, the complex structure of living creatures, from the whole organism down to the cells, is notoriously difficult to reduce to simple geometric descriptions, and functional processes have very often non-linear properties. Organisms develop and grow from small and simple units to gradually achieve their size and complexity. In living systems, design and performance commonly combine strict rules with some random variation which gives each individual its species characteristics and its individual traits. Furthermore, the wonderful variety observed in the plant and animal kingdoms is the result of stepwise «variation over a common theme» — they are similar but not alike with basic features preserved between related species but expressed in different size and proportions.

All this has some resemblance to the construction principles of fractal geometry where small changes in the algorithms yield a great variety of form so that one can expect considerable new insights from subjecting biological structures and processes to fractal analysis. And, finally, diseases appear often as modulations of basic design patterns so that fractal geometry may even have an impact on the study of disease in medical sciences. However, to introduce the concepts and tools of fractal geometry into biological and medical research is not easy. It requires skills that even a well-educated biologist cannot necessarily master.

*

The goal of the Symposium on *Fractals in Biology and Medicine* was first to explore the potential that fractal geometry offers for elucidating and explaining the complex make-up of biological organisms and then to develop the concepts, questions and methods required in research on fractal biology. In order to approach this task the conference brought together mathematicians, engineers, natural scientists, biologists and medical scientists, because to discuss fractal geometry in the perspective of living systems must be an interdisciplinary endeavour. One of the problems in this kind of enterprise is that interdisciplinary intercourse must be based on a very particular communication culture to which most scientists are unaccustomed. Most contributors to this symposium have met the majority of the other participants for the first time, and they were only partially conversant in the language of the fields represented by the others. One of the primary aims of this Conference therefore was an attempt to bridge the communication gap between various disciplines, chiefly between theoreticians and experimental or comparative biologists.

There is, of course, a limit to what can be achieved in three days of intense discussions of this kind. Thus, much of what has been presented — and is brought together in this volume — still stands side-by-side; it is not yet amalgamated into a coherent body of knowledge. And yet, some of the gaps could be bridged and the discussions resulted in a broadened perspective that may well lead to new insights when the concepts presented are developed further. Some of this process should be captured in these remarks which are based on a summary prepared at the end of the conference and should serve as an introduction to the relations between the different contributions.

In that sense I propose to review the insights on the role of fractals in biology and medicine, gained at this symposium, under three main topics: morphogenesis, structure-function relationships, and the assessment of structural complexity. I would then conclude with some remarks on the role of the eye and the problems related to the attempt to reduce shape perception to simple numbers or words.

1   The fractal nature of morphogenesis in biological systems

The formation of structures in living systems is a non-linear dynamic process, in general, if only because structures form into a limited space. Furthermore, neither animals nor plants can grow indefinitely but approach their final size asymptotically. This all imposes limits on the morphogenetic processes, and much of this becomes translated into structures that appear to have fractal properties. This has become evident in the morphogenesis of fungal structure, of lung, bone and blood vessels. It is particularly noteworthy that very similar structures appear in the course of diffusion limited aggregation (DLA). The rules of proportional growth also translate into Phyllotaxis, the prevailing pattern of structural order in many plants, based on related logarithmic spirals.

The considerations of the evolutionary interplay between mutation and selection also falls under the perspective of morphogenetic principles. The dichotomous growth of bacterial populations gives rise to tree-like structures within which mutations and gene exchanges become propagated. At first sight, these patterns are not fractal because the scale is invariant. However, carrying the principle of the variation generator from the level of the nucleotide to the levels of genes, cells, organisms, and populations may well reveal scale invariant variations eventually, and this may well introduce new views on how biological diversity evolved.

Finally, the development of an epidemic, such as that currently observed with respect to the human immune deficiency virus infection is related to «morphogenetic principles» since it determines the development of the structure of a population, but here again it was not readily apparent that fractals must be invoked in describing these developments.

2 Importance of fractal analysis for understanding structure-function relationships

A central concern of this Symposium has been to explore the role of fractal structure in relation to the biological functions served. The attempt to learn from a comparison of living systems with engineering constructions, such as the metabolism of living organisms compared to bioreactors, or the equivalence between electrodes in a battery and gas exchange in the lung have introduced a clearly mechanistic concept in which fractal design offers distinct advantages. Some of this is expressed in well known features of living organisms, such as allometry which describes design and function as non-linear relationships to size or scale.

Fractal design principles occur in a very large number of biological structures in relation to many functions. The design of the airway and vascular trees, of the cardiac conduction system and of neurons, and the structure of interfaces at various scales (cytoplasmic membranes, plasma membranes, gas exchange surface of the lung), are a few examples only. The question was raised why are fractals ubiquitous? Why is fractal design in biology important?

In that respect it is of interest to note that fractals occur in nature under a variety of conditions. They first of all result in many random phenomena where randomness or disorder shows similar patterns at various scales, such as in random walks or in percolation. Similarly, irregular objects such as coastlines and trees will show fractal features. On the other hand, fractal structure occurs in non-linear systems, such as in turbulence; it appears that vascular trees mimic the pattern of turbulence. Fractal structure also characterizes the interfaces in efficient exchangers such as electrodes and the lung. It has been noted that nature does not build in one scale but that living organisms in fact occur from microbes to the elephant. Similar design features are evident among species differing in size by several orders of magnitude, in mammals ranging from two grams to several tons, but most structural and functional parameters are not varying linearly with body size but rather follow logarithmic or power law regressions of allometry.

One of the interesting conclusions on the importance of fractals in biological design is that it imparts the organism with a considerable error tolerance. The lack of a preferred scale in biological evolution results in an important evolutionary advantage of fractal progeny. Also, the fractal nature of the heart beat, resulting in chaotic complexity of heart frequency, appears to be a safety measure, since the occurrence of regular periodicities may lead to cardiac arrest.

To consider the role of fractal design and fractal organization of time structure in dynamic events holds a great promise for the future. The discovery of similar patterns between biological structures and fractal models will allow the derivation of complex models that describe functional processes. This will allow the formulation of precise hypotheses on structure-function correlation which can be subjected to experimental test.

3 Assessing complexity of biostructures in health and disease

The fractal dimension of a structure is a measure of its complexity. Thus it is tempting to use the fractal dimension as a parameter to describe changes as they occur in pathological alterations of structure. For example, transformation of cells, such as leucocytes, or of tissues in malignancies in general cause a greater degree of disorder, particularly at the interface between the malignant cell or tissue and its surrounding. It would, therefore, appear that this should be expressed in an increase of the fractal dimension. Several examples were presented where this indeed seems to be the case; however, the predictive value of the fractal dimension for the degree of malignant transformation is still ambiguous. It may well be that the fractal dimension alone is not a sufficiently sensitive parameter. Mandelbrot, in his concluding paper, has shown that structures with the same fractal dimension can have very different structure and that additional measures such as that of lacunarity are required for a better characterization of structure. In general it would, therefore, appear, that one would have to define a set of parameters to characterize malignant transformations, one of which would be the fractal dimension.

The same may be true in the attempt to define the structure of neurons and of neural networks where the estimation of fractal dimension alone cannot provide the discriminatory parameters of different types of neurons. This again will require a set of descriptors, fractal and also metric, to arrive at a full picture. Thus one of the important tasks for the future will be to find such parameter combinations which make sense, also from a functional perspective.

With respect to using fractal concepts in structural analysis a number of difficult technical problems must still be overcome. One of them is the rigorous development of methods of measurement based on the basic principles of stereology since microscopic measurements are done on sections, and it must be clearly established how measures of the fractal dimension or of lacunarity are translated into the threedimensional space. Also, we must beware that we are not simply measuring the fractal dimension of artifacts of the method. This is particularly dangerous when using methods of automatic image analysis where the finite pixel size and routine methods of smoothing and image segmentation may well have an important effect on these measurements.

4   The role of the eye in morphology

One of the important messages that Benoit Mandelbrot conveys in his lectures is that the development of science has led to an elimination of the eye from science. The development of biology and biological morphology is no exception to this rule. Early last century textbooks of anatomy and embryology were written as purely verbal descriptions without any illustrations. In recent decades the tendency has evolved to replace pictorial evidence by numbers also in microscopy — and I must confess guilty of having promoted this approach through my own work. Quite clearly, to attach numbers and precise size estimates to biological structures is important particularly in view of interpreting structure-function relationships. However, we must realize that every attempt at reducing the image of structure to a few numbers or words results in a wilful loss of over 99 % of the information contained in the picture. A student has once sprayed a graffito on the wall of my Institute which says: «One word = one millipicture». This is the only graffito I have not had erased because I believe it conveys an important message.

Any measurement of biostructures only makes sense if it is interpreted critically in the framework of knowledge on the integral nature of the structures concerned.

The lecture by Benoit Mandelbrot at the end of the Symposium made it very clear that the application of the principles of fractal geometry in biology and medicine is only at the beginning. He impressively showed that by estimating the fractal dimension from the slope of a log-log regression one is casting away a vast and rich amount of information that is contained in the structure. At least two new parameters must be taken into consideration. One of them is lacunarity which is a measure of texture of the fractal structures, or of the distribution of local fractal properties over the entire structure. It is also evident in the study of many natural structures that the fractal dimension need not be constant throughout the structure, particularly at different levels of organization, and that the multifractal properties are a most important feature of structural design.

5   Conclusion

This Symposium has brought together on Monte Verità, the «Mountain of Truth», biologists, medical scientists, physicists and mathematicians to explore the potential of fractal geometry in describing biological phenomena in a very broad perspective. I would conclude that the undertaking was very worth-while and productive in the sense that it has not only shown the potentials of this concept, but also the pitfalls of a too simplistic application of these principles. It has, however, also become very clear that the further development of the principles of fractal geometry holds much in store for improving our approaches to understand the organization of biological organisms and its role in maintaining health and causing disease.

Fractal Geometry
and
Biomedical Sciences

A Fractal's Lacunarity, and how it can be Tuned and Measured

Benoit B. Mandelbrot
Yale University
New Haven CT 06520, USA
IBM Fellow Emeritus, IBM Yorktown Heights, NY 10598, USA

Abstract. The main exhibit in this paper is a stack of Cantor dusts that have identical fractal dimensions but differ violently from each other. Some look clearly fractal, while others look to the unassisted eye as filled intervals (they are said to be of low lacunarity), and others seem to reduce to the end points of a hollowed interval (they are said to be of high lacunarity). One of several quantitative measures of lacunarity is put forward, and the impact of low lacunarity fractals on modeling of nature is discussed.

1 Introduction

This brief paper is mostly devoted to fractal lacunarity, but the introduction itself begins with brief but rambling general comments, and the last two sections also digress.

The readers of this *Proceedings* need not be told that, today, the basic ideas of fractal geometry are viewed as being astonishingly simple. It is becoming hard to believe that, until I introduced computer pictures in a broad and deep fashion as a tool of science, and used old and new fractals to model nature, those shapes were few in numbers, were known to few scientists, and were viewed as «bizarre» and «pathological.» Today, to the contrary, the fractals keep increasing in numbers and variety, and they attract the attention of everyone — including children. This sharp change due to picturing and applications deserves special mention in this *Proceedings* devoted to medicine and biology, because it unavoidably raises a very important question of anatomy and/or of physiology: Could it be that, somehow, human vision is «wired for fractals,» as we know it is wired for line and color?

A second look shows, however, that the apparent simplicity of fractal geometry is altogether deceptive in many ways. After a field has sufficiently matured, an «intuition» becomes formed. One acquires a «feeling» for those cases where «cookbook» techniques can be trusted blindly, as contrasted with the cases where those techniques do not work or must be used with caution. Fractal geometry, however, has not yet moved beyond the early stage where even the most skilled users often grope for the proper tool.

To continue the preceding comment in somewhat self-serving style, all too many readers of *FGN (The Fractal Geometry of Nature* [4]) seem to recall my book as devoted to the notion of self-similarity, and to Julia and Mandelbrot sets. But this is an oversimplification. In fact, my book also deals with other topics, including self-affine fractals, multifractal measures and lacunarity. The first two of these topics, which were relegated to Chapter 39 of *FGN*, have by now been taken up by many scientists, spawning a vast literature, and the best I can do here, in Sections 7 and 8, to direct the readers among the books and papers that concern these topics and followed *FGN*.

Lacunarity, to the contrary, was allowed several chapters in my book. The stress was on qualitative considerations. True, page 315 of *FGN* hinted that the key to a quantitative approach resides in the properties of diverse prefactors — mostly the average and their variability, but this hint was not developed. Scattered uses ensued (e.g., [3]); but, to my surprise, lacunarity was slow to develop beyond the treatment in *FGN*. Now I have taken the topic up again, using prefactors as hinted in *FGN*. Several papers are on the way, and this paper's goal is to provide an advance «announcements» of material to appear. Figure 1 and the accompanying comments merely present pictorially simplest form of the considerations and illustrations one can find in *FGN*, Chapters 34 and 35. But this paper's Section 3 incorporates comments that are new and important. They will make lacunarity more complicated, but also, I hope, easier to understand and less controversial. Section 4 is the first presentation in print of one of the basic numerical measures for lacunarity.

2 Examples that show that, for Cantor dusts, a given fractal dimension is compatible with a wide range of different textures that seem to range from an interval to two points

Let us examine Figure 1, beginning with the «midline,» which is the line to which a hand points. The bottom part of this line represents the generator of a standard Cantor dust of fractal dimension $D = 1/2$. To construct this dust, one starts with the full closed interval $[0, 1]$. First, one erases the open middle *half*, leaving $N = 2$ closed intervals, each of length $r = 1/4 = N^{-2}$. Just as in *FGN*, Plate 60 (which illustrates the original removal of middle *thirds*), those intervals are shown as bars, because otherwise they would be invisible. Second, one erases the middle halves of each of the above remaining intervals. The same process, if continued without end, erases most of the interval $[0, 1]$, except for the fractal dust that is pictured in the top part of the midline of Figure 1.

A fractal dust with $D = 1/2$ can be increasingly «homogenous-looking.» Consider the line that is second down from the midline on Figure 1. The bottom part represents a generator made of $N = 4$ intervals, with $r = 1/16 = N^{-2}$. The novelty is that they are uniformly spaced. Similarly, on the bottom part of the k-th line from the midline is a generator with uniformly spaced intervals, such that $N = 2^k$ and $r = N^{-2}$. Each of these generators can be used to construct the Cantor dust pictured on the top of the corresponding line. The corresponding fractal dimensions are

$$D = \frac{\log N}{\log 1/r} = \frac{\log N}{\log N^2} = \frac{\log N}{2 \log N} = \frac{1}{2}. \tag{1}$$

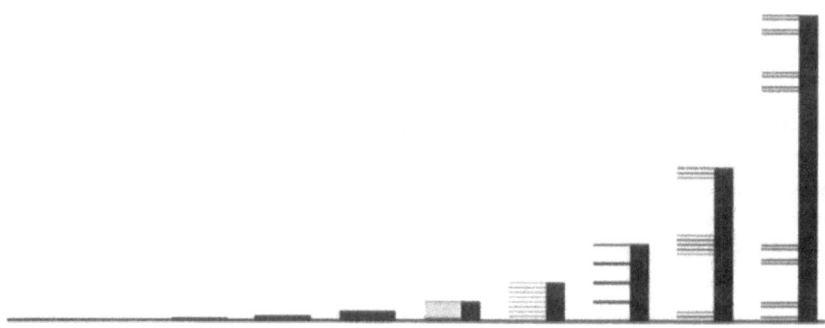

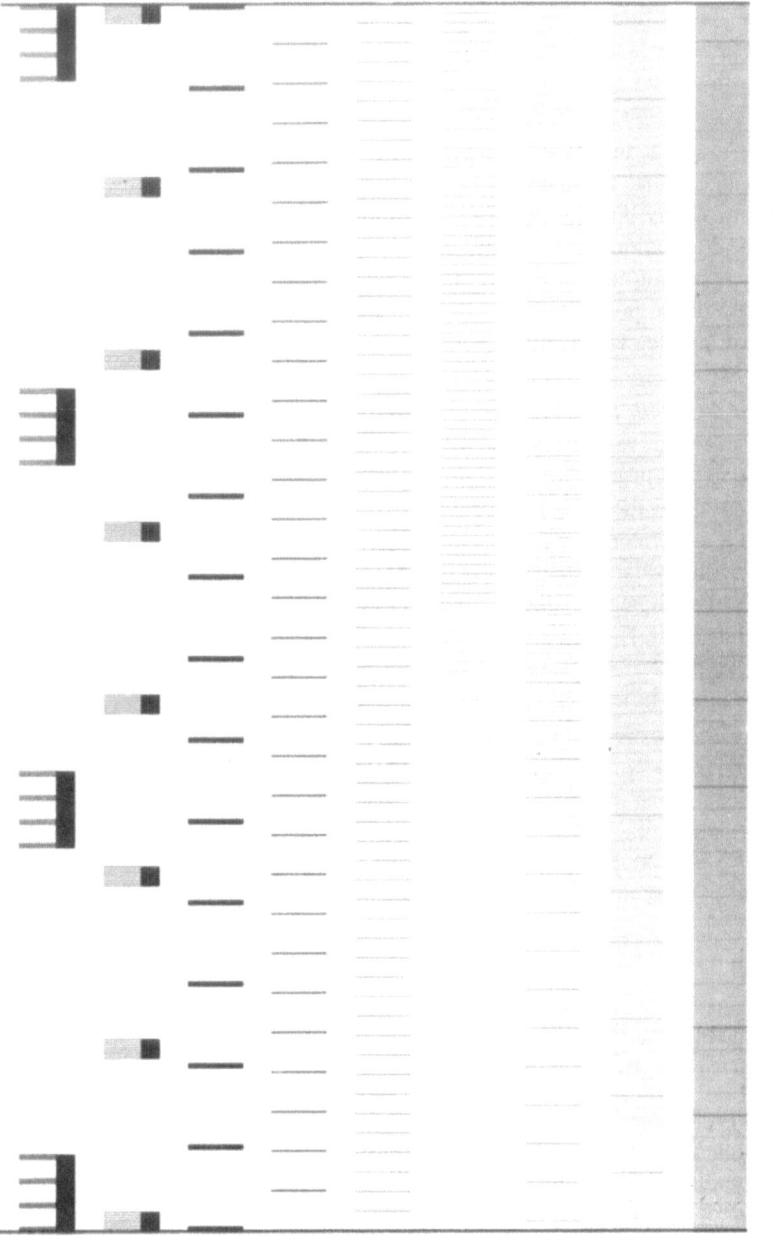

Fig. 1 A stack of Cantor dusts. By construction, all these sets are self-similar, and have the same fractal dimension $D = 1/2$. Yet, as one moves from bottom to top, one seems to move from a full unit interval with no holes, through a number of sets clearly recognizable as fractals, up to a unit interval that has been hollowed out, so that only the endpoints are left in. Hole is *lacuna* in Latin, and this paper proposes a measure of a fractal's *lacunarity* (= tendency to have holes) that varies from 0 at the bottom to 1 at the top.

In other words, every Cantor dust in the lower half of Figure 1 (midline or below) has precisely the same fractal dimension. Yet, these dusts differ violently from each other. In particular, the dusts near the middle of the whole stack are very far from being homogeneous, i.e., they are far from being translationally invariant. Despite the fact that few stages of construction can be shown, everyone acquainted with fractal geometry immediately recognizes the outcome as being fractal.

But the fractal at the bottom of the stack is sharply different. As a matter of fact, it is likely that the unavoidable imperfection of printing shows it as a solid bar; one can say that the bottom line mimics a homogeneous medium; i.e., it seems to be translationally invariant. More generally, the shapes one meets as one goes down starting in the middle of the stack «look less and less fractal.» If we had plotted all the possible values of N instead of only those of the form 2^k, the transition from clearly fractal to seemingly homogeneous would have become very progressive.

A fractal dust with $D = 1/2$ can be increasingly «discrete-looking.» This is achieved by a second way of playing with the generator in the midline of Figure 1. This new way is illustrated in the upper half of Figure 1. The k-th line above the midline, just like the k-th line below, involves $N = 2^k$ intervals of length $r = N^2$, but the generator they form is altogether different. It divides into 2 equal groups of $N/2$ intervals, each group being squeezed together near an end point of $[0, 1]$. (One is tempted to squeeze them into a single group — all to the right or all to the left of $[0, 1]$. However, it happens that this would be self-defeating). Now, the unavoidable imperfection of printing has a different effect: the fractals near the top of the whole stack seem to reduce to 2 isolated points.

Generalization to $D \neq 1/2$. It is obvious that one can modify Figure 1 by varying N, while $r = N^{-1/D}$, with $0 < D < 1$: it follows that $\log N / \log(1/r) = D$ for every fractal dust in the stack. Now change D but use the same laser printer or the same printing press. When $D > 1/2$ (e.g., in the removal of middle thirds, which yields $D = .6309$), apparent homogeneity is achieved higher in the stack, i.e., for a lower value of k. When $D < 1/2$, apparent homogeneity is achieved lower in the stack.

3   Implications of the examples in Section 2; the loose concept of qualitative lacunarity

Despite the striking differences seen on Figure 1, all the sets it illustrates are perfectly self-similar, hence unquestionably fractal by every definition. The existence of such fractals is a prime illustration of an assertion made in the Introduction, that the apparent simplicity of fractal geometry is altogether deceptive. To face the obvious need to distinguish between those fractals, I was led (in the 1970s), to introduce the new notion of *lacunarity*. *Lacuna* is Latin for lack or hole, hence the term is self-explanatory: Lacunarity concerns the tendency to have holes. It is as high as can be for the fractals at infinity above the top of Figure 1, and as low as can be for the fractals at infinity below the bottom.

A troubling conclusion. Figure 1 is extraordinarily troubling. Years ago, while introducing fractals into the sciences, I noted that their non-integers dimensions made them geometric «chimeras,» that is, shapes «in between» the Euclidean shapes like the line and the plane. To this «in-betweenness,» lacunarity adds new depth (I almost said a new dimension). As already mentioned, lacunarity was touched upon in *FGN*, but not completely enough. Lately, I have (at long last) advanced numerical measures, of which an example will be described in Section 4, and moved forward in the study of the implications of lacunarity, first in theory, and then in practice. Here is a sketch of some of these implications.

In the theory that holds after infinite interpolation, the answer is simple: Euclidean shapes differ from all fractals. As $k \to \infty$, many subtle issues of mathematics are bound to arise, but this is not the place to even raise this sort of question.

In practice, the answer is very delicate. Let us proceed from the essential point of view of physical problems that involve a positive smallest atomic size ϵ and a largest size L. This double cutoff confirms the point of view of what the eye can actually see.

In a Euclidean universe cut off at ϵ and L, physical reality is embodied in L/ϵ atoms. A continuous interval of length L is nothing but an approximation to those L/ϵ atoms, but continuity deserves attention because it happens to be far easier to investigate mathematically.

Now let us turn to the fractal universe. Far enough down the stack in Figure 1, the values of k and N are so large that $r = N^{-1/D} \leq \epsilon$. If so, a further fractal interpolation of the generator has no physical meaning. It follows that the opposites meet: a collection of L/ϵ atoms is a common discrete model for a) a low lacunarity fractal and b) an interval. Up the stack in Figure 1, similarly, one cannot distinguish between a) a high lacunarity fractal b) the endpoints of an interval.

Two conclusions. On the line a fractal can «mimic» an interval or two points, or any set of dimension different from its true dimension. The effect of lacunarity on our perception of dimension may be illustrated by a well-known metaphor: that of the «wind chill» factor expressing the effect of wind speed on the perception of temperature.

The distinction between fractal and Euclidean is only obvious for «middling» values of lacunarity. As lacunarity decreases or increases in a concretely meaning-ful environment, the distinction between fractal and Euclidean becomes increas-ingly blurred.

The case of the distribution of galaxies. From the above second conclusion, one must expect, given a a set of empirical data, that some statistical tests or other methods of analysis will declare this set to be fractal, while other tests declare it to be homogeneous.

Such a source of possible controversy is by no means hypothetical. As a matter of fact, the context which compelled me (in the 1970s) to develop the qualitative notion of lacunarity was the study of the large scale structure of the universe, which continues to this day to be full of controversy. Very recently, thanks to the comments summarized in the preceding few paragraphs, I have gained a better understanding of the points in dispute.

The conventional viewpoint [6] derives from the following facts. On the one hand, there is an overwhelming visual impression that the maps of galaxies are more or less uniform. This explains the conventional view that the overall distribution of galaxies is homogeneous. But, on the other hand, there is overwhelming numerical evidence that the large scale distribution of galaxies is fractal in a «local range» up to a crossover at 5-20 Mpc (depending on the definition of the crossover; the abbreviation pc stands for parsec, equal to a few light years). However, I saw no reasons to believe that the distribution fails to remain fractal throughout. The fact that this scenario is conceivable, is established by the material in *FGN*, Chapters 33 to 35. Defenders of the conventional view did not even design to discuss the low lacunarity scenario, and later they tried to rough up L. Pietronero when his analysis of the actual data failed to produce any crossover. This analysis is summarized in [7]. Were the discussion intrinsically limited to galaxy maps projected on the sky, I would fear that a controversy could continue forever. But evidence that clinched the fractal model of galaxies came only with the advent of three-dimensional evidence about the distribution of galaxies. The widely advertised «big voids» contradict homogeneity but are a standard characteristic of fractals.

The preceding case story yields a disquieting conclusion: the same difficulties are likely to be repeated whenever a possible fractal model is «atypical,» i.e., like those up or down the stack in Figure 1.

From the fact that the value of D does not suffice to characterize a fractal numerically, an upbeat conclusion is that there is work to do. I have been challenged to proceed from qualitative to quantitative lacunarity. At long last, I have responded to this challenge, as will be seen in the following section.

4   A possible measure of lacunarity: via the filling rate of ϵ-neighborhoods

FGN largely left lacunarity as a loose visual notion. The next vital task was to make it quantitative. Having now faced this task, I soon discovered that it can be achieved in several separate ways. Each must be studied on its own, and they face us with the new task of studying the relation between separate measures of lacunarity (see Section 6).

It will suffice here to sketch the measure of lacunarity that is based upon a set's ϵ-neighborhood (or *Minkowski sausage*). Behind its complicated name, there is a very simple notion, to which we have already alluded in words. The basic idea is that infinitely interpolated fractals are neither visible nor physically realistic. Observational or experimental science necessarily views any set as being a little out of focus. Thus, «point» is the proper name for «a little ball of radius ϵ.» If each point in a set is replaced by such a small ball, one obtains the set's ϵ-neighborhood. On the line, each point is really an interval of length 2ϵ; neighboring points in a fractal dust smear out together to form intervals, separated by other intervals, named *gaps*, which are of length $> 2\epsilon$.

Let us now evaluate the total length $L(\epsilon)$ of the ϵ-neighborhood of a Cantor dusts in Figure 1, and determine the dependence of $L(\epsilon)$ on ϵ. Part of the answer is well-known and reported in *FGN*, p.358: in the 1920s, Bouligand showed that

$$L(\epsilon) \propto \epsilon^{1-D}; \tag{2}$$

here the symbol $\propto$ means that

$$\lim_{\epsilon \to 0} [\log L(\epsilon)/\log \epsilon] = 1 - D. \tag{3}$$

On the line, $1 - D$, is a set's codimension C. This result of Bouligand holds for *all* fractal dusts, and yields one of several alternative paths to introduce D as a fractal dimension. But the very generality of this result means that (2) does not take account of the special structure of the Cantor dust. In particular, for some dusts in Figure 1, the covering by intervals of length 2ϵ may be efficient, i.e., with relatively little overlap; for other dusts, the covering may be very inefficient, i.e., with extreme overlap. In order to compare degrees of overlap, I observed that in certain cases that include those of Figure 1, one has the following far less general but far stronger property:

$$0 \le \min\{L(\epsilon)\epsilon^{D-1}\} \le L(\epsilon)\epsilon^{D-1} \le \max\{L(\epsilon)\epsilon^{D-1}\} < \infty. \tag{4}$$

·Circumstantial evidence suggests that in the case of a fractal it is impossible for $L(\epsilon)\epsilon^{D-1}$ to be a constant.

For the lower half of the stack in Figure 1, the calculation was performed by J. Klenk. Using (2ϵ) for reasons that will transpire soon,

$$1/\max\{L(\epsilon)(2\epsilon)^{D-1}\} = \frac{(r^{-D} - 1)^D r^D (1 - r^{1-D})^{1-D}}{1 - r}. \tag{5}$$

One can imagine using $\max\{L(\epsilon)(2\epsilon)^{D-1}\}$ to define $1/\Lambda_F$, and calling Λ_F the rate of lacunarity. This maximum is attained when ϵ is of the form $\epsilon = r^n(1 - Nr)/(N - 1)$ for integer n. Between two successive values of ϵ in this sequence, the variation of $L(\epsilon)$ is linear. Another calculation yields $\min\{L(\epsilon)(2\epsilon)^{D-1}\}$. One may want to introduce this min in the definition of Λ_F. Yet another possibility is to take a wide range of ϵ's and define $1/\Lambda_F$ as the average of $L(\epsilon)(2\epsilon)^{D-1}$, or better, define $-\log \Lambda_F$ as the average of $\log [L(\epsilon)(2\epsilon)^{D-1}]$. In the latter case,

$$\log L(\epsilon) - (1 - D) \log(2\epsilon) \tag{6}$$

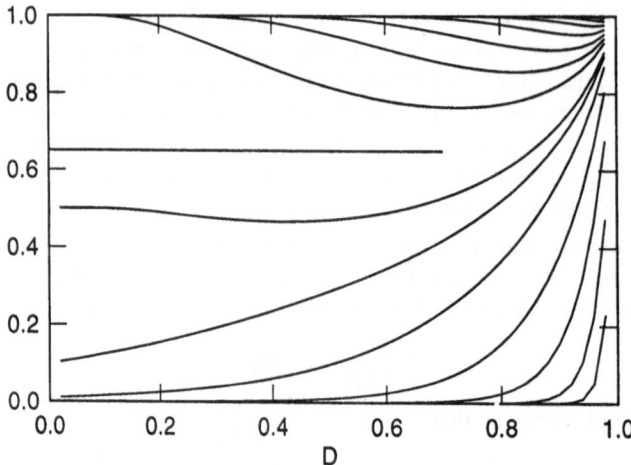

Fig. 2 The variation of $1/\max\{L(\epsilon)(2\epsilon)^{D-1}\}$ with D and N, in the case of a family of fractal dusts. This family includes the stack illustrated on Figure 1 and the corresponding stacks for other D's, and fills them for all values of N. Each vertical slice corresponds to the D given in the abscissa, and each curve corresponds to a given N, with r being obtained as $N^{-1/D}$. The horizontal line separates the high — and the low — lacunarity cases. The top 4 lines correspond to high lacunarity dusts; the bottom 4 lines correspond to low lacunarity dusts, all constructed as in Figure 1. In each case, as one moves away from the horizontal line, one has $N = 2, 10, 10^2, 10^4, 10^8, 10^{16}$ and 10^{32}.

will oscillate around its average $-\log \Lambda_F$. Before selecting Λ_F among these possibilities, I feel it is best to await further developments in the field.

The same argument holds for the upper half of the stack in Figure 1. The calculation (performed by I. Yekutieli) shows that

$$1/\max\{L(\epsilon)(2\epsilon)^{D-1}\} = \frac{(1 - r^{1-D})^{1-D}(1 - r^D)}{1 - r}. \tag{7}$$

The values of both expressions for $1/\max\{L(\epsilon)(2\epsilon)^{D-1}\}$ are plotted on Figure 2. As one goes down the stack on Figure 1, the value of Λ_F decreases from a maximum of 1, which holds in the most lacunar case (top) to a minimum of 0, which holds in the least lacunar case (bottom). The fact that this scale ranges from 0 to 1 is most satisfactory and came out without having to be forced upon the problem, except for the use of (2ϵ) instead of ϵ. (I do not know whether or not the inequality $0 < 1/\max\{L(\epsilon)(2\epsilon)^{D-1}\} < 1$ holds in other constructions).

After a choice is made between the possible definitions of Λ_F, I shall propose Λ_F as a measure of the lacunarity of a Cantor dust. Closely related definitions are possible in higher-dimensional spaces.

It may also be worthwhile to consider the variability of $L(\epsilon)(2\epsilon)^{D-1}$, as defined by max−min, or by the variance of either $L(\epsilon)(2\epsilon)^{D-1}$, of its inverse, or of its logarithm. This variability also contains lacunarity information.

5 Generalization, random dusts, and a connection between low lacunarity and a form of global statistical dependence

For exactly (recursively) self-similar fractals, the prefactor of the ϵ-neighborhood's length settles one aspect of the issue of lacunarity. But there are numerous other aspects: we have studied rates of lacunarity based on the prefactor of scaling relations other than (6), and began to examine the connections between various rates.

The original Cantorial example described in Section 2 is extremely over-simplified, and its systematic character is unrealistic, especially after it has been generalized in the plane and higher spaces. Hence the need for random fractal models. When dealing with them, one must face an additional problem; for example, normalization is very far from being completely solved. These problems cannot be studied here, and are only mentioned for an already-discussed reason: to avoid creating the false impression that the problem of lacunarity is already under full control.

I have devoted much attention to a special random fractal construction based on tremas. This simplest form consists in cutting out statistically a collection of independent filled-in intervals, discs (filled-in circles) or balls (filled-in spheres). Chapter 33 of *FGN* illustrates this procedure, and shows that the resulting sets are of high (but not extreme) lacunarity. An example is provided by Figure 3, which reproduces part of Plate 307 of *FGN*. The next obvious step was to replace spherical by non-spherical tremas. This change turns out to be sufficient to decrease lacunarity. In *FGN*, Chapter 34 and 35, this decrease was assessed visually and quantitatively. Major progress has recently been reached towards quantitative description, but a full description of what is already known would go beyond the bounds of this paper.

The inquisitive reader may wonder which feature of the non-spherical tremas makes them fit to accomplish the desired task. As I see it now, the basic underlying feature is an increase in a very original form of global statistical dependence.

6 The tedious issue of how to define fractals, fractal dimension and fractal lacunarity

At this stage, fractal lacunarity cannot be measured by a single well-defined number. It is like *bigness*, a broad notion in the process of being broken down into distinct strands that may well coincide in some cases and differ in other cases.

Will this process converge and, if it does, how rapidly? I am not optimistic about convergence, because of the increasing complications that accompany other broad notions, those of fractal and of fractal dimension. Many years ago, I observed that structural similarities no one had noticed should bring together several existing mathematical construction and concepts. They deserved to be put together, given a common name, and used as a base (I never said foundation) for a new discipline,

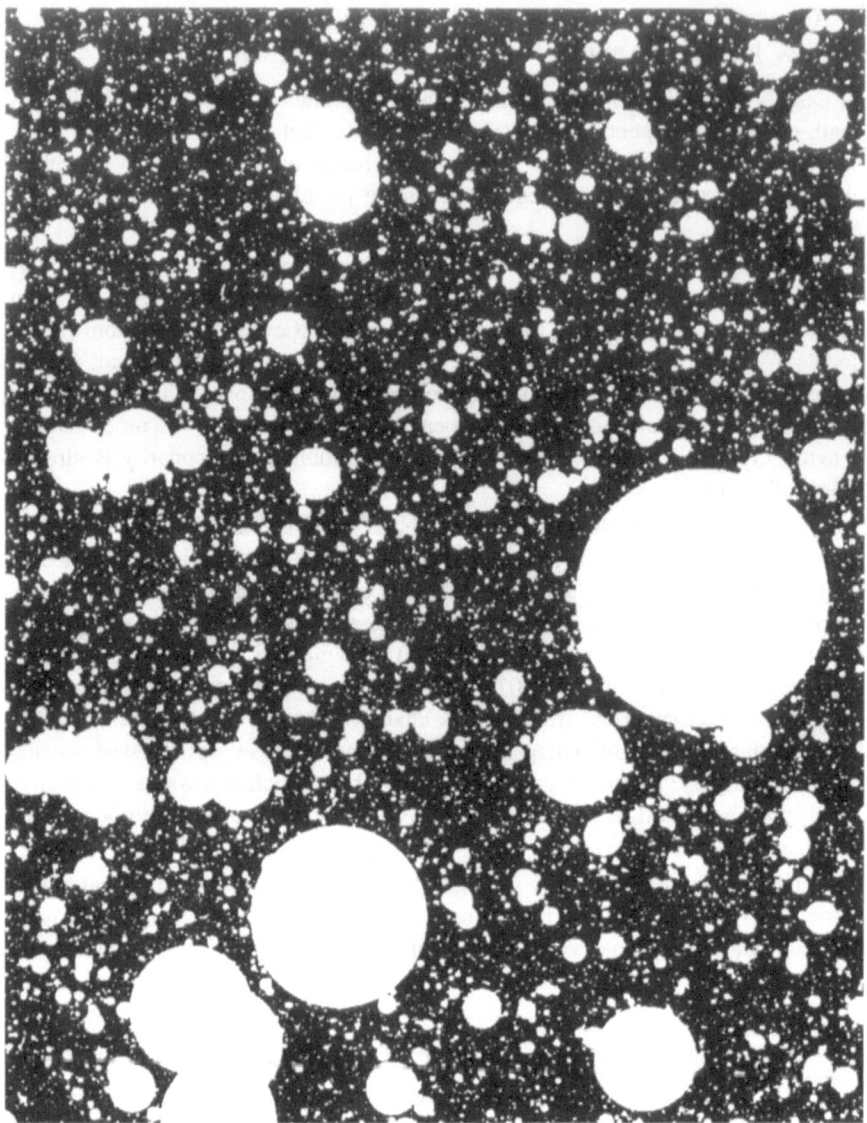

Fig. 3 A slice of Plate 307 of *FGN*, illustrating a trema set obtained by cutting off filled-in circles from the plane. (From: Benoit B. Mandelbrot, The Fractal Geometry of Nature; W.H. Freeman & Co., New York, and also from: Benoit B. Mandelbrot, Die fraktale Geometrie der Natur; Birkhäuser Verlag, Basel)

which I called fractal geometry. Did I succeed in defining fractals uniquely? Not only have I failed, but recent work on fractal aggregates (DLA) is revealing that shapes everyone wishes to call fractal can exhibit deep complexities, for which none of the previously advanced possible definitions would allow.

What about fractal dimension? Some of these notions I was able to borrow from existing mathematics literature were several distinct generalizations of dimension. The best known was due to Hausdorff and Besicovitch and purely tactical reasons (as I have kept repeating) led me at one time to give it top billing in fractal geometry. But this top billing has long since become inappropriate and not sustainable. It becomes increasingly clear that fractal dimension is necessarily a multi-faceted concept.

Against this background, I have little hope of finding a unique measure of lacunarity.

7   Bibliographic guidance concerning self-affinity

To gain an understanding of self-affinity, the best source is the collection of papers put together by Family and Vicsek [2].

To preface a few words on the topic, observe that, Cartesian coordinate axes having become second nature to every scientist, there is little awareness of the extremely fundamental distinction between two kinds of planes: the *isotropic* Euclidean plane and the *affine* planes. Let us state this distinction on two sharply different cases.

For a clearcut example of isotropic plane, consider a map of a piece on Earth's surface that is so very small that Earth's curvature does not matter. The custom is to place the coordinate axes along the direction NS and the orthogonal direction EW. But all other orthogonal lines would do equally well. In particular, distance can be measured in any direction.

For a clearcut example of self-affinity, begin with a graph of volume versus pressure. Distances along each axis, i.e., differences of volume ΔV or of pressure Δp are meaningful and can be measured. But what about $\sqrt{(\Delta V)^2 + (\Delta p)^2}$, which is a formal distance along a straight line between two points on the diagram. This formal expression depends on the units chosen for V and p. But those units can be chosen independently of each other; therefore, this formal distance is meaningless. The same command applies to «noises» that are voltages as function of time. Turning a knob changes their appearance. The same comment also applies to the Brownian motion.

The reader must be warned that there exist examples that do not fall neatly in the above contrast. As obvious example is a vertical cut across a mountain. In that case, vertical and horizontal coordinates are both measured in meters, and $\sqrt{(\Delta X)^2 + (\Delta h)^2}$ is well defined as the «distance as the crow flies» between the two points. However, horizontal and vertical distances have distinct meanings because of gravity.

Be this as it may, self-affine fractals have by now spawned a large literature. Once again, the early papers are collected in [2].

8   Bibliographic guidance concerning multifractals

To gain an understanding of multifractals, I recommend a text I wrote with Carl Evertsz [1].

To preface a few words on this topic note that the bulk of elementary fractal geometry is devoted to *fractal sets*. A set's visual expression is a region drawn in colored ink against white paper or board (or in chalk against a blackboard). A set's defining relation is an *indicator function* $I(P)$, which can only take two values:

$$I(P) = 1, \quad \text{or} \quad I(P) = \text{«true,»} \text{ if the point } P \text{ belongs to the set } S;$$

and

$$I(P) = 0, \quad \text{or} \quad I(P) = \text{«false,»} \text{ if } P \text{ does not belong to } S.$$

However, most facts about nature cannot be expressed in terms of the contrast between «black and white», «true and false», or «1 and 0». I faced this issue early on, and the opening lines of chapter IX of my first book on the subject (written in French) translate as follows:

»Before we generalize [fractal sets to measures], it may be recalled that, among our uses of fractal sets [to describe nature], several involve an approximation. While discussing clusters errors, we repressed our conviction that, between the errors, the underlying noise weakens, but does not stop. While discussing the distribution of stars, we repressed our knowledge of the existence of interstellar matter, which is also likely to have a very irregular distribution. While discussing turbulence, we approximated it as having [nonfractal] laminar inserts. In addition, no new concept would have been needed to deal with the distribution of minerals. Between the regions where the abundance of a metal like copper justifies commercial mining, the density of this metal is low, even very low, but one does not expect any region of the world to be totally without copper. All these voids [within fractals sets] must now be filled — without, it is hoped, inordinately modifying the mental pictures we have achieved thus far.

«Those aspects cannot be illustrated by sets; they demand more general mathematical objects that succeed to embody the idea of «shades of grey.» Those more general objects are called *measures.»*

It is most fortunate that the idea of self-similarity is readily extended from sets to measures; self-similar measures are usually called *multifractals*. The original approach to multifractals, which is probabilistic, was first described in two papers I published in 1974. However, multifractals became best known thanks to formal and heuristic papers that came out in the mid-1980s. Against this background of my approach, the nature of various heuristic arguments in those papers becomes clear, their limitations and proneness to error become obvious, and the unavoidable generalizations demanded by both logic and the data become easy.

Once again, as a simple introduction to multifractals (with a bibliography), I recommend [1] to the readers.

Acknowledgements

Figure 1 was drawn by Lewis N. Siegel, and Figure 2 was drawn by Juergen Klenk to illustrate the formulas he had obtained. The details of the theory of fractal lacunarity are being worked out with several collaborators, including J. Klenk and also D. Fracchia, M. Frame, I. Yekutieli, R. Cioczek-Georges, and D. Gatzouras. The mathematics is related to the so far little-used notion of *Minkowski content* as generalized from integer to non-integer dimensions.

References

[1] Evertsz, C.J.C. and Mandelbrot, B.B: Multifractal Measures. Appendix B to *Chaos and Fractals* by H-O. Peitgen, H. Jurgens and D. Saupe. New York, Springer, 1992, pp 849–881.

[2] Family, F. and Vicsek, T.: Dynamics of Fractal Surfaces, Singapore: World Scientific, 1991.

[3] Gefen, Y., Meir, Y., Mandelbrot, B.B., Aharony, A: Geometric Implementation of Hypercubic Lattices with Noninteger Dimensionality by Use of Low Lacunarity Fractal Lattices. Phy. Rev. Lett 50: 145–148, 1983.

[4] Mandelbrot, B.B.: The Fractal Geometry of Nature. New York, W. H. Freeman, 1982.

[5] Mandelbrot, B.B.: Plane DLA is not self-similar; is it a fractal that becomes increasingly compact as it grows? Physica **A 191** 95–107, 1992.

[6] Peebles, P.J.E.: Principles of Physical Cosmology. Princeton University Press, 1983.

[7] Pietronero, L.: The Fractal Structure of the Universe, Physics Reports 213 (No.6) 311–89, 1992.

Spatial and Temporal Fractal Patterns in Cell and Molecular Biology

Theo F. Nonnenmacher
Department of Mathematical Physics
Albert-Einstein-Allee 11
University of Ulm
D-89069 Ulm

Abstract. When discussing spatial and temporal fractal patterns in cell biology, we critically investigate various methods for a practical determination of the fractal dimension D. Asymptotic fractal formulas are presented and applied to describe self-similar cell profiles or — in the temporal case — scale invariant protein dynamics. However, because cells are of finite size, and due to the existence of a largest time scale for dynamical processes, such geometrical objects or dynamical processes are not fractal even in an asymptotic sense, instead they may be only «fractal between limits». A method of finding these scaling limits (crossover points) and their mathematical characteristics will be proposed and applied.

1 Introduction

Many objects in biology and medicine like tissues, cells or subcellular organelles display irregular shapes and discontinuous morphogenetic patterns in support and in connection with their functional diversity. To capture all this richness of complex structure and function into a theoretical model is one of the major challenges of modern theoretical biology. In recent years a wide range of irregular structures observed in physical and chemical sciences has been quantitatively characterized by using Mandelbrot's concept of fractal geometry [13]. However, applications of this concept to study self-similar patterns are rather scarce in cell and tissue biology despite the fact that it can also be used as a design principle for morphological complexity and function in living organisms [28].

In this paper, we will first apply and critically discuss different methods of digital image analysis to measure the fractal dimension D for geometric structures of cellular profiles. As prototype examples we investigate two samples of cells: (i) 19 human T-lymphocytes from various normal donors and (ii) 18 hairy leukemic cells from two different patients.

Second, in order to discuss scale invariant protein dynamics, we need to deal with the concept of time-fractals. Just as geometric fractals do not have a characteristic length scale, a fractal time process generates fluctuations which do not have a single or characteristic time scale and consequently, such fluctuations cannot be adequately designed with standard statistics based on mean and variance. Lévy statistics [11] [14] which can appear both in time and space is an example for such a non-standard statistics that brings into play power law behavior $f(t) \sim t^{-q}$ for a fractal interpretation of certain stochastic, chaotic-like processes. More recently, some interest has been focused on the question of an integral or differential equation that leads to power-law solutions t^{-q} where q is not necessarily an integer. Such an analytical approach is based on fractional operators like $d^{\nu}f(t)/dt^{\nu}$ with $0 < \nu < 1$. Here, we will present the basic ideas of such a fractional operator design. We will relate the scaling dimension q to the fractional order ν of the operator d^{ν}/dt^{ν}, and we will discuss scale invariant kinetic mechanisms as an application to protein dynamics in Myoglobin and in ion channel gating kinetics.

2 Quantitative Structural Analysis of Self-Similar Cell Profiles

In this section, we will critically discuss and apply different experimental methods of digital image analysis to measure the fractal dimension D of cellular profiles. The pioneering work using the concept of fractal geometry in cell biology was done by Paumgartner et al [22]. They estimated mitochondrial and endoplasmic reticulum membrane surface densities stereologically and found fractal behavior up to a critical magnification. Using computerized image analysis, Rigaut [23] observed that most cell contours are ideally fractal only asymptotically at low resolutions and at higher resolutions the cell perimeter tends to a maximum. Methods of digital image analysis are also adapted to measure the fractal dimension for human granulocytes [15], lymphocytes [10] [12] and for neurons [26], for example.

The mathematical basis for an understanding of fractal geometric structures, their measurement and interpretation is the Richardson-Mandelbrot equation [13] $L(\epsilon) = N(\epsilon)\epsilon$, commonly used in applications, where $L(\epsilon)$ is the length (perimeter) of the closed cell contour in the image plane, ϵ is the unit length to measure the perimeter $L(\epsilon)$, and

$$N(\epsilon) \sim 1/\epsilon^D \tag{1}$$

is the number of unit lengths ϵ necessary to cover the perimeter $L(\epsilon)$. D is the fractal contour dimension which increases from 1 to 2 at increasing «wiggliness». Inserting (1) into the Richardson-Mandelbrot equation, one obtains the perimeter scaling law

$$L(\epsilon) \sim \epsilon^{1-D}. \tag{2}$$

Mathematically generated fractals (Koch curve, Sierpinski triangle, etc.) show the property of exact self-similarity over an infinite range of ϵ-scales, i.e. ϵ takes all values between zero and infinity. Self-similar geometric objects observed in nature, however, are usually random fractals with a limited range of ϵ-values (i.e. $\epsilon_1 < \epsilon < \epsilon_2$) in which statistical self-similarity can be observed and to which the scaling laws (1) and (2) are applicable. Self-similarity is an important defining characteristic of fractal geometry. Hence, strategies for measuring the fractal dimension D cannot ignore the scaling limits ϵ_1 and ϵ_2. Natural fractal objects like cells that have been analyzed up to now manifest self-similarity over a scale length between one [10] and almost three [12] orders of magnitude depending on the range of resolution,

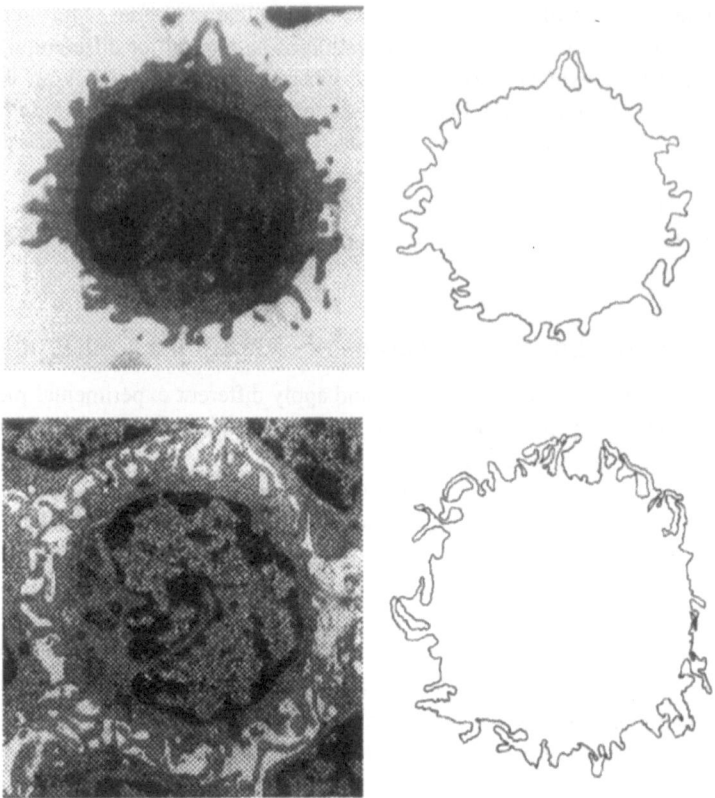

Fig. 1 Normal lymphocyte (upper row) and hairy leukemic cell (lower row) in a grey value representation and the extracted contours.

on the experimental device involved and on magnification as we will show in this investigation. Besides these «technical» limits, there exists a natural lower bound ϵ_L, defined by a critical magnification limit beyond which the surface appears as two physical sheets of finite thickness and finite surface area [28].

2.1 Measurements of the Fractal Dimension

In our analysis of cell boundaries, the contour profiles of the cells visible on the electron micrographs (see fig.1) were picked up by a scanner. The data set was stored into the memory of an HP-9000/835 and a NEC SX-3 computer and subsequently analyzed by different methods: a) yardstick method; b) box counting method; and c) probabilistic method. The analysis reported here used the entire perimeter (closed contour) and not just pieces of it; i.e., we determined the contour dimension D for the entire cell profile. Pieces of it may have different values for D as pointed out by Mandelbrot [13]. According to the various methods a) to c) different estimates of D are obtained.

By making use of the *Yardstick Method* (YM), the computer program selects a yardstick size of length ϵ, selects a starting point on the contour and then begins to count the number $N(\epsilon)$ of yardsticks of length ϵ necessary to cover the contour profile completely. This procedure is repeated for each point on the contour being the starting point and for many different choices of ϵ. The results are plotted in a $\log N(\epsilon)$ versus $\log \epsilon$ graph, giving – if self-similarity is detected — a straight line. According to (1), the slope of the linear region detected by this procedure directly yields the contour dimension $D = -\log N(\epsilon)/\log \epsilon$. Throughout the paper, we will denote by log the logarithm to base 10 ($\log = \log_{10}$).

In order to apply the *Box Counting Method* (BCM), the image plane is divided into small square grids, each of edge length ϵ. Counting the number $N(\epsilon)$ of occupied squares, one again obtains a measure for the contour length since the number $N(\epsilon)$ of squares intersected by the curve is roughly proportional to the number of steps N needed by YM to cover the contour. Starting with a smallest ϵ-scale (one pixel), the grid-length ϵ was then increased successively up to 1300 pixels depending on the size of the EM-picture. In a $\log N(\epsilon)$ versus $\log \epsilon$ plot, one obtains the fractal (box) dimension D for a self-similar structure. Shifting the grid-lattice (for each ϵ) several times within the plane, an averaged value for $N(\epsilon)$ is obtained. A verification of the BCM is the modified box counting method (MBCM) which we also used for our analysis and which is described in detail by Barth et al [1]. Besides YM and BCM, we further applied a probabilistic method (PM) which is discussed in detail by Baumann et al [2].

2.2 Estimation of Scaling Limits

In order to find the scaling limits ϵ_1 and ϵ_2, we developed and used an automatic fitting procedure. In contrast to a manual fit (see, for example, [10]) where that proportion of the curve which shows a linear dependence of $\log L(\epsilon)$ upon $\log \epsilon$ is selected subjectively, i.e. by visual inspection of curves like those given in fig. 2 or fig. 3, our automatic procedure which is based on a least square fit algorithm searches for the widest interval $[\log \epsilon_1, \log \epsilon_2]$ within which the standard deviation (*s.d.*) of the estimated slope does not exceed a given limit which we required arbitrarily to be *s.d.* = 0.0086. This corresponds to a 95% confidence interval of the slope estimate with a width of 0.0337. By this method, the dimension D is estimated at a fixed *s.d.* = 0.0086 for each contour data set of an individual cell.

2.3 Estimation of the Fractal Dimension by Asymptotic Fractal Formulas

Fitting data points which start from a plateau for small ϵ-values and approach a straight line for large ϵ-values (in a $\log L(\epsilon) - \log \epsilon$ plot, see fig. 3), one can use asymptotic fractal formulas like

$$L(\epsilon) = \left[a^8 + (b\epsilon^{1-D})^8 \right]^{1/8} \tag{3}$$

or

$$L(\epsilon) = L_o \beta(\epsilon/\epsilon_0)^{-\beta} \gamma(\beta, \epsilon/\epsilon_0). \tag{4}$$

Formula (3) has been suggested by Barth (private communication). Taking $g = -1$ in eq. (3), one just recovers Rigaut's asymptotic fractal formula being frequently used for the data fitting procedure. The additional parameter g that is introduced in (3) essentially guarantees a more accurate fit than Rigaut's formula for ϵ-values on the plateau and in the intermediate region between the plateau and the straight line (a, b, g and D are fit parameters).

The asymptotic formula (4) was *derived* by making use of Bernoulli scaling ideas and renormalization techniques (see Eq. (10)). γ is the incomplete gamma function and β is related to the fractal dimension D via the relation $D = \beta + 1$. In the limit $\epsilon \to 0$ one finds $\lim_{\epsilon \to 0} L(\epsilon) = L_0$ and, consequently, L_0 represents the plateau-value for $\epsilon \to 0$. For large ϵ-values, one obtains asymptotically $L(\epsilon) \sim \epsilon^{-\beta}$, i.e. inverse power-law scaling.

Some warnings: Using asymptotic formulas for the data fitting procedure, one will always get some value for D even if the data set is restricted to the plateau. In such a case, however, one has not observed a fractal structure! To capture a fractal structure in the data, you need to continue taking measurements into the asymptotic region (at least over one order of magnitude).

2.4 Results

First, we tested the computer methods (YM and BCM) as to their accuracy and reliability and applied them to determine the fractal dimension D for an exact self-similar mathematical construct. YM was tested with a Koch curve. The result is that at increasing resolution R (number of pixels) YM approaches the exact value ($D_{exact} = 1.261...$) from below. For instance, taking $R = 1200 \times 1200$ pixels, we obtain $D = 1.22$. For $R = 2400 \times 2400$ pixels, we get $D = 1.24$. BCM was tested with a Sierpinski triangle ($D_{exact} = 1.584...$). The results are: (i) the higher the resolution R, the more accurate the value of D; (ii) BCM and YM approach the exact value for D from below, PM approaches D from above (see Ref [2]).

Figure 2 demonstrates (in a $\log N(\epsilon)$ versus $\log(\epsilon)$ plot) how the self-similar (fractal) behavior indicated by a straight line is approached by using (a) YM (*) and (b) BCM (+). Fig. 2 showing the data of a hairy leukemic cell contains even more information: (i) according to eq. (1), the slope of the straight line yields the fractal contour dimension $D = 1.35$, and (ii) the data points that follow the straight line are restricted to a finite ϵ-domain ($\epsilon_1 < \epsilon < \epsilon_2$) which defines self-similar (fractal) structure of the cell profile. The upper limit ϵ_2 where the data points break out of the linear behavior is found to be approximately $\epsilon_2 \approx 4$ microns for all contours of the same cell line under analysis. We interpret the existence of this upper limit as an indication of the finite macroscopic size of the cell contour ($L(\epsilon_2) \approx 24.8\mu m$) if it is approximated by a circle with radius ϵ_2. Indeed, using an Euclidean measure for the cell perimeter $L(\epsilon_2) = 2\pi\epsilon_2 = 25.1$ microns, one discovers that ϵ_2 is closely related to the classical radius of a circle. Depending on the method used for the data analysis, the existence of a lower bound ϵ_1 is observed by inspection of fig. 2. In case of BCM, we estimate $\epsilon_1^B \approx 0.02$ microns and by using YM, we find $\epsilon_1^Y \approx 0.2$ microns. In this case, BCM detects self-

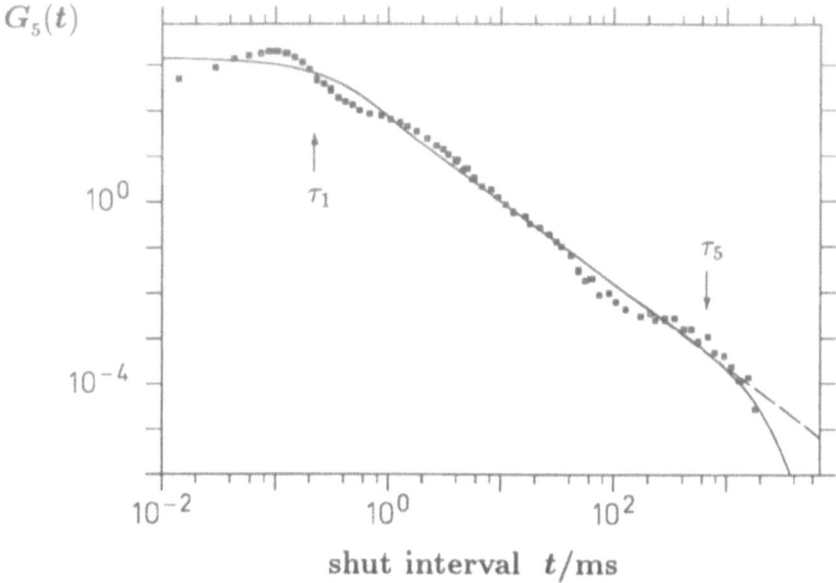

$G_5(t)$

10^0

10^{-4}

T_1

T_5

10^{-2} 10^0 10^2

shut interval t/ms

Fig. 2   Contour data of a hairy leukemic cell determined by YM ($\star$) and BCM (+). ϵ_1 and ϵ_2 denote the scaling limits (see text in Sec. 2.4).

similarity over a domain ϵ ranging between $\epsilon_1^B = 0.02 \ \mu m < \epsilon < \epsilon_2 = 4 \ \mu m$ which covers a factor $\epsilon_2/\epsilon_1^B = 200$, i.e. more than two orders of magnitude for the hairy cell under consideration. In case of YM, the self-similar domain shows a ratio of $\epsilon_2/\epsilon_1^Y = 20$ which is also an acceptable result because it is beyond the «standard norm» at a factor of 10 which is conventionally required in order to talk about «fractal structures».

Applications of the fractal concept to stereology are traditionally based on the perimeter scaling law (2) which yields — besides the fractal dimension D — additional structural information. Fig. 3 shows eq. (2) in a $\log L(\epsilon)$ versus $\log(\epsilon)$ plot for two different magnifications of one and the same hairy leukemic cell. The slope α of the straight line indicating self-similar structure is related to D by $D = 1 - \alpha$. The data points present our numerical results for the perimeter $L(\epsilon)$ of one and the same contour of a hairy leukemic cell taken at two different magnifications (8 700x, 18 400x). The slope α of the straight line is found to be $\alpha = -0.34$ leading to $D = 1.34$ which is in good agreement with the value 1.35 estimated via fig 2. This dimension is far beyond the dimension of normal T-lymphocytes as expected: Just by visual inspection of EM-pictures (see fig.1), more wiggliness or roughness of a hairy leukemic cell contour can be seen with respect to the cell boundary of a normal T-lymphocyte. However, computerized image analysis yields a reproducible and consistent number for the parameter D free of any subjective influence. Consequently, D can be used as a quantitative measure

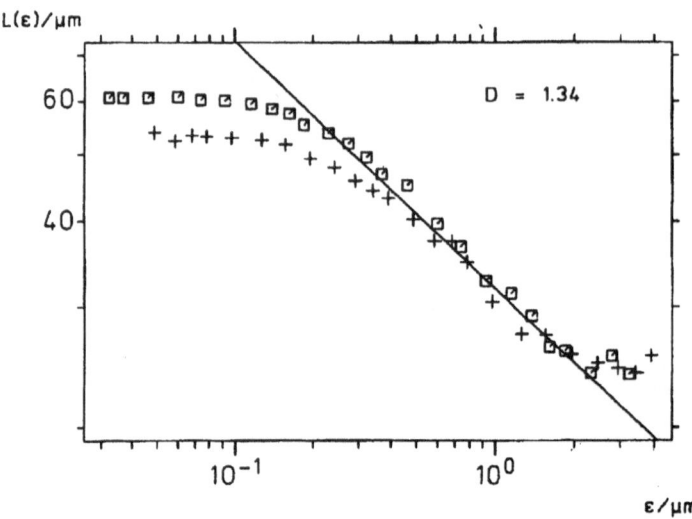

Fig. 3   Data points in a $\log L(\epsilon)$ *vs* $\log \epsilon$ plot of one and the same hairy cell determined by YM at two different magnifications: $M_1 = 18400 \times$ ($\square$) and $M_2 = 8700 \times$ ($+$). The straight line indicates ideal scaling behavior (D=1.34).

for the structural complexity of the cell contour. Besides the dimension D, fig.3 in addition shows that for small ϵ-values, the perimeter $L(\epsilon)$ reaches a saturated value (plateau) L_{max}. The plateau value depends on the magnification M. For $M = M_1$ = 18 400x the measured maximum value is given by $L_{max} = L(M_1) = 60.8$ μm (upper curve) and for $M = M_2 = 8700$ x we find $L_{max} = L(M_2) = 53.1 \mu m$ (lower curve). Additionally, we observe that the higher the magnification, the larger the ϵ-span that follows the straight line. In other words, the higher the magnification, the more structural details of the contour are observable and thus one observes a larger domain of self-similar structures.

Following the way just described, we determined the contour dimension D for 19 T-lymphocytes and for 18 hairy leukemic cells by using BCM. A crucial point in finding a reliable D-value for each cell may be seen in determining the scaling limits ϵ_1 and ϵ_2. We attacked this problem by using an automatic fit procedure as discussed in subsection 2.2. All data points between the limits ϵ_1 and ϵ_2 were fitted by a least square fit procedure resulting in a straight line. The slope of this line gives just an individual value for D with $s.d. = 0.0086$. In case of T -lymphocytes, we found individual D-values ranging between 1.12 and 1.23 for all cells of this type. In case of hairy leukemic cells, we found values for D ranging between 1.29 and 1.37. In all cases the ratio ϵ_2/ϵ_1 is larger than 10, i.e. the scaling region is larger than one order of magnitude. Thus we may talk about «fractal structures».

Having the D-values for several members of one and the same cell line (19 T-lymphocytes) at our disposal, we can prepare a histogram in order to study the underlying statistics. We found that a Gaussian distribution is compatible with the individual D-values. Therefore, we were able to calculate a mean value $\bar{D} = 1.15(s.d. = 0.034)$ representing the fractal dimension of the sample of 19 T-lymphocytes. In case of 18 hairy leukemic cells, we found $\bar{D} = 1.32(s.d. = 0.023)$ to be the mean fractal dimension of this cell line.

We now shall discuss estimates of D by making use of so-called asymptotic fractal formulas (eqs. (3) and (4)). Here, least square fit procedures take all data points into account which are shown in fig.3 (apart from the last five points for large ϵ-values). Thus, one has available a large set of data in order to fit the curve. However, one should realize that the data points (see fig.3) which enter into and which form the plateau for small ϵ-values (high resolution) do not fall into the scaling region represented in fig.3 by the straight line. Nevertheless, such asymptotic scaling formulas can be used and frequently have been used in order to fit experimental data sets. To get some feeling for the accuracy of such an estimation of D, we used formulas (3) and (4) for a fit of the data (magnification 18 400x) shown in fig.3. Via formula (3), we estimated $D = 1.39$. From formula (4), we found $D = 1.38$. These values for D are in fairly good agreement with the estimate ($D = 1.34$) found by the method described in subsection 2.2 which takes only those data points into account that fall into the scaling region between ϵ_1 and ϵ_2. As a final comment on this section we notice that the fractal dimension D alone does not completely specify the fractal structure because objects having the same D may differ by other characteristics as, for instance, *lacunarity*.

3   Scale-Invariant Protein Dynamics

In this section, we will mainly investigate temporal aspects of scale-invariant dynamical processes. Usually, fractals are being considered as self-similar objects in space; i.e., the term *fractal* is intuitively based on a structural (geometric) concept that applies to a wide class of complex shapes whereby self-similarity is a key feature of such irregular geometric patterns. Just as a geometric fractal does not have a characteristic (dominant) scale of length, a scale-invariant dynamical process generates fluctuations that do not have a characteristic scale of time and thus cannot be adequately measured by statistics based on mean and variance [11] [14] [25]. If the average time of an event were finite, then this would provide a time scale. A scale-invariant process can be regarded to be a self-similar superposition of many processes with different time scales for which the term «fractal time process» was coined [25]. Processes which are scale-invariant can be modelled by Lévy statistics [14] [15] [20] [25] [29].

It is generally accepted that according to their biological function, proteins assume (during functional activity) a number of different (conformational) states. Transitions which lead from one of these states to another are called «functionally important motions» [5] and are usually formulated as a kinetic or relaxation process characterized by kinetic rate constants $k_i = 1/\tau_i$. The number N of rate constants k_i ($i = 1, ..., N$) indicates how many conformational states are taking part in the relaxation process. If there would exist just a single characteristic or dominant time scale $\tau_i = \tau_0$, one would expect an exponential Debye relaxation $\sim \exp(-t/\tau_0)$. Experimental observations, however, indicate that measured relaxation data sets cannot be fitted by a single exponential function. Instead, slow, non-exponential relaxation processes are observed giving evidence for scale-invariant underlying protein dynamics. Theoretical concepts to describe slow relaxation phenomena like Kohlrausch-Williams-Watts relaxation $\exp\{-(t/\tau)^\beta\}$ or inverse power-law relaxation $t^{-\alpha}$ have been developed over the last few years in order to understand non-standard relaxation processes in glasses [4], ion channel proteins [16] [19] [20] polymers [6] [17] and other disordered systems. Here, we will demonstrate how scaling ideas can be applied to self-similar, scale-invariant protein dynamics.

3.1 Bernoulli Scaling Ideas and Renormalization Technique
Patch clamp experiments [24] for studying ion channel gating kinetics provide important information on the molecular structure and function of ion channel proteins. Generally, ion channels are considered to be large integral membrane proteins which allow passive flux of ions through cell membranes. Currents recorded from ion channels show that the channels repeatedly open and close their pores during normal activity. Counting the number of channel events and plotting such a count versus duration time, one can construct a histogram which can be approximated by a probability function $f(t)$ for the life-time of the conducting state [24]. A typical plot is given in fig.4 showing the data of a fast Cl^- channel [3].

When modelling patch clamp data theoretically, one usually assumes that channel gating is associated with transitions among a number of conformational protein states with amplitudes a_i and time constants τ_i. Data from the duration histograms are fitted to an N-component probability density function of the form

$$f(t) = \sum_{i=1}^{N} \frac{a_i}{\tau_i} e^{-t/\tau_i} \tag{5}$$

where the amplitudes a_i and the corresponding time constants τ_i are usually presumed (for the fitting procedure) to be independent. The number of free parameters for such a discrete Markov chain is $2N-1$ because one of the $2N$ constants (a_i, τ_i) is fixed by the normalization condition [19] [20].

As already mentioned, proteins may have a large number of conformational states and thus many terms in eq. (5) should be taken into account. For instance, in the case of a fast Cl^- channel, Blatz & Magleby [3] conclude N to be five. For this example, one has at least nine parameters which must — and do — fit the data.

In order to model such a situation with less than $(2N - 1)$ parameters, we are considering a process where the time between events is a random variable. A prominent random (no memory) process is the Poisson process $g(t) = 1/\tau \exp(-t/\tau)$. The average time between events is defined by $\langle t \rangle = \int_0^\infty tg(t)dt$.

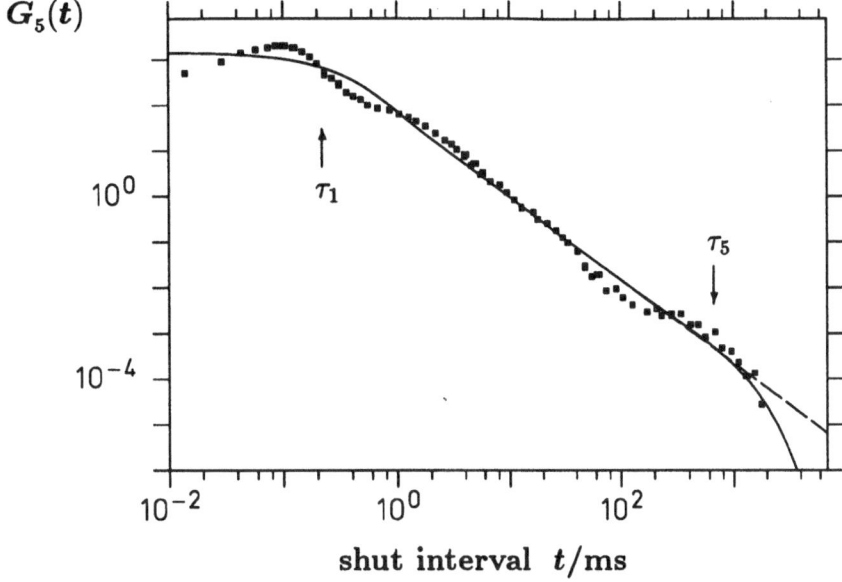

shut interval t/ms

Fig. 4 Measured data points [3] of a Cl^- channel are oscillating along the power-law trend (dashed line, eq. (10)). The full curve (eq. (11)) approaches the power-law trend between the limits τ_1 and τ_5, and crosses over to exponential decay for $t \geq \tau_5$ (see text, Sec. 3.1).

If $\langle t \rangle$ is finite ($\langle t \rangle = \tau$ in the Poisson process), then a natural time scale $\langle t \rangle$ exists in which to measure time. If one waits long enough, i.e. if the observation time t_{obs} is much longer than $\langle t \rangle$, events occur at an average rate $k = 1/\langle t \rangle$. However, if $\langle t \rangle$ is infinite (actually much longer than the time of observation t_{obs}), then no natural time scale exists in which to gauge measurements. In this case, one expects to find temporal scaling laws and the concept of fractal time emerges [25]. Now the questions arise: How can we model such a situation and which are the physical processes behind all that? Several mechanisms can be mentioned which generate a fractal time distribution of events. A good example is thermally activated hopping over a distribution of energy barriers: Let $G_N(t)$ be a sum of N Poisson processes with time constants τ_i $(i = 1, \ldots, N)$ each weighted by a probability p_i of occurrence, i.e.

$$G_N(t) = \sum_{i=1}^{N} \left(\frac{p_i}{\tau_i} \right) e^{-t/\tau_i}. \qquad (6)$$

This time series looks similar to the standard Markov-chain series (5) in terms of «amplitudes» $p_i = a_i$ and time constants τ_i. However, $G_N(t)$ is still not in a «renormalized» form because renormalization implies the concept of scaling. A frequently used scaling concept is Bernoulli scaling. The basic idea [14] is this: A duration between events which is longer than an order-of-magnitude occurs with an order-of-magnitude less probability. This scaling condition can be described mathematically by taking $\tau_i = \tau\lambda^i$ and $p_i = a_N p^i$ with $0 < p < 1$ and $\lambda > 1$, leading to

$$G_N(\xi) = \frac{a_N}{\tau} \sum_{i=1}^{N} \left(\frac{p}{\lambda}\right)^i e^{-\xi\lambda^{-i}}, \tag{7}$$

where a_N [20] is a normalization constant and $\xi = t/\tau$.

This way, a distribution $G_N(t)$ having no single characteristic time scale is generated (it incorporates many time scales τ_i in form of a discrete geometric spectrum λ^i of which no time-scale is dominant). For $N \to \infty$ the average time $\langle t \rangle$ to gauge measurements $\langle t \rangle = \int_0^\infty tG_N(t)dt$ tends to infinity as required for fractal time processes. Here, we note that any finite number N would lead, of course, to a finite $\langle t \rangle$ increasing with N. However, once N and λ are large enough, one can find an average time $\langle t \rangle$ that becomes much larger than t_{obs}. Hence, the concept of fractal time can still be applied, but it is restricted by the limits $\tau_1 = \tau\lambda$ and $\tau_N = \tau\lambda^N$ (note: $\tau_N \to \infty$ for $N \to \infty$).

Our result (7) in which the Bernoulli scaling condition is built into $G_N(t)$ represents a first stage of renormalization. The second stage of renormalization utilizes the idea of evaluating the discrete series (7) for $G_N(t)$ by replacing the sum with a series of integrals using the Poisson summation formula; i.e., going to a continuum model that brings into play the power-law representation of $G_N(t) \sim t^{-1-\mu}$ [16] [19] [20] [29].

Here, we would like to point out that the discrete series (7) represent no more (and no less) than the original Markov chain series (5) subjected to the constraint requiring Bernoulli scaling in terms of $\tau_i \propto \lambda^i$ and $a_i \propto p^i$. Therefore, our model does not contradict, but rather is a consequence of the Markov chain model which incorporates many more possibilities and parameters for modeling patch clamp data than our renormalized (scaling) model that filters out of the Markov model just those gating processes which are consistent with Bernoulli scaling.

Applying the Poisson summation formula to the discrete series (7), one is coming up with the power-law representation [19] [20]

$$G_N(\xi) = \frac{a_N}{\tau \ln \lambda} \sum_{m=-\infty}^{\infty} \xi^{-\nu_m} I_N(\nu_m, \xi/\lambda) + \frac{a_N}{2\tau}(\lambda^{-\nu_0}e^{-\xi/\lambda} + \lambda^{-N\nu_0}e^{-\xi/\lambda^N}) \tag{8}$$

where $\nu_m = 1 - (\ln p + 2\pi im)/\ln \lambda$ is a complex-valued exponent and I_N denotes the integral

$$I_N(\nu_m, x) = \int_{x\lambda^{-N+1}}^{x} dw\, w^{\nu_m-1} e^{-w} = \gamma(\nu_m, x) - \gamma(\nu_m, x\lambda^{-N+1}). \tag{9}$$

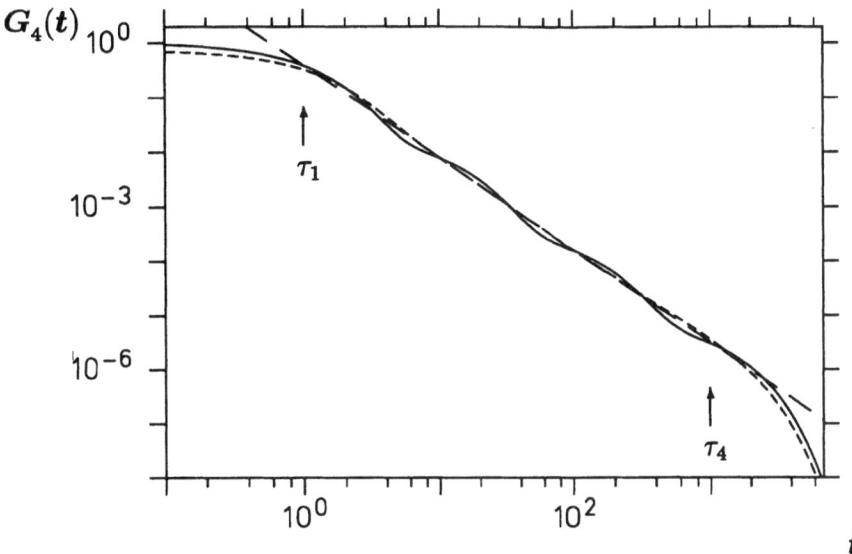

Fig. 5   This plot demonstrates how the scaling region between the limits τ_1 and $\tau_N(N = 4)$ emerges out of eq. (8): The dashed curve (- - -) is calculated from eq. (11) taking into account just the term $m = 0$ (i.e. neglecting oscillations in eq. (8)); the full curve takes into account many m-terms of the sum in eq. (8), i.e. $m = 0, \pm 1, \ldots, \pm 12$. The straight line (——) indicates ideal fractal behavior without limits. Parameters are: $\tau = 0.1, N = 4, \lambda = 10$ and $\mu = 0.7$.

In the limit $N \to \infty$, I_∞ given by $\lim_{N\to\infty} I_N (\nu_m, x) = \gamma(\nu_m, x)$ approaches the incomplete gamma function. Introducing the positive number $\mu = -\ln p/\ln \lambda$, one may write $\nu_m = \nu_0 - 2\pi i m/\ln \lambda$ with $\nu_0 = 1 + \mu$, and thus, taking into account only the term ($m = 0$) in (8) one obtains in the limit $N \to \infty$ the asymptotic fractal formula

$$G_\infty(\xi) = \frac{a_\infty}{\tau \ln \lambda} \xi^{-\nu_0} \gamma(\nu_0, \xi/\lambda) + \frac{a_\infty}{2\tau} \lambda^{-\nu_0} e^{-\xi/\lambda}. \tag{10}$$

A re-interpretation of the parameters $a_\infty/\tau \ln \lambda = \beta L_0$, $\mu + 1 = \beta$, $\xi/\lambda = \epsilon/\epsilon_0$, and neglecting the last term in (10), brings us back to Eq. (4) which we already offered as a three-parameter (L_0, β, ϵ_0) formula for an estimation of fractal dimensions.

In fig. 4, Eq. (10) is tested with measurements of ion-channel data which oscillate around the power-law trend. Similar oscillations following a power-law trend were also observed in the lateral diffusion coefficient [18], in the QRS-spectrum [9] and (if ξ is interpreted as generation number) in the dichotomous branching system of the bronchial tree [15] [29].

Fig. 5 shows an illustrative plot that demonstrates how the property «fractal between limits» emerges out of the formula

$$G_N(\xi) = a_N/(\tau ln\lambda)\xi^{-\nu_0} I_N(\nu_0, \xi/\lambda) + \frac{a_N}{2\tau}(\lambda^{-\nu_0} e^{-\xi/\lambda} + \lambda^{-N\nu_0} e^{-\xi/\lambda^N}), \tag{11}$$

which represents just the first term ($m = 0$) of eq. (8) and approaches the asymptotic formula (10) for $N \to \infty$. Thus, if there is a finite number N of terms in the sum (7) and consequently in the integral I_N, one has two crossover points ($\tau_1 = \tau\lambda$ and $\tau_N = \tau\lambda^N = \tau_1\lambda^{N-1}$). For $t < \tau_1$, the relaxation kinetics is dominated by the initial value $G_N(0)$ and for $t > \tau_N$, the decay becomes exponential. *Scale invariant processes* are restricted to the time interval $\tau_1 < t < \tau_N$. When interpreting measurements, it is essential to get that range, i.e. $\tau_N/\tau_1 = \lambda^{N-1}$, included in the data. Of course, if you would «see» in the data just a single conformational state (i.e. $N = 1$), then the scaling range λ^{N-1} would collapse to a single point and you would not have observed a fractal process. In Fig. 4, we have re-analyzed the data set [3] for a Cl^- channel with $N = 5$; i.e., five shoulders have been identified [3]. Our fit is based on eq. (11) with $\tau = 0.033$, $N = 5$, $\mu = 0.82$ and $\lambda = 7$, leading to a scaling range $\tau_5/\tau_1 = 7^4 \approx 2400$; i.e., three orders of magnitude before the curve crosses over to exponential decay (for $t > \tau_5$). Thus, the way to tell what is going on in the dynamics is to look over as many orders of magnitude as possible.

3.2 Fractional Operator Design

Investigating kinetic or relaxation processes in proteins, polymers, glasses or other disordered systems, one observes slow decay of the relaxation functions. In a series of experiments, Frauenfelder [5] studied the function and dynamics of Myoglobin (Mb). It is a typical heme protein and its central function is the storage of oxygen (O_2). Actually, Frauenfelder [5] studied the reaction $Mb + CO \longleftrightarrow MbCO$. A laser flash breaks the bond between the iron atom in the heme pocket and the ligand. Then, the ligand molecule moves away from the binding side and later rebinds. If $N(t)$ denotes the fraction of Mb molecules which have not yet rebound a CO molecule at time t after photodissociation, then Frauenfelder's findings are that the rebinding is not exponential in time but is closer to the asymptotic fractal formula $N(t) = N(0)(1 + t/\tau)^{-\alpha}$ indicating power-law kinetics for the underlying processes. Clearly, if $t/\tau \gg 1$, $N(t)$ approaches asymptotically inverse power-law behavior $N(t) \sim t^{-\alpha}$. Theoretical models borrowed from the theory of polymers, glasses and other disordered systems have already been applied to protein dynamics.

In the current investigation, however, we will not go into a discussion of such well-known theoretical concepts. Instead, we alternatively propose a method based on fractional integral and differential operators that enables us to generalize the standard relaxation equation

$$dN(t)/dt = -(1/\tau)N(t) \tag{12}$$

with the exponential solution $N(t) = N(0)\exp(-t/\tau)$ to a fractional differential equation representing slow relaxation phenomena.

Our procedure is as follows: The integrated form of (12) reads $N(t) - N(0) = -(1/\tau)_0D_t^{-1}N(t)$, where $N(0)$ is the initial value $N(t = 0)$ and the integral operator is defined by $_0D_t^{-1}N(t) = \int_0^t N(t')\,dt'$. Introducing now the fractional

Liouville-Riemann integral operator $_0D_t^{-\beta}$, defined by [21] [27]

$$_0D_t^{-\beta}N(t) = \int_0^t \frac{(t-t')^{\beta-1}}{\Gamma(\beta)} N(t')\, dt' \tag{13}$$

with $0 < \beta < 1$ and replacing $(1/\tau)_0D_t^{-1}N(t)$ by $(1/\tau^\beta)_0D_t^{-\beta}N(t)$, the integrated version of (12) takes the form

$$N(t) - N(0) = -(1/\tau^\beta)_0D_t^{-\beta}N(t) \tag{14}$$

representing a fractional integral equation with an incorporated initial value of $N(0)$. Eq. (14) can be inverted by introducing the fractional differential operator

$$_0D_t^\nu N(t) = \left(\frac{d}{dt}\right)^n {}_0D_t^{\nu-n}N(t), \qquad 0 < \nu < n \tag{15}$$

into the fractional differential equation [7] [17]

$$_0D_t^\beta N(t) - N(0)\frac{t^{-\beta}}{\Gamma(1-\beta)} = -(1/\tau^\beta)N(t). \tag{16}$$

The solution of either (14) or (16) is given by [6] [7]

$$N(t) = N(0)\sum_{k=0}^\infty \frac{(-1)^k}{\Gamma(1+\beta k)}(t/\tau)^{\beta k} = N(0)E_\beta(-(t/\tau)^\beta), \tag{17}$$

where $E_\beta(x)$ is a Mittag-Leffler function.

One recognizes that in the limit $\beta \to 1$, the exponential solution of the standard relaxation equation (12) is rediscovered. For large t-values, solution (17) approaches the asymptotic fractal formula [7]

$$N(t) \sim N(0)\sum_{k=0}^\infty \frac{(-1)^k}{\Gamma(1-\beta(k+1))}\left(\frac{\tau}{t}\right)^{\beta(k+1)}. \tag{18}$$

Just as a byproduct of our fractional analysis, we comment that the leading order term of (18), i.e. $N(t) \sim t^{-\beta}$ exhibits the same inverse power-law exponent β which defines the fractional order of the Liouville-Riemann integral operator (13).

We notice that a re-interpretation of Frauenfelder's Myoglobin data is given by Glöckle and Nonnenmacher [8] in terms of such a fractional relaxation model.

My final *comments and conclusions* are: self-similar geometric structures and scale-invariant dynamical processes in biomedical systems are not fractal on all length and time scales. Instead, they are fractal only within limits. Estimating the fractal contour dimension D we observed that contour profiles of hairy leukemic cells have D-values *significantly larger* than those of healthy T-lymphocytes. However, investigating profiles of blast cells of acute lymphoblastic leukemia (not discussed here) we estimated D-values that are *close to but smaller* than those of normal T-lymphocytes (see Losa, this Volume). To confirm this and to get a more complete understanding of this small change in smoothing down the structural contour irregularities when T-lymphocytes are forced to transform into lymphoblasts requires further investigations and, possibly, more information than being stored in the fractal contour dimension D. Currently we are searching for a unifying principle correlating both spatial and temporal fractal properties.

Acknowledgments:
This work has been supported in part by Deutsche Forschungsgemeinschaft (SFB 239, Project C 8) and by the Swiss National Science Foundation, grant 31-25702.88. I gratefully acknowledge stimulating discussions with Dr. Gerd Baumann, Andreas Barth and Walter Glöckle. I also acknowledge helpful comments from Professors Albrecht Kleinschmidt and Torsten Mattfeldt while preparing this manuscript.

References

[1] Barth A, Baumann G & Nonnenmacher TF: J. Phys. A: Math. Gen. **25**, 381 (1992).

[2] Baumann G, Barth A & Nonnenmacher TF: This volume.

[3] Blatz AL & Magleby KL: J. Physiol. (London) **378**, 141 (1986).

[4] Blumen A, Klafter J & Zumofen G: In: Optical spectroscopy of glasses, Zschokke I, ed. Reidel, Dordrecht (1986).

[5] Frauenfelder H: Ann NY Acad. Sci. **504**, 151 (1987).

[6] Glöckle WG & Nonnenmacher TF: Macromolecules **24**, 6426 (1991).

[7] Glöckle WG & Nonnenmacher TF: J. Stat. Phys. **71**, 741 (1993).

[8] Glöckle WG & Nonnenmacher TF: Fractional relaxation equations for protein dynamics. This volume.

[9] Goldberger AL, Bhargava V, West BJ & Mandell AJ: Biophys. J. **48**, 525 (1985).

[10] Keough KMW, Hyam P, Pink DA & Quinn B: J. Microsc. **163**, 95 (1991).

[11] Lipsitz LA & Goldberger AL: Loss of complexity and aging. JAMA **267**, 1806 (1992).

[12] Losa GA, Baumann G & Nonnenmacher TF: Path. Res. and Pract. **188**, 680 (1992).

[13] Mandelbrot BB: The Fractal Geometry of Nature. Freeman, San Francisco (1983).

[14] Montroll EW & Shlesinger MF: Proc. Natl. Acad. Sci. USA **79**, 3380 (1982).

[15] Nonnenmacher TF: In: Thermodynamics and Pattern Formation in Biology, Lamprecht I & Zotin, AI, eds. Walter de Gruyter, Berlin, 371–394 (1988).

[16] Nonnenmacher TF: J. Colloid. Polym. Sci. **267**, 753 (1989).

[17] Nonnenmacher TF: In: Rheological Modeling: Thermodynamical and Statistical Approaches, Casas-Vázquez J and Jou D, eds. Springer, Berlin (1991).

[18] Nonnenmacher TF: Eur. Biophys. J. **16**, 375 (1989).

[19] Nonnenmacher TF & Nonnenmacher DJF: Phys. Lett A **140**, 323 (1989).

[20] Nonnenmacher TF & Nonnenmacher DJF: In: Stochastic Processes, Physics and Geometry, Albeverio S, Casati G, Cattaneo U, Merlini D and Moresi R, eds. World Scientific, Singapore (1990).

[21] Oldham KB, & Spanier J: The fractional calculus. Academic Press, New York (1974).

[22] Paumgartner D, Losa G & Weibel ER: J. Microsc. **121**, 51–63 (1981).

[23] Rigaut JP: J. Microsc. **133**, 41–54 (1984).

[24] Sakmann B & Neher E: Single channel recording, Plenum, New York (1983).

[25] Shlesinger MF: Ann Rev. Phys. Chem **39**, 629 (1988).

[26] Smith TG, Marks WB, Lange GD, Sheriff WH & Neale EA: J. Neurosc. Meth. **27**, 173-180 (1989).

[27] Srivastava HM & Bushman RG: Theory and applications of convolution integral equations. Kluver Academic Publishers, Dordrecht (1992).

[28] Weibel ER: Am. J. Physiol. **261**, L361-L369 (1991).

[29] West BJ, Bhargava V & Goldberger AL: J. Appl. Physiol. **60**, 1089 (1986).

Chaos, Noise and Biological Data

Bruce J. West, W. Zhang and H.J. Mackey
Physics Department
University of North Texas
Denton, Texas 76203

1   Introduction

In the recent past there has been a torrent of papers concerned with data processing, see e.g., Casdagli and Eubank [1]. In large part this flood of interest has been
associated with the realization that irregular time series, long thought to contain
signal and *noise*, may in fact be *chaotic*. Noise is produced by the coupling of
a signal to an infinite dimensional environment, e.g., the billions of neurons in
the brain producing the erratic EEG signal. Chaos on the other hand is produced
by the intrinsic deterministic dynamics of a low-dimensional nonlinear system.
Difficulties arise when one attempts to discriminate between colored noise and
chaos in erratic time series. The difficulties are associated with how one interprets
the erratic nature of the time series. If it is noise, the random character of the
data masks the underlying signal. If it is chaos, the statistical fluctuations contain
information and can not be conveniently suppressed as one would noise. However, since quantities such as the correlation or information dimension of the time
series can not distinguish between the two, the development of new measures are
important and new methods of data processing required.

Herein we are concerned with the construction of measures that can distinguish between chaos and noise in time series. Osborne and Provenzale [2] have
shown that obtaining a finite fractal dimension D from the processing of a time series is not in itself sufficient to infer the existence of a low-dimensional dynamical
system. A random fractal., e.g., a Gaussian random process with an inverse power-
law spectrum $P(f) \propto 1/f^\alpha$, has a fractal dimension given by $D = 2/(\alpha - 1)$. This
theoretical result has been tested [2,3] using the attractor reconstruction technique
(ART) that is usually applied to chaotic time series, i.e., «random» time series
generated by deterministic nonlinear dynamical systems. Therefore the fractal dimension alone can not distinguish between a chaotic signal and colored noise, i.e.,
D is not an unambiguous measure of chaos.

One measure that has been proposed for discriminating between chaos and
noise is the power-law form of the spectrum itself. Frisch and Morf [4] presented
an intuitive argument that systems of coupled nonlinear rate equations that have
strange attractor solutions give rise to chaotic time series whose power spectra decay more rapidly than algebraic, e.g., *exponential* in frequency at high frequencies,
i.e., $P(f) \propto e^{-\lambda f}/f^\alpha$ where λ and α are positive constants. This argument was

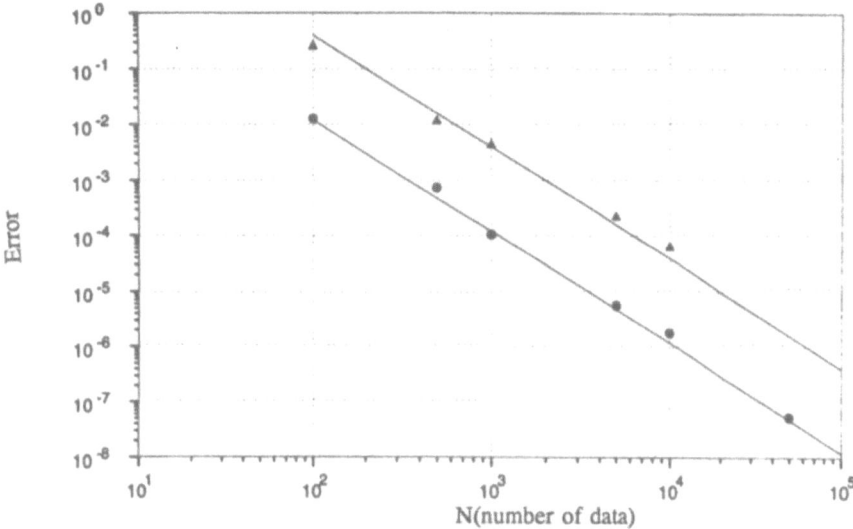

Fig. 1 The average prediction error is plotted versus the number of data points in the time series for the logistic map (•) and the Hénon map (△). The local approximation of FS was used with a linear polynomial approximation and a sample size of 100. The solid curve is the theoretical prediction from (7).

numerically verified by Greenside, et al. [5] using systems with known strange attractor solutions. Sigeti and Horsthemke [6] give a general proof establishing the power-law form of the spectrum for stochastic systems driven by additive or multiplicative noise. No such proof exists for the exponential decay of spectra for chaotic system although mathematical investigations have been initiated to establish such a proof, e.g. see Ruelle [7]. Contrary to this expectation inverse power-law spectra have been found for certain deterministic dynamical systems that have chaotic solutions, e.g., the driven Duffing oscillator [8] and the driven Helmholtz oscillator [9] in which low frequency inverse power-law spectra are observed in both analog simulations and numerical integration of the equations of motion. Inverse power-law spectra also appear in discrete dynamical systems with intermittent chaos, i.e., systems that manifest randomly spaced bursts of activity that interrupt the normal regular behavior [10]. Therefore an inverse power-law spectrum is also not an unambiguous measure of *colored noise* nor of chaos.

The technique that we adopt herein is based on the deterministic generator of chaos, which is to say that since chaos is deterministically generated one ought to be able to predict its evolution. At least for short intervals of time. In fact even beyond where one can faithfully forecast a chaotic signal, the deviation of the forecast from the data (error) behaves in a characteristic way. Therefore one can even use the growth of error in forecasts to determine if a given time series is chaos or noise.

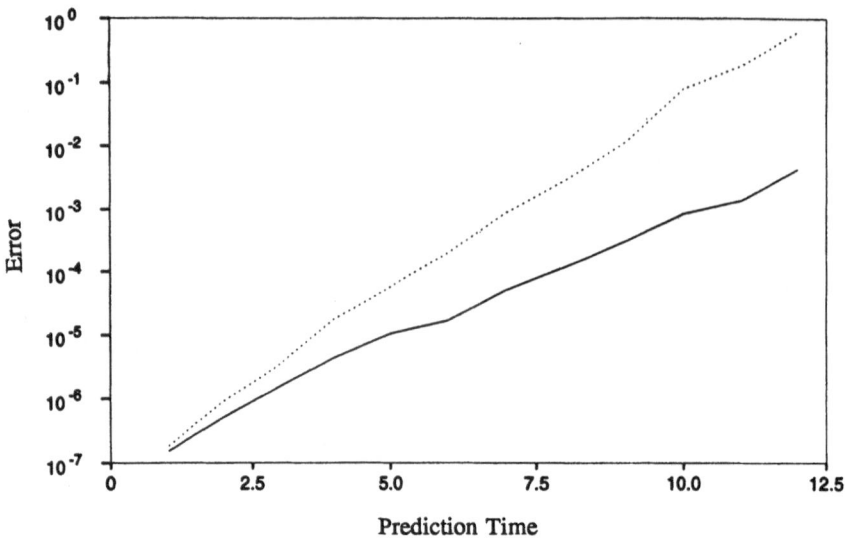

Fig. 2   The prediction error is graphed as a function of the prediction time T for the logistic equation using two versions of the local approximation (LA): no updating of neighbors (· · ·) and neighbors are updated at each time step (–).

In Section 2 we review some of the latest techniques for predicting chaos from chaos. Herein we recalculate the predictions made by a number of authors and test how the prediction-error grows with the prediction time and with the number of data points. These methods are then applied to the problem of discriminating chaos from colored noise and a number of valid indicators are obtained. Included in this review are the results from a neural network (NN) prediction of chaos from chaos using a back propagation technique. The NN method is determined to be more robust to contamination by noise in its predictions of chaos from chaos than are polynomial methods.

2   Forecasting Chaos from Chaos

We have indicated that «random» time series may really be produced by deterministic equations and may therefore be chaos rather than noise. When this is the case the system dynamics may be modelled by: (i) finding a state space with minimum embedding dimension (maximum determinism), and (ii) fitting a nonlinear function to the map that transforms the present states into future states. In this way a data stream (time series) can be used to determine not only the number of variables necessary to model the underlying dynamical system, but it can also provide an approximate functional form for nonlinear dynamics governing the evolution of the system [1,11].

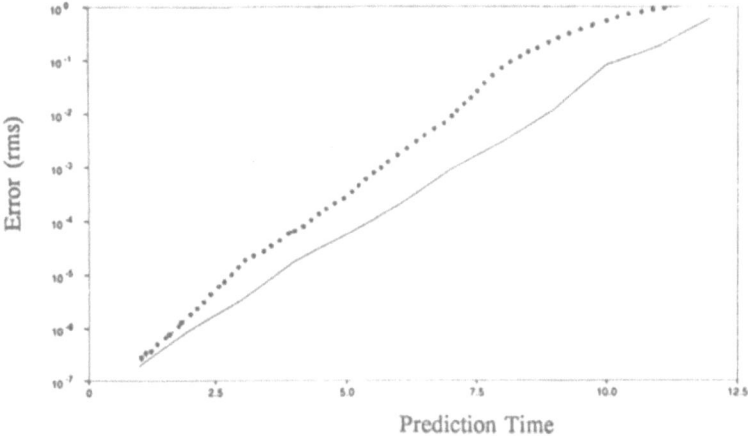

Prediction Time

Fig. 3 The prediction error using the local approximation (LA) for logistic data is graphed versus prediction time for a one-dimensional data vector (solid curve) and a two-dimensional data vector (dotted curve).

2.1 Scaling Error

The great promise of chaos is that although it puts limits on long term predictions, it implies predictability over the short term. How well this can be done in a biological context remains to be seen.

As Farmer and Sidorowich [12,13] point out, most forecasting is currently done with linear methods, and since linear dynamics cannot produce chaos, these methods cannot produce good forecasts for chaotic time series. Although the development of nonlinear forecasting techniques is an active area of research, with few exceptions, the full exploitation of the concept of chaos has not as yet been made. Preliminary investigations have been done to compare the quality of forecasts using linear and nonlinear techniques and we discuss some of these here.

Finding the state space representation is accomplished using the attractor reconstruction methods discussed by many researchers. To recapitulate let us consider a discrete time series $[\xi(t_j)]$, $j = 0, 1, \ldots, M$ where $\xi(t)$ is the observable in a biological system, e.g., an EEG, ECG, blood pressure, etc. We assume that $[\xi(t_j)]$ is stationary, as it would be if the data points are generated by a mapping on an attractor. We construct a state vector $\vec{X}(t) = [X_1, (t), X_2(t), \ldots, X_m(t)]$ by assigning the coordinates

$$X_1(t) = \xi(t)$$
$$X_2(t) = \xi(t + \tau)$$
$$\ldots \tag{1}$$
$$X_m(t) = \xi[t + (m - 1)\tau]$$

where τ is a delay time. If the dynamics take place on an attractor of dimensions d, then a necessary condition for determinism is $m \geq d$. If D is the dimension of the

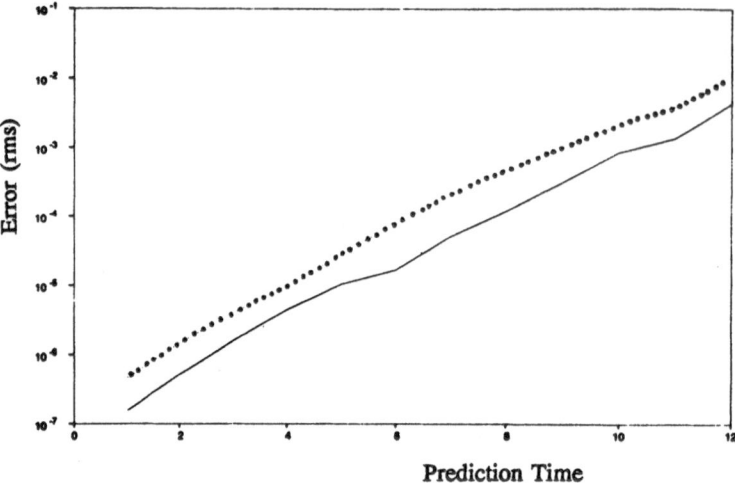

Prediction Time

Fig. 4 The prediction error using the local approximation (LA) for logistic data is plotted against the prediction time using two nearest updated neighbor (solid curve) and using four updated nearest neighbors (dotted curve).

manifold containing the attractor, Takens [14] showed rigorously that $m = 2D + 1$ is sufficient, in principle, to determine the dimension of the attractor. Note the ambiguity in determining the delay time used in (1).

Once the phase space representation is found, we construct a model to fit the data. We assume time to be discrete and express the dynamics by the map

$$\vec{X}(t + T) = \vec{f}_T[\vec{X}(t)] \tag{2}$$

for the present state $\vec{X}(t)$ and future state $\vec{X}(t + T)$. Here both $\vec{f}_T$ and $\vec{X}$ are m-dimensional vectors and $\vec{f}_T$ is the nonlinear map that propagates $\vec{X}$ T-units forward in time. To estimate $\vec{X}(t + T)$, the problem of interest, we approximate the dynamics of the map $\vec{f}_T$ by

$$\hat{\vec{X}}(t, T) = \hat{\vec{f}}_T[\vec{X}(t)] \tag{3}$$

where the carets denote approximants. The accuracy of the forecast is measured here using

$$E = \frac{|\hat{\vec{X}}(t, T) - \hat{\vec{X}}(t + T)|}{\sigma_x} \tag{4}$$

where $\langle E \rangle = 0$ for perfect forecasts and $\langle E \rangle = 1$ for forecasts that are no better than using $\hat{\vec{X}}(t, T) = \langle \vec{X}(t) \rangle$. Note that (4) measures how well $\hat{\vec{f}}_T$ approximates $\vec{f}_T$. This error estimate requires a data sample $\vec{X}(t)$ and its prediction. This error is

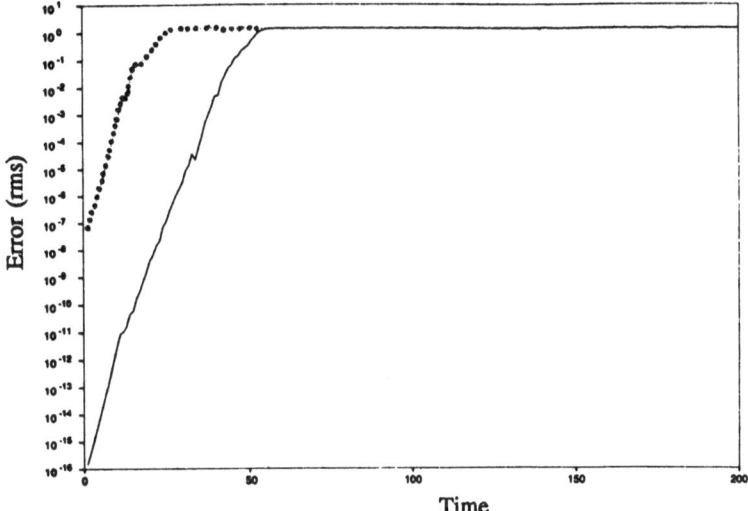

Fig. 5   The prediction error is graphed as a function of prediction time for single precision (dotted curve) and double precision (solid curve) for logistic data to determine the effect of rounding error.

defined relative to the standard deviation of the entire data stream σ_x. The average of E over a sample $\vec{X}(t)$ approximates $\langle E \rangle$ in (4).

Of course to construct the nonlinear functions $\vec{f}_T$ one must choose a representation since there are an *infinite* number of ways to select the nonlinear mapping function. As Farmer and Sidorowich point out, in the absence of any theoretical understanding, finding a good representation is largely a matter of trial and error. They themselves use a polynomial for the nonlinear map. Due to the deterministic nature of chaos, locally a smooth map can be approximated by a polynomial through a Taylor expansion. The parameters in such a representation can be linearly fit to minimize the least square sequence deviations, i.e., yields a linear least square problem. Nearly 50 years ago Wiener [15] suggested the use of polynomials for forecasting residuals in random time series, i.e., the differences between the data and the prediction $\vec{X}(t) - \hat{\vec{X}}(t)$ referred to as the moving average (MA) model. The most general form of the nth degree m-dimensional polynomial is

$$A_n(x_1, \ldots, x_m) = \sum_{j_1=0,\ldots,j_m=0} a_{j_1,j_2,\ldots,j_m} x_1^{j_1} x_2^{j_2} \ldots x_m^{j_m}, \qquad (5)$$

where $j_1 + j_2 + \ldots + j_m \leq n$. The number of parameters $a_{j_1,j_2,\ldots,j_m}$ is

$$\frac{(m+n)!}{m!n!} \approx m^n. \qquad (6)$$

Fitting these many parameters with a data set rapidly becomes impractical when m and n are large, but is convenient when they are not.

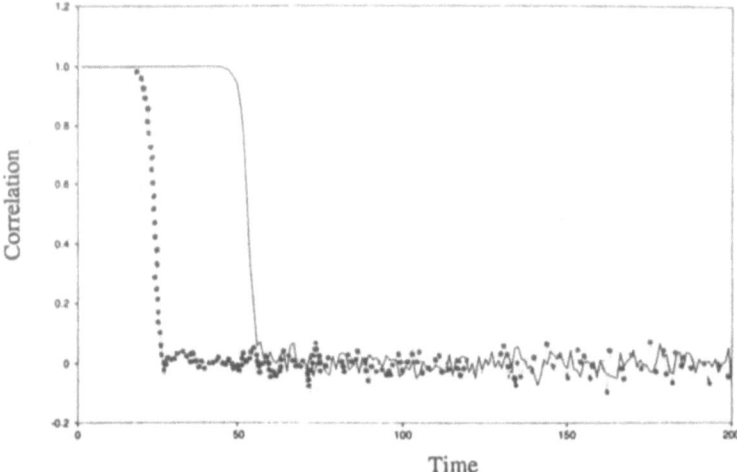

Fig. 6   The correlation of the generated logistic data as a function of the prediciton time is graphed for both the single precision (dotted curve) and double precision (solid curve) data.

A well-behaved function $\vec{f}_T$ can be easily modelled using any of a number of representations provided the data stream is sufficiently long. For a complicated function, one having many variations in a short time interval, it is unclear that any representation can provide an adequate approximation to $\vec{f}_T$. The dependence on representation may be reduced by means of a *local approximation* in which the domain is segmented into local neighbourhoods and the parameters, in a polynomial representation say, are fit in each neighborhood separately. Local approximations are found by Farmer and Sidorowich to produce better fits for a given number of data points than global approximations, particularly for larger data sets. This result has been verified by Casdagli [16].

Here we applied the *local approximation* (LA) developed by Farmer and Sidorowich (FS) [12] to a number of mappings. As shown by FS the scaling behavior of the prediction error $\langle E \rangle$ with the number of data points shown in Fig. 1 is predictable. Consider an nth order polynomial approximant. Ideally the error then depends on the $(n+1)$st derivative of $\vec{f}_T(\vec{x})$, so that if ϵ is the spacing between data points, the error is proportional to ϵ^{n+1}. If the data points are uniformly distributed throughout the volume then $\epsilon \sim N^{-1/d}$ as they do for a fractal where d is the fractal dimension of the space and N is the number of data points. Thus, the rms prediction error grows as

$$\langle E \rangle \sim N^{-q/d} \tag{7}$$

where $q = n+1$. In Fig. 1 $\log \langle E \rangle$ is graphed against $\log N$, so that the slope of the curve should be $-(n+1)/d$. It is found that this is indeed the case. Thus the scaling of the prediction error in itself provides a technique for determining whether a

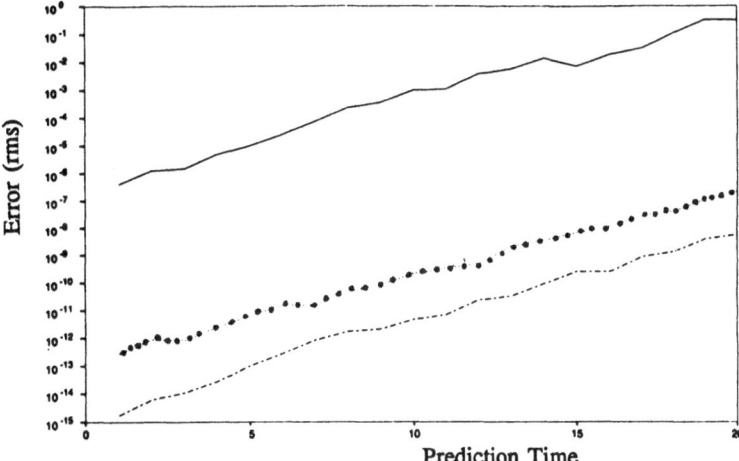

Fig. 7 The prediction error is graphed as a function of prediction time using the local approximation with updated neighbors for a linear polynomial (solid curve), a quadratic polynomial (dotted curve) and a cubic polynomial (dot-dash curve) for logistic data.

given time series is noise or chaos. Again this scaling result was independently verified [16].

There is a second important dependence of the prediction error on the forecasting parameters, this is the prediction time T. The rate at which the errors grow depends on the manner in which we make predictions; either *directly* or *iteratively*. A direct forecast is made by fitting a new model for each individual T. An iterative one is made by fitting a model to $T = 1$ and iterating to make predictions for $T = 2, 3 \ldots$. Intuitively one would guess that the direct model would be more accurate, since each model is «tailored» for the time it is supposed to predict, and there is no accumulation of errors due to iterations. FS show that this is in fact not always the case. They estimate that for direct forecasting $\langle E \rangle \sim N^{-q/d} e^{qT\lambda_{max}}$ whereas for iterative forecasting $\langle E \rangle \sim N^{-q/d} e^{T\lambda_{max}}$ where λ_{max} is the largest Lyapunov exponent. Thus the exponential growth of errors is a factor q larger in the former case over the latter. They argue that the superiority of iterative estimates comes from the fact that this scheme makes use of the regular structure of the higher iterates. In addition, they also give a numerical example showing the superiority of the direct over the iterative forecasting.

2.2 Local Approximation Results

We implement the local approximation (LA) technique and forecast the data sequence generated by the logistic map:

$$f(X_n) = 4X_n(1 - X_n). \tag{8}$$

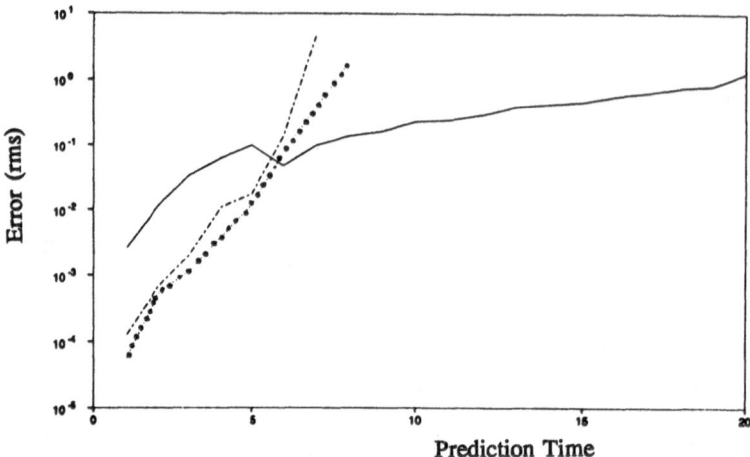

Fig. 8 The prediction error is graphed as a function of prediction time for the Hénon map. The forecast uses the local approximation with updated neighbors with a linear polynomial in two dimensions (solid curve), three dimensions (dotted curve) and four dimensions (dot-dash curve).

We wish to examine a number of properties of the forecasting schemes using the logistic map. Firstly, we check the prediction (8) and verify in Figure 2 that the error does grow exponentially with time, when a linear LA is used. We implement two ways of forecasting the data in time (here time is given by the discrete integers of the iterations). The first procedure for predicting all $X_{n+1}, X_{n+2}, \ldots$, etc., is based on the neighbors of X_n. The second method is to predict X_{n+j} based on the neighbors of X_{n+j-1}, i.e., we update the neighbors along with the prediction. We call the latter the LA with updated neighbors (LAU). Our numerical results shown in Figure 2 indicate that updating neighbors reduces the prediction error significantly. This is indicated by the reduction in the slope of the prediction error curve.

In Figure 3 the error for LA is again determined as a function of prediction time T, and the growth of error for two different dimensions for the reconstructed data vector is compared. Here it is clear that increasing the dimension from one to two increases the rate at which the error grows. If the forecast is updated at each time step the dependence on the number of nearest neighbors used in update does not appear to have a strong effect on the rate of error growth [cf. Figure 4]. There is also a significant dependence of the error growth rate on the precision of the calculation. In Figure 5 we compare the effects of rounding errors in generating the data from the map using both single and double precision. This gives quantitatively the typical length of the prediction time of the logistic map. It is apparent that double precision increases the predictable time by nearly an order of magnitude. However, the error is not the only measure available to assess the quality of the

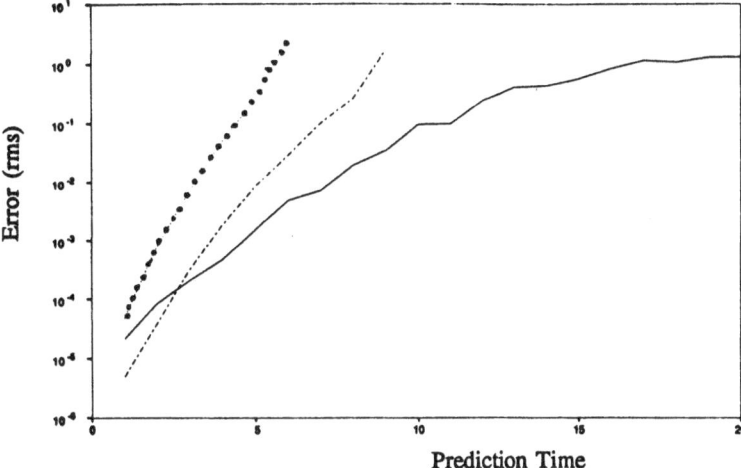

Fig. 9 The prediction error is graphed as a function of prediction time for the logistic map. The forecast uses a local approximation with updated neighbors with a linear polynomial in one dimension (solid curve), two dimensions (dot-dash curve) and three dimensions (dotted curve).

forecast. The correlation between the data and the forecast is also a good measure:

$$\rho(T) = \frac{\langle(\hat{f}_T - < \hat{f}_T >)(f_T - < f_T >)\rangle}{\left[\langle \hat{f}_T - < \hat{f}_T >^2\rangle\langle f_T - < f_T >^2\rangle\right]^{1/2}} \tag{9}$$

This procedure is equivalent to an auto-regressive (AR) model which mimics the original time series in terms of its mean, variance, and the autocorrelation function for multiple time delays.

In Figure 6 the correlation function plotted as a function of prediction time for both the single precision and double precision data. It is clear that the suppression of numerical error increases the prediction time by nearly a factor of two. The precipitous drop in the correlation function after a prolonged interval of nearly perfect correlation ($\rho = 1$) is characteristic of chaotic time series. This observation was also made by Shugihara and May [17], who also point out that the correlation for white noise is flat as a function of prediction time.

Here again we use the logistic map as the generator of our «experimental data» and determine the effects of modeling the forecasting function with polynomials of various degrees. In Figure 7 we show the result of using a linear, a quadratic and a cubic polynomial for $\hat{f}(X_n)$. It is clear that the latter two polynomials do about 10^6 better in error level than does the linear forecast. Further the quadratic function is 10^2 better than the cubic, as it should be since the logistic map is a quadratic polynomial. Indeed we observe the error in quadratic forecasting results is of the same order as that of iterating the logistic map with double precision

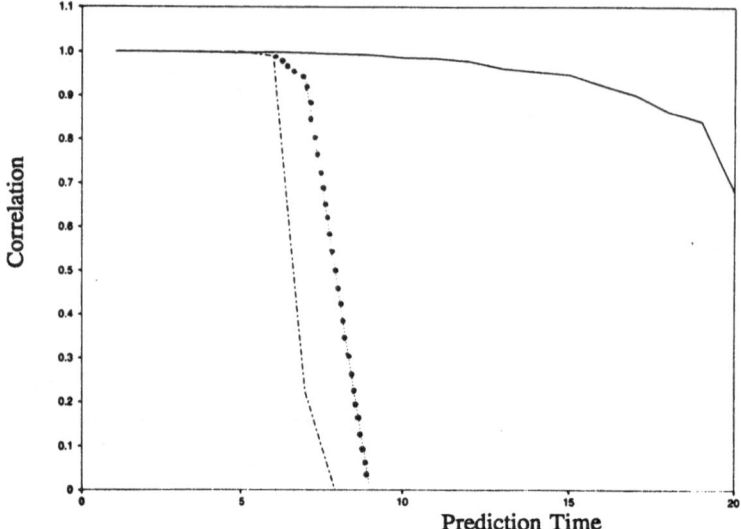

Fig. 10 The correlation between the forecast and the Hénon data as a function of the prediction time is graphed for the reconstructed data vector in two dimensions (solid curve), three dimensions (dotted curve) and four dimensions (dot-dash curve). A local approximation with updated neighbors is used for the prediction.

shown in Figure 5. Thus we should be able to reproduce these data exactly with the quadratic forecast function.

Let us now consider a data stream generated by the Hénon map

$$y_{n+1} = 1 - ay_n^2 + by_{n-1} \tag{10}$$

with the parameter values $a = 1.4$ and $b = 0.3$. Note that the Hénon map can be written either as a two-dimensional map with the mapping function relating (y_n, x_n) to (y_{n+1}, x_{n+1}) or as the one-dimensional map (10) with y_{n+1} related to the two values y_n and y_{n-1}.

In Figure 8 we depict the growth of error for reconstructed Hénon data vectors of two, three and four dimensions. Here we use a linear forecast with the minimum number of neighbors. It is clear from this figure that a linear forecast with updates at each prediction time gives the best results for two dimension, compared with the error for a higher dimension forecast which is greater than unity. A similar behavior is observed in Figure 9 where the error growth for the logistic data vectors of one, two and three dimensions. Here it is clear that the higher dimensions eventually yield arbitrarily large errors, whereas the error for the one-dimensional vector stays below unity. The correlation function for the predictions using the Hénon data shown in Figure 10 has the sudden decrease observed earlier for the logistic data. Here the two-dimensional forecast remains above .75 for 20 time steps into the future, whereas the three and four dimensional forecasts both fall

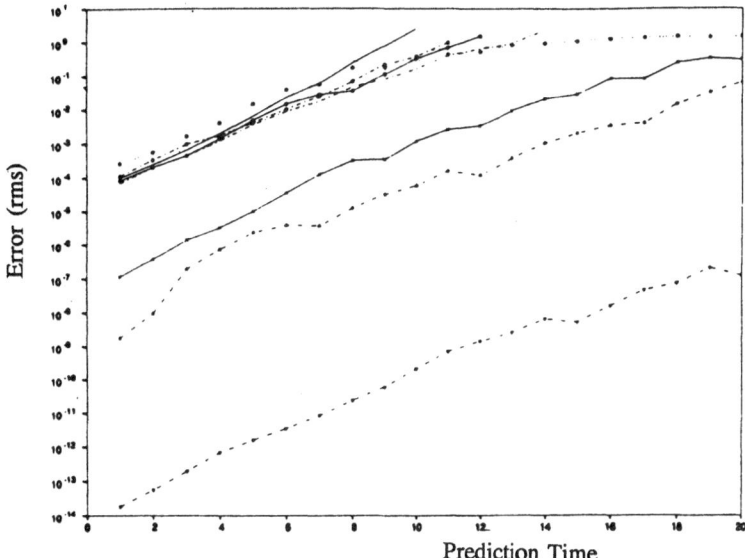

Prediction Time

Fig. 11 The root mean square prediction error is plotted as a function of prediction time for logistic
data. The most significant curves are those using the NN (x), the LA for a quadratic function
using the nonlinear least squares (– △ –) and the LA for a quadratic function using linear
least square (– · – – – –). The upper cluster of curves are for various combinations of linear
and nonlinear solvers using the local approximation.

sharply before $T = 9$, thereby indicating the two-dimensional nature of the Hénon
map.

2.3 Neural Network Results

In this section we compare the results obtained by means of determining the
forecasting function using a neural net [18,19] and that using the polynomial
techniques of the preceding section. It is useful to bear in mind that the neural
network technique always involves a least squares determination of parameters in
the numerical algorithm. Although it is true that a linear least square problem can
be solved exactly to within machine precision, a neural net is a nonlinear least
squares problem which can not be solved exactly in general. In Figure 11 the
rms prediction error is compared between forecasts of logistics data using LA and
those using a neural net (NN). We see that all the forecasts using either linear
or nonlinear least square solvers for a linear forecast are about equally bad and
substantially worse than that produced by the NN. However, when the quadratic
approximation is implemented in the forecast function we see that linear least
square solver does substantially better than either the NN or the nonlinear least
square solver, the latter two being about the same with NN or a little worse.

A similar result is obtained for the Hénon data as we show in Figure 12. Here
the rms error is seen to be significantly reduced from the LA linear polynomial

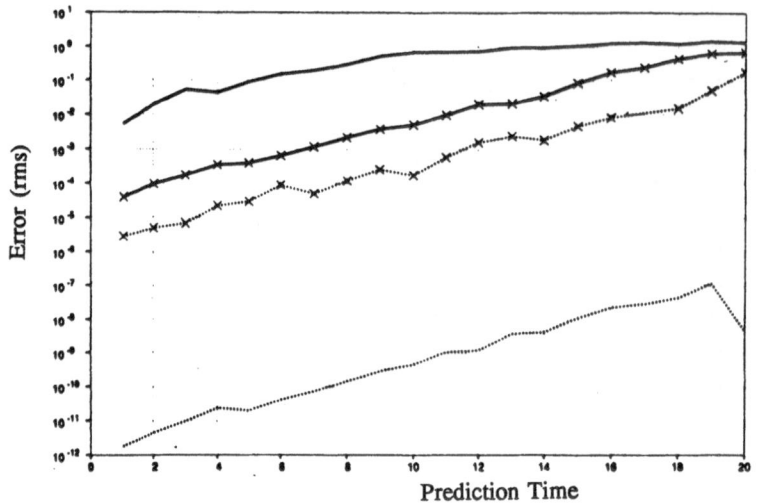

Fig. 12 The root mean square prediction error is plotted as a function of the prediction time for data
from the Hénon map using LA and a linear polynomial (—), LA and a quadratic polynomial
(...), NN with a tolerance in the nonlinear solver 10^{-3} (– × –) and NN with a tolerance
10^{-6} in the nonlinear solver (··· × ···).

approximation to the NN result, and reduced again from the NN result to the LA
with a quadratic polynomial. The two neural net curves result from increasing the
required tolerance from 10^{-3} to 10^{-6} in the NN training algorithm, i.e. nonlinear least square solver. This decreases the rms error by an order of magnitude,
however the NN error remains six orders of magnitude above that of the LA
quadratic polynomial. This might suggest that the NN does not work as well as
the polynomial techniques. However such a conclusion would be premature since
we have only tested its utility on polynomial maps and we must broaden the base
of application before deciding.

Another way of determining the utility of a given method is its robustness
in the presence of noise. The question is whether the NN techniques are more
sensitive to noise than are the LA procedure. The figure of merit we use here is
the signal-to-noise ratio (SNR) which is twenty times the logarithm to the base ten
of the ratio of rms signal strength to the rms noise strength i.e., $10\log_{10}(\sigma_s^2/\sigma_N^2)$
where σ_s^2 and σ_N^2 are the variances of the signal and noise, respectively. In Figure 13 we show the average relative error on an ensemble of ten members for
the logistic data with zero-centered Gaussian white noise added of the indicated
strength. In this figure the forecasts are made with the LA method, whereas in
Figure 14 the same data is forecast using the NN. It is clear that both procedures
are equally poor when the noise has the same rms strength as the signal. However not only is the rate of growth of the average relative error faster for the LA
method than for the neural net, but the overall level of the error is two orders of

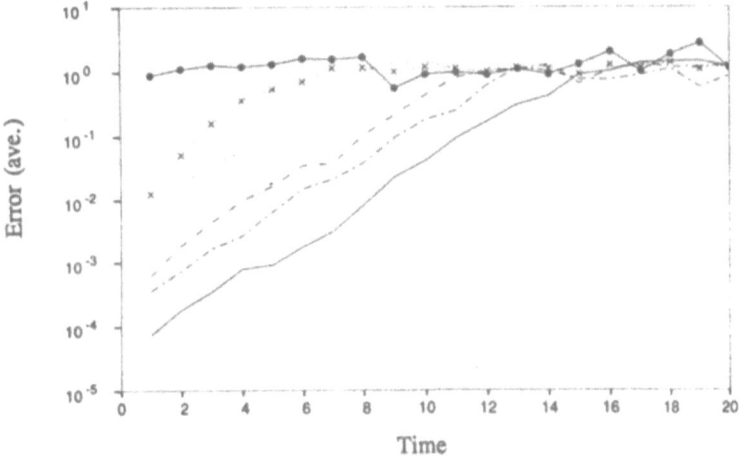

Fig. 13 The average relative prediction error is plotted as a function of prediction time for the LA forecasts of logistic data with Gaussian white noise. The SNR is: 90 (—), 80 (– - –), 70 (– · –); 50 (· · ·); 40 (· · × · ·) and 0 (- - · - -). Here we need 10 members in the ensemble each of which had 10^3 data points in the time series.

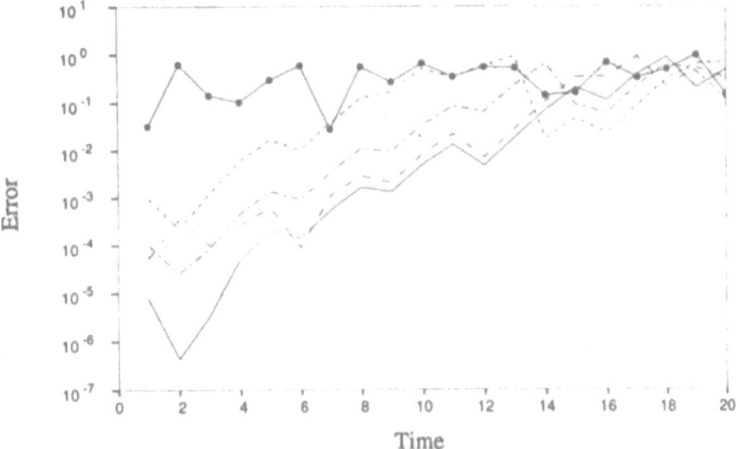

Fig. 14 Same data as for Figure 13 but using the NN forecasts and SNR's: 85 (- - -); 80 (– - –); 65 (– · –); 60 (- - -), 45 (. . .) and 0 (– · –).

magnitude smaller in the latter than in the former for high noise. From this and other calculations we conclude that the NN is less sensitive to the noise than are the polynomial forecasting techniques.

The correlation dimension of these two dynamic processes can be obtained using the Grassberger-Procaccia method. In Figure 15a we plot the slope of the

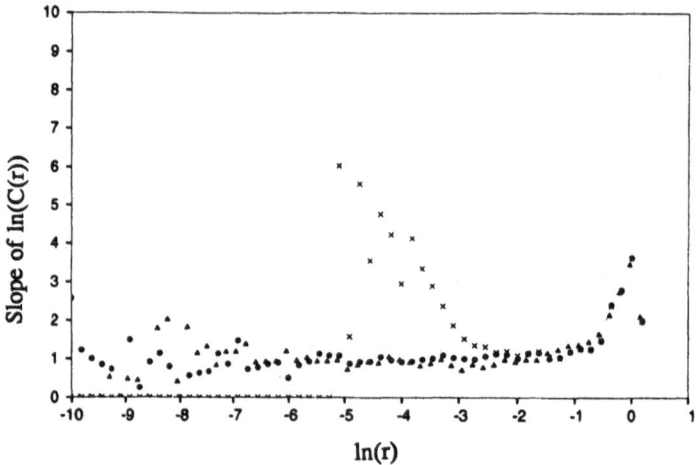

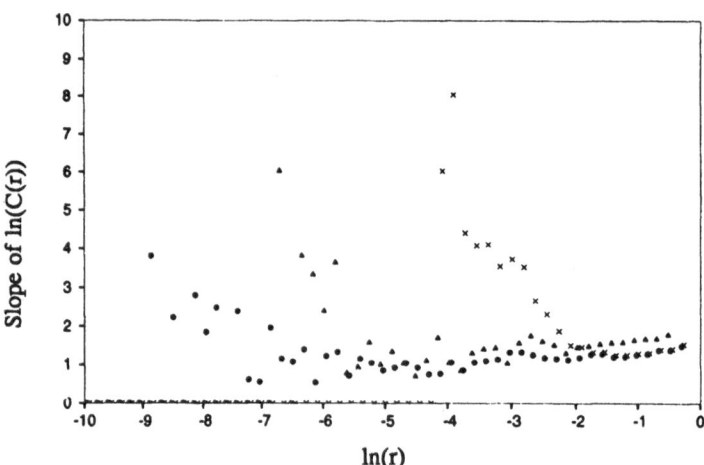

Fig. 15 The slope of the logarithm of the correlation function [ln $C(r)$] is plotted versus the logarithm of the separation between trajectories on the attractor [ln r]. (a) logistic data with Gaussian white noise superimposed having an SNR of 30 dB is given by the crosses; the noise-free logistic data is given by the filled circles and the noise data processed through a neural net is given by the triangles. (b) The same as (a) using a Hénon data stream.

logarithm of the GP-correlation function versus the logarithm of the separation (r) between trajectories on the reconstructed attractor. The flat region $-5 \leq \ln r \leq -1$ indicates the correlation dimension of the noise-free logistic data. The dimension obtained from the logistic data plus noise with a 30 dB SNR using the neural net coincides with the noise free case in this region of r. It is clear that one can not obtain the dimension of the time series directly from the data. This result carries down to a SNR of 10 dB without change. A SNR below 10 dB begins to restrict the region of lnr over which the NN yields the proper correlation dimension.

We again follow the above procedure using a data stream generated by the Hénon map. In Figure 15b the slope of $\ln C(r)$ is plotted versus lnr and we again see a relatively flat region for $-5 \leq \ln r \leq -1$ for the noise-free data. The noise contaminated signal again does not have a well defined dimension and the signal plus noise as processed by the NN yields a correlation dimension slightly above that obtained from the noise-free data. These results agree with the more detailed analyses of Albaro et al. [19]

Acknowledgement:
The support of this work by grant number N6-2269-92-C-0548 from the Naval Air Warfare Center is gratefully acknowledged.

References

[1] M. Casdagli and S. Eubank, editors, Nonlinear Modeling and Forecasting, Addison-Wesley, Redwood City, CA (1992).

[2] A.R. Osborne and A. Provenzale, Physica 35D, 357–381 (1989).

[3] B.J. West and H.J. Mackey, J. Appl. Phys. 69(9), 6747–6749 (1991).

[4] U. Frisch and R. Morf, Phys. Rev. 23A, 2673 (1981).

[5] H.S. Greenside, G. Ahlers, P.C. Hohenberg and R.W. Walden, Physica 5D, 322 (1982).

[6] D. Sigeti and W. Horsthemke, Phys. Rev. 35A, 2276 (1987).

[7] D. Ruelle, Phys. Rev. Lett. 56, 405 (1986).

[8] F.T. Arecchi and F. Lisi, Phys. Rev. Lett 49, 94 (1982).

[9] M.A. Rubio, M. De La Torre and J.C. Antoranz, Physica 36D, 92 (1989).

[10] H. Frijisaka and T. Yamada, Prog. Theor, Phys. 74, 918 (1985).

[11] B.J. West, H.J. Mackey and D. Chen, «Methods for distinguishing chaos from colored noise», in Patterns, Information and Chaos in Neuronal Systems, ed. B.J. West, World Scientific, Singapore (1993).

[12] J.D. Farmer and J.J. Sidorowich, Phys. Rev. Lett. 59, 845 (1987).

[13] J.D. Farmer and J.J. Sidorowich, Los Alamos preprint (1988).

[14] F. Takens, in Lecture Notes in Mathematics 898, eds. D.A. Rand and L.S. Young, 366–381 (1981).

[15] N. Wiener, Time Series, MIT Press (1949).

[16] M. Casdagli, Physica 35D, 335 (1989).

[17] G. Sugihara and R.M. May, «Nonlinear Forecasting as a way of distinguishing chaos from measurement error in time series,» Nature 344, 734 (1990).

[18] A.S. Lapedes and R. Farber, Los Alamos Technical Report LA-UR-87-2662.

[19] A.M. Albano, A. Passamente, T. Hediger and M.E. Farrell, «Using neural nets to look for chaos,» Physica D 58, 1–9 (1992).

Fractal Landscapes in Physiology & Medicine: Long-Range Correlations in DNA Sequences and Heart Rate Intervals

C.-K. Peng[1),2)], S.V. Buldyrev[2)], J.M. Hausdorff[1)],
S. Havlin[2),3)], J.E. Mietus[1)], M. Simons[1),4)],
H.E. Stanley[2)] and A.L. Goldberger[1)]

1) Cardiovascular Division, Harvard Medical School, Beth Israel Hospital, Boston, MA 02215, USA

2) Center for Polymer Studies, Boston University, Boston, MA 02215, USA

3) Department of Physics, Bar-Ilan University, Ramat-Gan 52100, Israel

4) Biology Department, MIT, Cambridge, MA 02139, USA

Abstract. Healthy systems in physiology and medicine are remarkable for their structural variability and dynamical complexity. The concept of fractal growth and form offers novel approaches to understanding morphogenesis and function from the level of the gene to the organism. For example, scale-invariance and long-range power-law correlations are features of non-coding DNA sequences as well as of healthy heartbeat dynamics. For cardiac regulation, perturbation of the control mechanisms by disease or ageing may lead to a breakdown of these long-range correlations that normally extend over thousands of heartbeats. Quantification of such scaling alterations are providing new approaches to problems ranging from molecular evolution to monitoring patients at high risk of sudden death.

We briefly review recent work from our laboratory concerning the application of fractals to two apparently unrelated problems: DNA organization and beat-to-beat heart rate variability. We show how the measurement of long-range power-law correlations may provide new understanding of nucleotide organization as well as of the complex fluctuations of the heartbeat under normal and pathologic conditions.

1 Long-Range Correlations in Nucleotide Sequences

Genomic sequences contain numerous «layers» of information. While the means of encoding some of these instructions is understood (for example, the codes directing amino acid assembly and intron/exon splicing, etc.), relatively little is known about other kinds of information encrypted in the DNA molecule. In higher eukaryotic organisms, only a small portion of the total genome length is actually used for protein coding. The role of introns and the intergenomic sequences that constitute a large portion of these DNA polymers remains unknown.

Recently we [1] proposed a novel method for studying the global organizational properties of genomic sequences by constructing a 1:1 map of the sequence onto a «DNA walk». Consider a one-dimensional walker [2] dictated by the sequential order of nucleotides. The walker steps up $[u(i) = +1]$ if a pyrimidine occurs at position a linear distance i along the DNA chain, while the walker steps down $[u(i) = -1]$ if a purine occurs at position i. The question we ask is whether such a walk displays only short-range correlations (as in an n-step Markov chain)

[3] or long-range correlations (as in critical phenomena and other scale-free «fractal» phenomena).

This DNA walk provides a novel graphical representation for each DNA sequence and permits the degree of correlation in the nucleotide sequence to be directly visualized (Fig. 1). A useful quantity that measures the degree of the correlation is obtained by calculating the «net displacement» $y(n)$ of the walker after n steps, which is the sum of the unit steps $u(i)$ for each step i,

$$y(n) \equiv \sum_{i=1}^{n} u(i). \tag{1}$$

A useful statistical description of any «landscape» can then be derived by considering a sliding window of size ℓ through the landscape and measuring the change of the «altitude» across this window, i.e.,

$$\Delta y_\ell(n) = y(n + \ell) - y(n), \tag{2}$$

where n indicates the starting position of the window. We define the fluctuation measurement, $F(\ell)$, as the standard deviation of the quantity Δy_ℓ.

The calculation of $F(\ell)$ can distinguish three possible types of behavior. (i) If the nucleotide sequence were random, then the landscape has the same statistical properties as that generated by a *normal* random walk, i.e., $F(\ell) \sim \ell^{1/2}$ (ii) If there were a local correlation extending up to a characteristic range (such as in Markov chains), then *the behavior $F(\ell) \sim \ell^{1/2}$ would be unchanged from the purely random case* (for $\ell \gg 1$). (iii) If there is no characteristic length (i.e., if the correlation is «infinite-range»), then the fluctuations will be described by a power law

$$F(\ell) \sim \ell^\alpha, \tag{3}$$

with $\alpha \neq 1/2$ [4]. If $\alpha > 1/2$, it indicates persistent correlation, i.e., one type of nucleotide (purine or pyrimidine) is likely to be close to another of the same type. In contrast, $\alpha < 1/2$ indicates that the nucleotides are organized such that purines and pyrimidines are more likely to alternate («anti-correlation») [5].

The power-law form of Eq. (3) implies a self-affine (fractal) property in the DNA walk landscape. To visualize this finding, one can magnify a segment of the DNA walk to see if it resembles (in a statistical sense) the overall pattern. Fig. 1(a) shows the DNA walk representation of a gene and Fig. 1(b) shows a magnification of the central portion. Fig. 1(c) is the further magnification of a sub-region of Fig. 1b. Note the similar fluctuation behavior on all three different length scales.

We calculate α from the slope of double logarithmic plots of the mean square fluctuation $F(\ell)$ versus ℓ (Fig. 2). Measurement of this exponent for a broad range of representative genomic and cDNA sequences across the phylogenetic spectrum reveals that long-range correlations ($\alpha > 1/2$) are characteristic of intron-containing genes and non-transcribed genomic regulatory elements [1,6,7]. The

SELF-SIMILARITY OF DNA WALKS

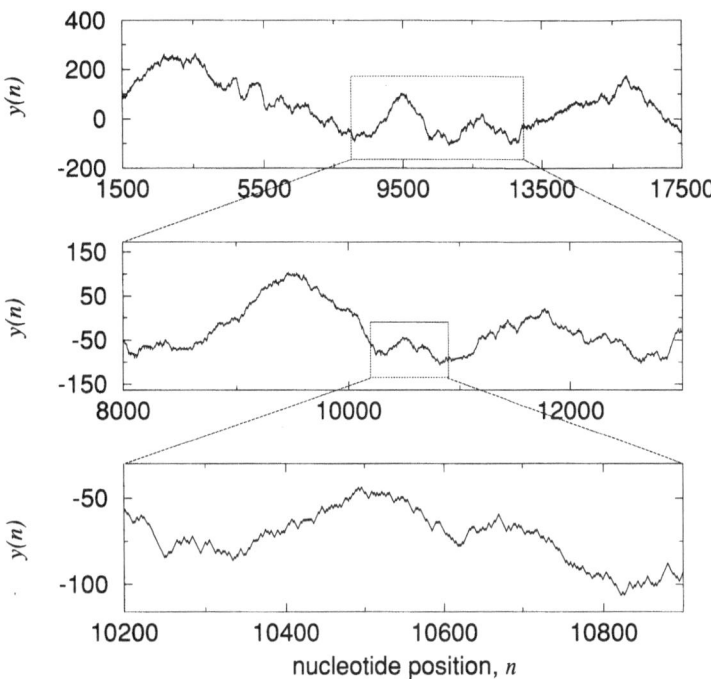

Fig. 1   The DNA walk representation for the rat embryonic skeletal myosin heavy chain gene. (a) The entire sequence. (b) The magnification of the solid box in (a). (c) The magnification of the solid box in (b). The statistical self-affinity of these plots is consistent with the existence of a scale-free or fractal phenomenon termed a fractal landscape. In order to observe statistically similar fluctuations within successive enlargement, the magnification factor along the vertical direction ($M_\perp$) and horizontal direction ($M_\parallel$) follows a simple relation: $\log M_\perp / \log M_\parallel = \alpha$. Here $\alpha = 0.63$. Note that these DNA walk representations are plotted so that the end point has the same vertical displacement as the starting point.

finding of long-range correlations in intron-containing genes appears to be independent of the particular gene or the encoded protein — it is observed in genomic sequences as disparate as myosin heavy chain, beta globin, adenovirus and yeast chromosome III [1,8].

In contrast, for cDNA sequences (i.e., the spliced together coding sequences) and genes without introns, we find that $\alpha \cong 1/2$, indicating no long-range correlation (Fig. 2). In fact, the *lack* of long-range correlations in coding regions is not very surprising considering that an uncorrelated sequence can carry more information than a correlated sequence [1]. On the other hand, the existence of long-range correlations in the non-coding regions is paradoxical and suggests a new organizational role for so-called «junk DNA». Ongoing investigations are directed at studying the implications of these correlations for DNA structure and

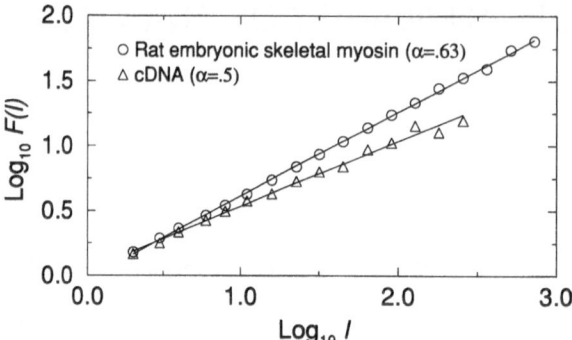

Fig. 2   Double logarithmic plot of $F(\ell)$ versus ℓ for rat embryonic skeletal myosin heavy chain gene shown in Fig. 1 (75% non-coding regions) and its cDNA. Note that the slope ($\alpha = 0.63$) for the intron-containing sequence is $> 1/2$ indicating the presence of long-range correlations. In contrast, the slope is 0.5 for the coding sequence (cDNA) indicating the absence of long-range correlations.

function, as well as for molecular evolution. Since power-law behavior represents a scale-invariant (fractal) property of DNA, it cannot be attributed simply to the occurrence of nucleotide periodicities such as those associated with nucleosome packaging. Whether these long-range correlations are related to higher order DNA/chromatin structure or to DNA bending and looping remains speculative.

A complementary approach to interpreting this correlation behavior is to relate it to the dynamic processes that modify nucleotide sequences over time. Buldyrev *et al.* [9] recently proposed a generalized Lévy walk model to account for the genesis of these correlations, as well as a plausible evolutionary mechanism based on nucleotide insertion and deletion [10].

From a practical viewpoint, the calculation of $F(\ell)$ for the DNA walk representation provides a new, *quantitative* method to distinguish genes with multiple introns from intron-less genes and cDNAs based solely on their statistical properties. The fundamental difference in correlation properties between coding and non-coding sequences also suggests a new approach to rapidly screening long DNA sequences for the identification of introns and exons [11].

2   Long-Range Correlations in Heart Beat Intervals

The healthy heartbeat is generally thought to be regulated according to the classical principle of homeostasis whereby physiologic systems operate to reduce variability and achieve an equilibrium-like state [12]. However, our recent findings [13] indicate that under normal conditions, beat-to-beat fluctuations in heart rate display the kind of long-range correlations typically exhibited by dynamical systems far from equilibrium. Since the heartbeat is under neuroautonomic control, our findings also imply that this feedback system is operating in a non-equilibrium state. Our results demonstrate that such power-law correlations extend over thousands of heart beats in healthy subjects. In contrast, heart rate time series from patients with severe congestive heart failure show a breakdown of this long-range correlation behavior, with the emergence of a characteristic short-range time scale. Similar alterations in correlation behavior may be important in modeling the transition from health to disease in a wide variety of pathologic conditions.

Clinicians traditionally describe the normal activity of the heart as «regular sinus rhythm». But in fact, rather than being metronomically regular, cardiac interbeat intervals normally fluctuate in a complex, unpredictable manner. Much of the analysis of heart rate variability has focused on short-term oscillations associated with respiration (0.15–0.40 Hz) and blood pressure control (0.01–0.15 Hz). Fourier analysis of lengthy heart rate data sets from healthy individuals typically reveals a $1/f$-like spectrum for lower frequencies (< 0.01 Hz), and some alterations in spectral features have been reported with a variety of pathologies. However, the long-range correlation properties of physiologic and pathologic heart rate time series had not been systematically described.

The mechanism underlying complex heart rate variability is related to competing neuroautonomic inputs. Parasympathetic stimulation decreases the firing rate of pacemaker cells in the heart's sinus node. Sympathetic stimulation has the opposite effect. The nonlinear interaction between these two branches of the nervous system is the postulated mechanism for the type of erratic heart rate variability recorded in healthy subjects (even during resting or sleeping hours), although non-autonomic factors may also be important.

Our analysis is based on the digitized electrocardiograms of beat-to-beat heart rate fluctuations over long time intervals (up to 24 h $\approx 10^5$ beats) recorded with an ambulatory (Holter) monitor. The time series obtained by plotting the sequential intervals between beat n and beat $n + 1$, denoted by $B(n)$, typically reveals a complicated type of variability. To quantitatively study these dynamics over large time scales, we pass the time series through a digital filter that removes fluctuations of frequencies > 0.005 beat^{-1}, and plot the result, denoted by $B_L(n)$, in Fig. 3. We observe a more complex pattern of fluctuations for a representative healthy adult (Fig. 3a) compared to the pattern of interbeat intervals for a subject with severe heart disease associated with congestive heart failure (Fig. 3b). These heartbeat time series produce a contour reminiscent of the irregular «landscapes» of DNA walks (Fig. 1).

To apply the previous fractal landscape analysis, we can make a simple mapping such that $B_L(n)$ for the heartbeat is equivalent to $y(n)$ for DNA. Thus we can measure the heartbeat fluctuation $F(\ell)$ the same way as in DNA walk, where ℓ indicates the size of the observational window (number of beats). Figure 4 is a log-log plot of $F(\ell)$ vs ℓ for the data in Figs. 3. This plot is approximately linear over a broad physiologically-relevant time scale ($\ell \sim 200$ to 4000 beats) implying that $F(\ell) \sim \ell^\alpha$.

We find that the scaling exponent α is markedly different for the healthy and diseased states: for the healthy heartbeat data, α is close to 0, while α is close to

CARDIAC INTERBEAT INTERVAL LANDSCAPES

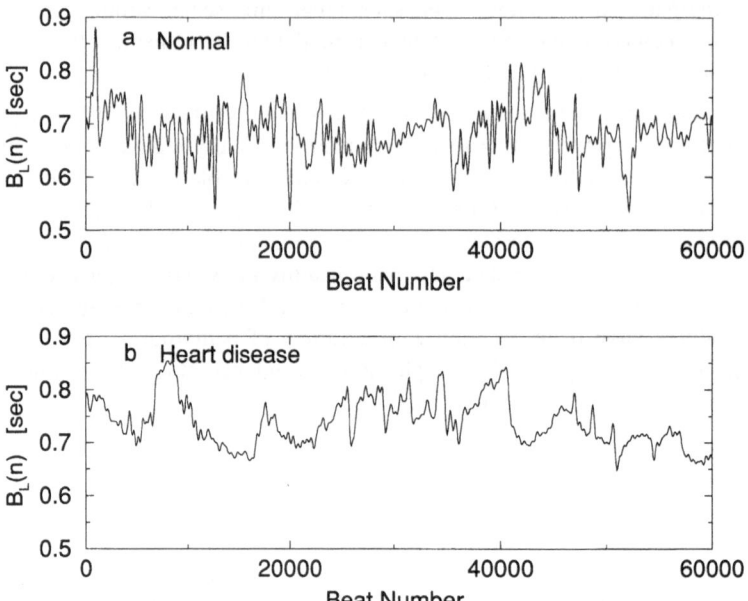

Fig. 3   The interbeat interval $B_L(n)$ after low-pass filtering for (a) a healthy subject and (b) a patient
with severe cardiac disease (dilated cardiomyopathy). The healthy heartbeat time series shows
more complex fluctuations compared to the diseased heart rate fluctuation pattern that is close
to random walk («brown») noise. The low-pass filter removes all Fourier components for
$f \geq f_c$. The results shown here correspond to $f_c = 0.005$ beat^{-1}, but similar findings are
obtained for other choices of $f_c \leq 0.005$. This cut-off frequency f_c is selected to remove
components of heart rate variability associated with physiologic respiration or pathologic
Cheyne-Stokes breathing as well as oscillations associated with baroreflex activation (Mayer
waves). After Ref. 13.

0.5 for the diseased case in this example. As we discussed previously, $\alpha = 0.5$
corresponds to a *random walk* (Brownian motion). Thus the low-frequency heartbeat fluctuations for the diseased state can be interpreted as a stochastic process,
in which case the interbeat increments $I(n) \equiv B(n+1) - B(n)$ (corresponding to
$u(n)$ in the DNA case) are uncorrelated for $\ell > 200$. For the healthy subject, the
interbeat increments are *anti-correlated* ($\alpha < 0.5$).

To study further the correlation properties of the time series, we choose
to study $I(n)$. Since $I(n)$ is stationary, we can apply standard spectral analysis
techniques [10]. Figures 5a and 5b show the power spectra $S_I(f)$, the square of
the Fourier transform amplitudes for $I(n)$, derived from the same data sets (without
filtering) used in Fig. 3. The fact that the log-log plot of $S_I(f)$ vs f is linear implies

$$S_I(f) \sim f^\beta. \tag{6}$$

The exponent β is related to α by $\beta = 1 - 2\alpha$ [5].

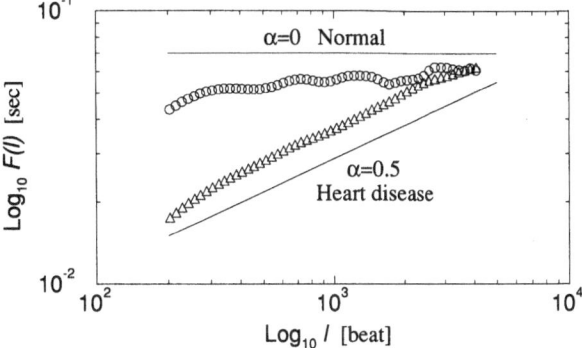

Fig. 4 Double logarithmic plot of $F(\ell)$ vs n. The circles represent $F(\ell)$ calculated from data in Fig. 3(a) and the triangles from data in Fig. 3(b). The two best-fit lines have slopes $\alpha = 0.07$ and $\alpha = 0.49$ (fit from 200 to 4000 beats). The two lines with slopes $\alpha = 0$ and $\alpha = 0.5$ correspond to «$1/f$ noise» and «brown noise,» respectively. We observe that $F(\ell)$ saturates for large ℓ (of the order of 5000 beats), because the heartbeat interval are subjected to physiological constraints that cannot be arbitrarily large or small.

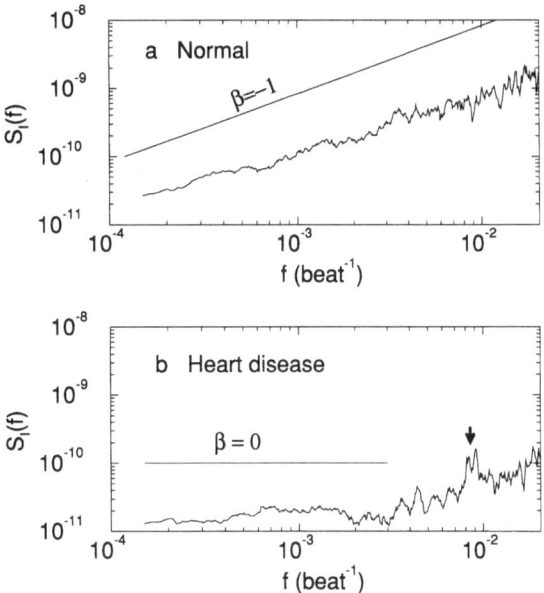

Fig. 5 The power spectra $S_I(f)$ for the interbeat interval increment sequences over ~ 24 hours for the same subjects in Fig. 3. (a) Data from a healthy adult. The best-fit line for the low frequency region has a slope $\beta = 0.93$. The heart rate spectrum is plotted as a function of «inverse beat number» (beat^{-1}) rather than frequency (time^{-1}) to obviate the need to interpolate data points. The spectral data are smoothed by averaging over 50 values. (b) Data from a patient with severe heart failure. The best-fit line has slope 0.14 for the low frequency region, $f < f_c = 0.005$ beat^{-1}. The appearance of a pathologic, characteristic time scale is associated with a spectral peak (arrow) at about 10^{-2} beat^{-1} (corresponding to Cheyne-Stokes respiration). After Ref. 13.

For the data set from the patient with severe heart disease, we observe a flat spectrum ($\beta \simeq 0$) in the low frequency region (Fig. 5b) confirming that $I(n)$ are not correlated over long time scales (low frequencies). Therefore, $I(n)$, the first derivative of $B(n)$, can be interpreted as being analogous to the *velocity* of a random walker, which is uncorrelated on long time scales, while $B(n)$ — corresponding to the *position* of the random walker — are correlated. However, this correlation is of a trivial nature since it is simply due to the summation of uncorrelated random variables.

In contrast, for the data set from the healthy subject (Fig. 5a), we obtain $\beta \simeq 1$, indicating *non-trivial* long-range correlations in $B(n)$ — these correlations are not the consequence of summation over random variables or artifacts of non-stationarity. Furthermore, the «anti-correlation» properties of $I(n)$, indicated by the positive β value, are consistent with a nonlinear feedback system that «kicks» the heart rate away from extremely high or low values. This tendency, however, does not only operate on a beat-to-beat basis (local effect) but over a wide range of time scales, a fractal property of cardiac regulation.

We [13] analyzed data from two different groups of subjects: 10 adults without clinical evidence of heart disease (age range: 32–64 years, mean 44) and 10 adults with severe heart failure (age range: 22–63 years; mean 54). Data from patients with heart failure due to severe left ventricular dysfunction are likely to be particularly informative in analyzing correlations under pathologic conditions since these individuals have well-defined abnormalities in both the sympathetic and parasympathetic control mechanisms that regulate beat-to-beat variability. Furthermore, such patients are at very high risk for sudden death. Both exponents (α and β) were significantly different between the diseased and normal groups [13].

Previous studies have demonstrated marked changes in short-range heart rate dynamics in heart failure compared to healthy function, including the emergence of intermittent relatively low frequency (~ 1 cycle/minute) heart rate oscillations associated with the well-described syndrome of periodic (Cheyne-Stokes) respiration, an abnormal waxing and wanning breathing pattern often associated with low cardiac output. This pathologic, *characteristic time scale* is indicated by a vertical arrow in Fig. 5b.

The long-range power-law correlations in healthy heart rate dynamics may be adaptive for at least two reasons [14–16]: (i) the long-range correlations serve as an newly described organizing principle for highly complex, non-linear processes that generate fluctuations on a wide range of time scales, and (ii) the lack of a characteristic scale helps prevent excessive *mode-locking* that would restrict the functional responsiveness (plasticity) of the organism. Support for these two related conjectures is provided by observations from severely pathologic states such as heart failure where the *breakdown* of long-range correlations is often accompanied by the emergence of a dominant frequency mode (e.g., the Cheyne-Stokes frequency). Analogous transitions to highly periodic behavior have been observed in a wide range of other disease states including certain malignancies,

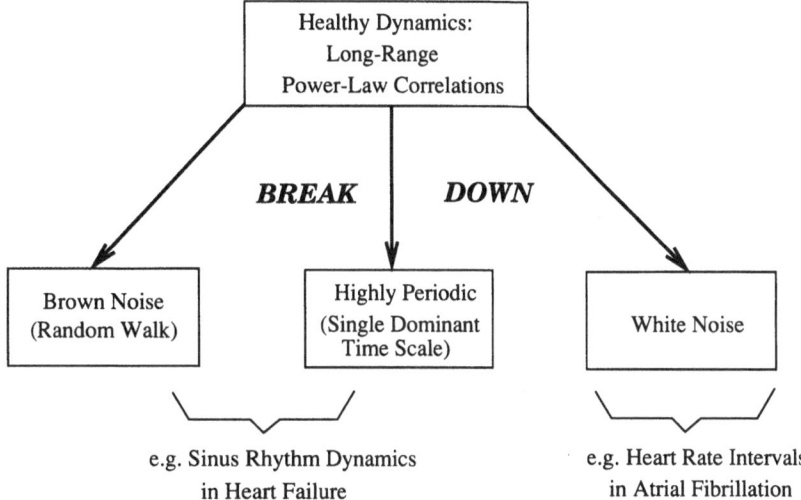

Fig. 6 The breakdown of long-range power law correlations may lead to any of three dynamical states: (i) a random walk («brown noise») as observed in low frequency heart rate fluctuations in certain cases of severe heart failure; (ii) highly periodic oscillations, as also observed in Cheyne-Stokes pathophysiology in heart failure, and (iii) completely uncorrelated behavior («white noise»), exemplified by the heart rate during atrial fibrillation.

sudden cardiac death, epilepsy, fetal distress syndromes and with certain drug toxicities [15, 16].

Important unanswered questions currently under study include: What are the physiological mechanisms underlying such long-range correlations in cardiac beat-to-beat intervals? Are these fluctuations entirely stochastic or do they represent the interplay of deterministic and stochastic mechanisms [17]? How do these findings relate to the suggestion that some features of normal heart rate variability are due to chaotic dynamics [17–20]?

From a practical viewpoint, these findings may have implications for physiological monitoring and in particular for cardiac rhythm analysis. The complete breakdown of normal long-range correlations in any physiological system could theoretically lead to three possible dynamical states (Fig. 6): (i) a random walk (brown noise), (ii) highly periodic behavior, or (iii) completely uncorrelated behavior (white noise). Cases (i) and (ii) both indicate only «trivial» long-range correlations of the types observed in severe heart failure. Case (iii) may correspond to certain cardiac arrythmias such as fibrillation. More subtle or intermittent degradation of long-range correlation properties may provide an early warning of incipient pathology, including an increased risk of sudden cardiac death. A breakdown of long-range correlations may also be an important marker of aging [21]. Finally, we observe that the long-range power-law correlations present in the healthy heartbeat imply that the underlying control mechanisms actually drive the system *away from* a single steady state. Therefore, the classical theory of homeostasis, according to which stable physiological processes seek to maintain «constancy» [12], and its more recently proposed modifications under the rubric of «homeodynamics», need to be revised and extended to account explicitly for this far from equilibrium behavior.

Acknowledgement:
We wish to thank C. DeLisi, F. Sciortino, M.M.E. Matsa and S.M. Ossadnik for help at various stages of this work, and AHA, CONACYT, Mathers Charitable Foundation, NIH, NIDA, NIMH, NSF, ONR and the US-Israel Binational Foundation for support.

References

[1] C.-K. Peng, S.V. Buldyrev, A.L. Goldberger, S. Havlin, F. Sciortino, M. Simons, and H.E. Stanley, *Nature* **356**, 168 (1992).

[2] E.W. Montroll and M.F. Shlesinger, «The Wonderful World of Random Walks», in *Nonequilibrium Phenomena II. From Stochastics to Hydrodynamics*, eds. J.L. Lebowitz and E.W. Montroll, pp. 1–121 (North-Holland, Amsterdam, 1984).

[3] S. Tavaré and B.W. Giddings, in *Mathematical Methods for DNA Sequences*, Eds. M.S. Waterman (CRC Press, Boca Raton, 1989), pp. 117–132.

[4] H.E. Stanley, *Introduction to Phase Transitions and Critical Phenomena*, pp. 120–121 (Oxford Univ. Press, London, 1971).

[5] S. Havlin, R.B. Selinger, M. Schwartz, H.E. Stanley and A. Bunde, *Phys. Rev. Lett.* **61**, 1438 (1988).

[6] C.-K. Peng, S.V. Buldyrev, A.L. Goldberger, S. Havlin, F. Sciortino, M. Simons, and H.E. Stanley, *Physica* **191**, 25 (1992).

[7] C.-K. Peng, S.V. Buldyrev, A.L. Goldberger, S. Havlin, M. Simons, and H.E. Stanley, *Phys. Rev. E* **47**, 3729 (1993).

[8] P.J. Munson. R.C. Taylor and G.S. Michaels, *Nature* **360**, 636 (1992).

[9] S.V. Buldyrev, A.L. Goldberger, S. Havlin, C.-K. Peng, H.E. Stanley, and M. Simons, preprint.

[10] S.V. Buldyrev, A.L. Goldberger, S. Havlin, C.-K. Peng, M. Simons, and H.E. Stanley, *Phys. Rev. E* **47**, 4514 (1993).

[11] H.E. Stanley *et al.*, preprint.

[12] W.B. Cannon, *Physiol. Rev.* **9**, 399 (1929).

[13] C.-K. Peng, J. Mietus, J.M. Hausdorff, S. Havlin, H.E. Stanley, and A.L. Goldberger, *Phys. Rev. Lett.* **70**, 1343 (1993), and references therein.

[14] B.J. West and A.L. Goldberger, *J. Appl. Physiol.*, **60**, 189 (1986).

[15] B.J. West and A.L. Goldberger, *Am. Sci.*, **75**, 354 (1987), and references therein.

[16] A.L. Goldberger, D.R. Rigney and B.J. West, *Sci. Am.* **262**, 42 (1990).

[17] D.R. Rigney *et al.*, preprint.

[18] D.R. Rigney, J.E. Mietus and A.L. Goldberger, *Circulation* **82** (Suppl. III), 236 (1990).

[19] J.E. Skinner, C. Carpeggiani, C.E. Landisman and K.W. Fulton, *Circ. Res.* **68**, 966 (1991).

[20] A.L. Goldberger, *NIPS (Int. Union Physiol. Sci./Am. Physiol. Soc.)* **6**, 87 (1991).

[21] L.A. Lipsitz and A.L. Goldberger, *J. Amer. Med. Assoc.* **267**, 1806 (1992).

Fractals in
Biological Design
and Morphogenesis

Design of Biological Organisms and Fractal Geometry

Ewald R. Weibel
Department of Anatomy
University of Bern
Bühlstrasse 26
CH-3000 Bern 9, Switzerland

Abstract. This paper explores the question whether fractal geometry forms an important basis for determining the design of biological organisms during morphogenesis. It is first observed that functionally important internal surfaces such as the intracellular membranes of the liver or the gas exchange surface of the lung have characteristics of fractal surfaces in that the surface area measured depends on the microscopic resolution. However, these surfaces are fractal within bounds and the sequential occurrence of different «generators» of surface texture may require several fractal regressions. It is secondly noted that airways and vascular trees show the properties of fractal trees and that this may offer significant functional advantages. The question is finally raised whether and to what extent fractal constructive algorithms may form part of genetic programming.

1 Introduction

It has become commonplace to note that many pictures generated by fractal algorithms are reminiscent of some of the beautiful structures found in biological organisms: trees, flowers, or animal structures alike. And it is easily noted that many of these structures are derived by growth from smaller and often simpler units, just as fractal algorithms generate progressively more complex structures. The fundamental underlying principle in fractal geometry is self-similarity and scale invariance of form as the structure develops. And clearly this is very often also an underlying principle in the development and growth of biological forms, as noted already by D'Arcy Thompson in 1942. Two well-known examples are evidence for this notion:

(1) the Nautilus, a large sea mollusc whose progression of chamber size follows a logarithmic growth law (Fig. 1); all chambers are self-similar, i.e. have the same proportions, because, with every growth step, the animal needs to add an air-filled floating chamber which is proportional to the weight gain of the animal. Since growth is proportional to attained size the growth curve is logarithmic and, in turn, becomes imprinted in the shell structure. This kind of structure also appears in many fractal patterns, e.g. at the periphery of the Mandelbrot set;

(2) the growth of trees, such as a beech, starts from small saplings whose stem grows in length and thickness and becomes enlarged by the addition of a new generation of twigs with each growth step. Fig. 2 compares a three years old sapling with a terminal twig taken from a branch of its parent tree, a beech about 100 years old; the similarity between sapling and twig is evident so that the growth of such a tree is clearly governed by self-similarity.

Fig. 1 Mid-plane section of a Nautilus shell showing the logarithmic spiral resulting from proportional growth of the air chambers.

One could easily come up with numerous examples of this kind. However, this is anecdotal evidence that seems to document the fractal nature of biological organisms; and one can, as so often, easily come up with anecdotal evidence that proves the contrary. Can we be more rigorous and more systematic in exploring whether fractal geometry serves an important template function in determining the design of living organisms?

2   The hypothesis of fractal morphogenesis of biostructures

Fractal geometry is conceived as a set of constructive algorithms that determine design by self-similarity of the structure at various scales, in other words by scale invariance of form. Such constructive algorithms lend themselves to the development of a theory of morphogenesis by proportional growth from which one can derive models that permit to test the *hypothesis* that *morphogenesis of biostructures follows fractal principles*. To approach this let us first examine two cases of morphogenesis in higher organisms which are of central importance for their good functioning:

(1) the formation of large internal surfaces in confined spaces, such as cellular membranes or the gas exchange surface of the lung;

(2) the design of branched vessel systems by which such large internal surfaces are efficiently accessed from central sources, examples being the vascular trees that distribute blood and blood-borne substances to the cells, or the airways which distribute oxygen to the large gas exchange area of the lung.

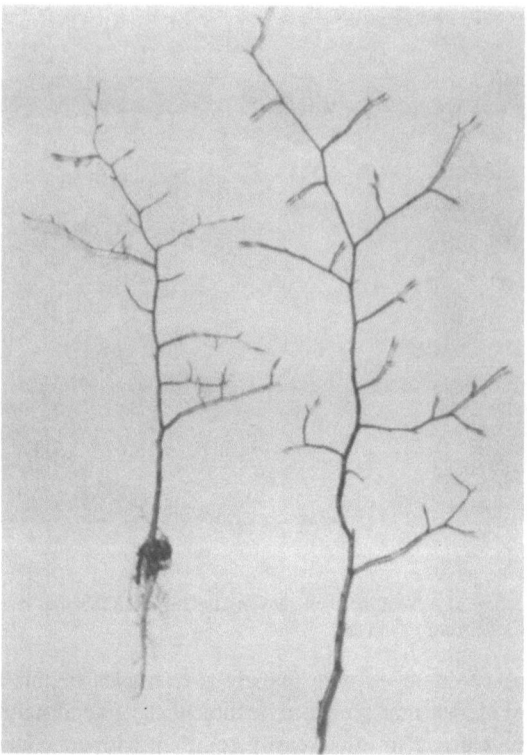

Fig. 2   Three years old sapling of a beech tree compared to an end twig of its nearly 100 years old parent tree.

3   Fractal nature of internal surfaces

3.1 First case study: cellular membranes

The discovery that cellular membrane systems have fractal properties arose from the uncertainty of observation about the extent of such membranes. In 1968 methods for measuring the surface area of cellular membranes on electron micrographs had been established (Weibel et al. 1966; Weibel 1979). The methods were simple as they were based on statistical sampling procedures derived from geometric probability of intersection of some test line probe with the membrane images observed on sections. They could be applied in studies in cell biology that wished to establish a relationship between some enzyme functions and the system of endoplasmic membranes, for example in liver cells (Fig. 3). When the first studies on the morphometry of liver cell membranes were published (Loud 1968; Weibel et al. 1969) the results did not match as we obtained much higher values than other groups (Table 1):

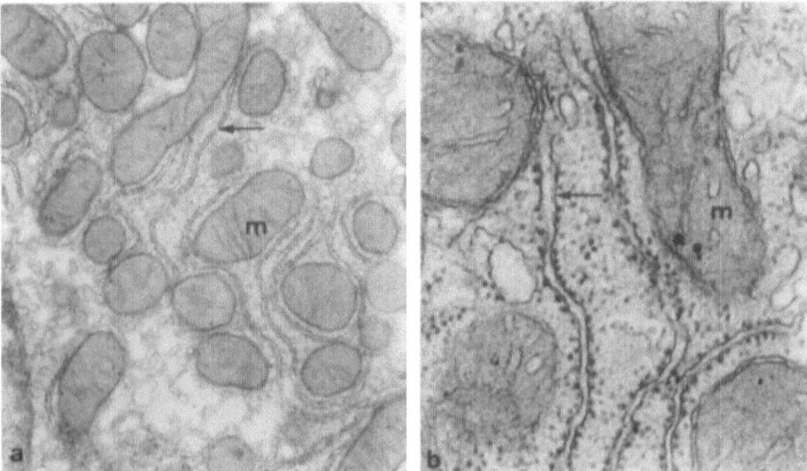

Fig. 3 Electron micrographs of liver cell cytoplasm at two different magnifications (a: 18'000 ×, b:58'000 ×) show complex of endoplasmic reticulum membranes (arrows) and mitochondria (m) with their membranes. Higher magnifications reveal greater details in the shape of membrane profiles.

Authors	Surface density of endoplasmic reticulum membranes $(m^2 \cdot cm^{-3})$	Magnification at which measurements were done
Loud (1968)	5.7	12'500
Weibel et al. (1969)	10.9	90'000

Table 1 Uncertainty in surface area estimates of endoplasmic membranes in rat liver measured by different authors.

Long debates followed about which of the estimates was correct, whether the liver cells contained 6 or 11 m^2 of membranes per cm^3, a quite significant difference, and whether the stereological methods used were reliable since it appeared possible that the same method may yield different results if the measurements were done at different magnifications of the electron micrographs (Table 1). The key event towards the resolution of this debate was the publication, in 1977, of Benoit Mandelbrot's book «Fractals: Form, Chance and Dimension». There he demonstrated, based on the work of Richardson, that the length of a coast line was indeterminate unless the fractal dimension and the yardstick were defined. Were we dealing with a «Coast of Wales Effect»? If so, measurements obtained at a higher magnification (Fig. 3) were bound to yield higher values than those done at lower magnification. We undertook a systematic study on measur-

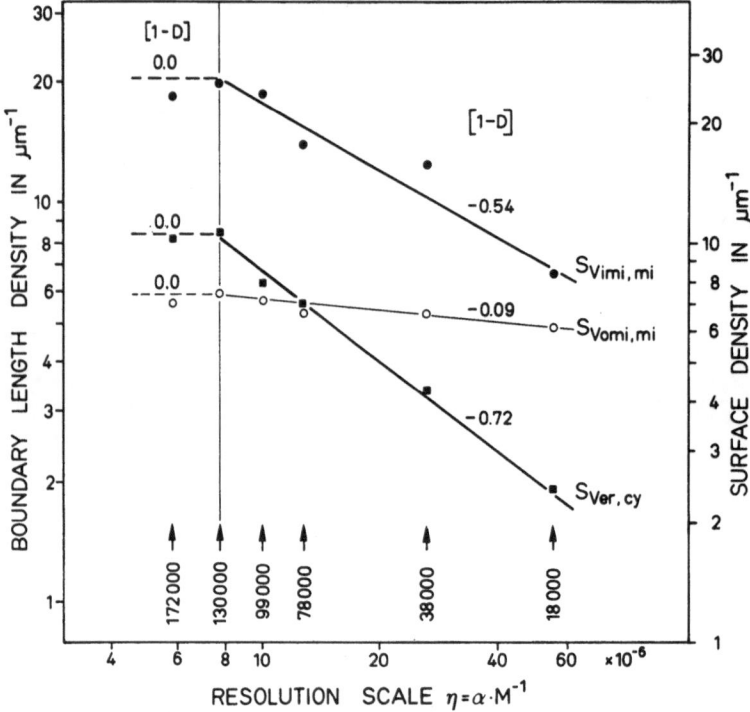

Fig. 4 Double logarithmic plot of boundary length density estimates (which stereology converts into surface density estimates) of membranes against resolution scale. Slope of regression is related to fractal dimension D. [From Paumgartner et al. 1981]

ing liver cell membranes at different magnifications and found that indeed the estimates of surface density increased with increasing resolution (Paumgartner et al., 1981). The test was the log-log plot of the data (Fig. 4) from which it followed that endoplasmic reticulum membranes were characterized by a fractal dimension of 2.7 (assuming that the slope for a membrane in 3D space is 2 - D), and that the fractal dimension of each type of cellular membranes was different. It was however interesting that in all cases there was an upper cut-off point at 130'000 × magnification beyond which the curve was flat. This suggested that there was a change in the nature of the membrane image at this resolution, and indeed this is the magnification at which the membrane becomes visible as a physical sheet of finite and measurable thickness. At this magnification the membranes change from fractal to deterministic structures, but if we could further increase resolution the surface of the sheets would again assume fractal properties and the surface area estimates would continue to rise, though possibly at a different slope, as we resolve surface molecules and their structure.

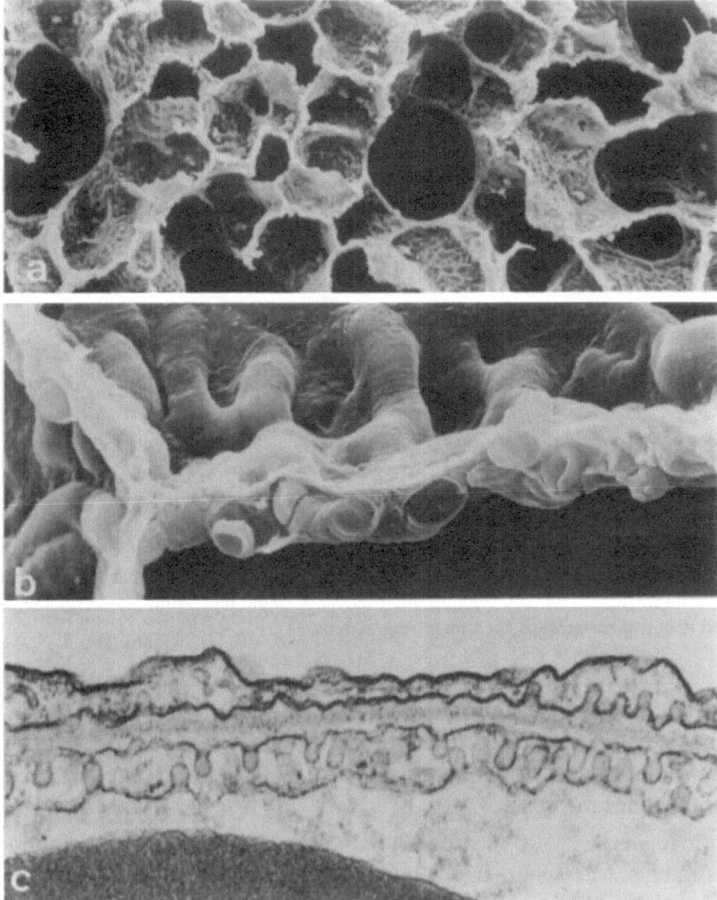

Fig. 5   Increasing microscopic resolution reveals a hierarchy of structures that form the inner lung surface: alveoli around the airways (A), capillaries in the alveolar walls with their imprints on the surface (B), and membrane folds of the epithelial cell forming the surface of the air-blood barrier. A and B are scanning electron micrographs magnified 75 × and 1'000 × respectively; C is a thin section electron micrograph magnified 46'000 ×.

3.2 Second case study: the gas exchange surface of the lung

A first look at the lung shows it to be similar to a foam made of some 300 million little bubbles, called alveoli, all open to the airways and densely covered by a network of blood capillaries (Fig. 5a + b). This is in the interest of efficient uptake of oxygen from the air which requires a large surface of contact to be established between air and blood. Just how large is this surface? Here again an uncertainty arose. The first estimates of alveolar surface area in the human lung, obtained around 1960 by light microscopy, yielded values of about 60 m^2 (Weibel and Gomez 1962; Weibel 1963), but when a few years later the elctron

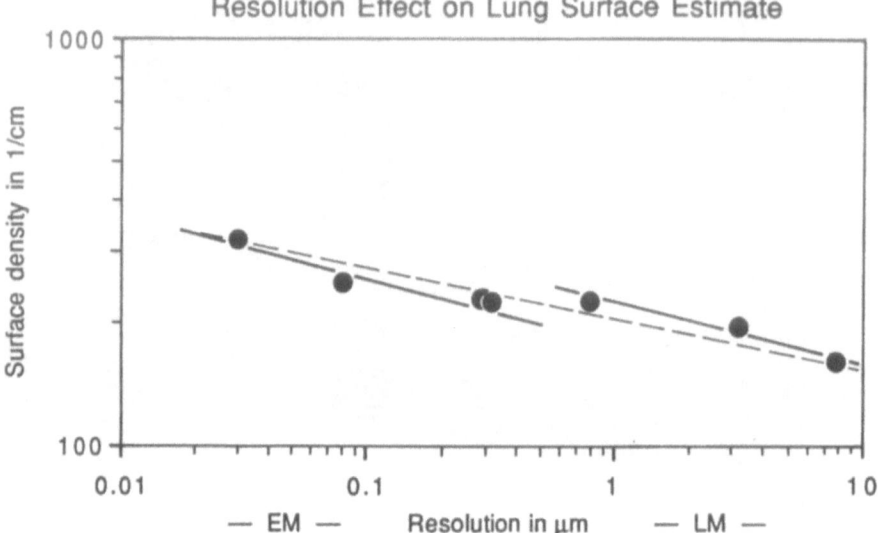

Fig. 6   Log-log plot of surface density estimates of inner lung surface obtained at different resolution
levels in the light (LM) and electron microscope. A single regression represents the data less
well than two (or perhaps even more) regressions.

microscope became available for such measurements I suspected that this must
be an underestimate of the «true» surface (Weibel 1964). Indeed, when in 1978
human lungs could be subjected to a thorough morphometric study we estimated
the alveolar surface area at 130 m^2 (Gehr et al. 1978). Is this again due to a fractal
nature of the inner lung surface?

The obvious test was to measure the alveolar surface area at increasing magnifications, and it was found that it increased with a slope corresponding to a
fractal dimension of 2.24 (Keller et al. 1978). However, there was a problem
because the successive structures that were resolved at increasing magnification
were of different nature (Fig. 5): at first we see «smooth» alveoli (Fig. 5a), then
we resolve the ridges that are imprinted by the capillaries in the thin alveolar
walls (Fig. 5b), and finally we see the fine wrinkles of the alveolar epithelial cell
membranes (Fig. 5c). Is it then justified to use a single fractal dimension as a
descriptor of surface complexity? We can hardly claim that alveoli, capillaries and
surface wrinkles are self-similar structures; each has its own «generator», is determined by its own «constructive algorithm» and, accordingly, we should search
for at least two or three self-similarity levels. This is indeed found if the log-log
regression is scrutinized (Fig. 6). We should note that Rigaut (1984) had found
the fractal dimension of the alveolar surface to change when he studied it by light
microscopy; his analysis was however limited to the lower half of the regressions
shown in Fig. 6.

In conclusion, we found that internal surfaces both at the level of cells and
organs have fractal properties, and that their full morphometric description is best

served by estimating their degree of complexity by determining a fractal dimension in addition to some metric estimates at a given yardstick or resolution. However, the fractal model must be used with caution and critically if the morphometric estimates are to be used in a functional context (Weibel 1991b). «Membranes» are physical sheets so that we will find a cut-off point on the log-log plots, and this may allow us to indicate a resolution at which a deterministic surface can be estimated, one which is related to specific structural entities and also makes functional sense. Furthermore, biological surfaces may be built of a hierarchy of non-selfsimilar elements, and it is of great functional significance to sort out the deterministic structures which serve as generators of surface texture. On the whole we can say that biological surfaces are fractal with bounds, and it may well be that the bounds are the critical points that the analysis has to reveal.

From the perspective of the cell or the organism we therefore find that functionally important interfaces show fractal structure although locally, in the functional domain, they are generated by the folding of deterministic sheets into a confined space. In the future, it may be heuristically very productive to ask the question whether the fractal design of interfaces brings metabolic or morphogenetic advantages to cells, organs, and organism.

4   Accessing large surfaces from a central source

The liver or the heart muscle contain a large number of cells which occupy the organ space rather homogeneously. They must all be supplied with oxygen and substrates to perform their functions and this is served by the blood vessels. Likewise, the alveolar surface is densely and quite homogeneously arranged within the lung, i.e. the area of 130 m^2 — nearly the size of a tennis court — is finely crumpled into a space of about 5 liters, forming some 300 million alveoli (Weibel 1963, 1984). The capillary network which is contained in the walls between alveoli (Fig. 5) must be perfused with blood as evenly as possible by pumping the blood from the heart through the arteries to the capillaries and then collecting it into veins that lead it back to the heart. Some alveolar surface elements are close to the heart, others are far away, so the functional problem is to divide the blood stream efficiently to allow blood cells to reach all points of the large alveolar surface evenly and in approximately the same time span. In addition, all alveoli must be efficiently supplied with fresh oxygen from each inspired volume of fresh air, and the goal must be to achieve a well-matched ventilation and perfusion of the large number of gas exchange units.

These problems are solved by designing the blood vessels and the airways as trees that have their roots in the heart or in the upper airways, nose and mouth, respectively, and extend with their twigs to the units that must be supplied, cells in the heart muscle or liver, and alveoli with their capillaries in the lung (Fig. 7).

We had noted above that many botanical trees have the properties of fractal structures and that the underlying self-similarity is the result of morphogenesis:

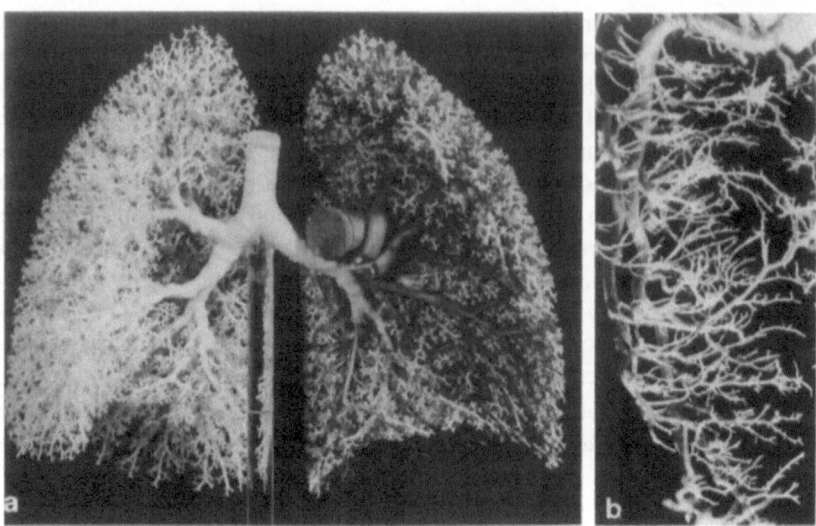

Fig. 7   Resin cast of the airways and blood vessels in the human lung (a) and of coronary arteries
in the heart muscle (b) reveal their tree structure based on dichotomous branching.

they grow and branch systematically from the stem to the twigs (Fig. 2). The
morphogenesis of the lung's airway tree follows much the same principles (Weibel
1984). It begins in the young fetus by formation of the stem, the trachea, and
two bronchial buds that extend sideways into a pad of mesenchyme, primitive
connective tissue, which contains a network of early blood vessels. The bronchial
buds are blind-ending epithelial tubes whose wall is a compact layer of cells. They
grow in length, and at a certain point the terminal cap of epithelium forms a Y-
shaped tip which initiates the formation of two branches which grow in length and
then split at their tips, each into two branches, and so on, a morphogenetic process
called dichotomy. Thus with each branching the number of end twigs is doubled:
2, 4, 8, 16, 32 ..., until, after 23 generations, some 8 million terminal airways are
formed, the total number found in the normal human lung (Weibel 1963). As this
branching goes on, all airway tubes grow proportionately in length and diameter,
just as in a beech tree (Fig. 2), so that in the final tree the dimensions of airways
decrease regularly from the central bronchi to the peripheral branches. Because
branching occurs in all directions the result is a tree whose end twigs fill the entire
space of the lung (Fig. 7a).

Fig. 8a shows a fragment of a cast of a human airway tree; it is quite evident
that, inspite of irregularities, the branching pattern is similar at all levels from
larger to smaller branches. Is this a fractal tree? In his books on *Fractal Objects*
Benoit Mandelbrot (1977; 1983) has constructed «Koch trees» which have the
properties to fill, with their tips, the space homogeneously. The generator was a T-
or Y-shaped element with fixed proportions in the length and diameter of the legs.

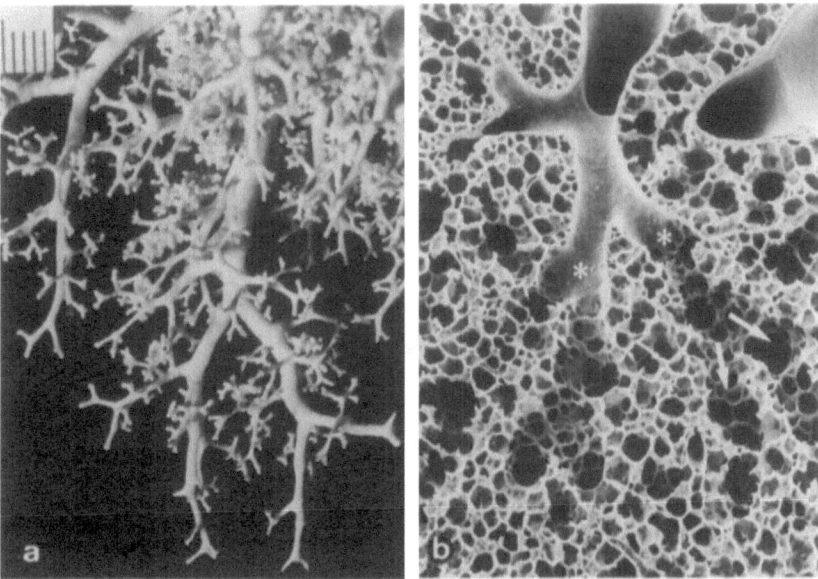

Fig. 8   Peripheral small airways shown in a resin cast (a) and in a scanning electron micrograph (b) which shows the transition of conducting airways (asterisk) to alveolar ducts (arrows). The largest airway in (b) corresponds about to the smallest end twigs in (a).

The tree was self-similar because these proportions were maintained throughout the branching process. These Koch trees resemble the pattern of airway branching. To test whether the airways of the human lung conform to such fractal models we need to test whether the proportions between length and diameter within one generation, and from generation to generation are maintained. Length and diameter of airways vary considerably (Fig. 8a); but when such measurements are averaged for each generation we find that the proportions are constant. The length-to-diameter ratio is invariant at 3.25, and the branching ratios (the reduction factor with each successive generation) are 0.86 for diameter and 0.62 for length, irrespective of the generation (Weibel 1963). These findings support the notion that the airway tree has the basic properties of self-similarity. The variation in the proportions of individual branches is the result of the necessity to reach all points in a space that is determined by the shape of the chest cavity (Nelson and Manchester 1988), and this also explains part of the differences in the branching pattern between different species (Nelson et al. 1990). Along another line of reasoning it has been argued that the airway tree is fractal if the progression of dimensional change, diameters or lengths, of the segments follows a power law when plotted against generation of branching (West et al. 1986). This assumes that the «yardstick» or scale is reduced by a constant factor from one generation to the next. Fig. 9 shows a plot of the average airway diameter against generation for the human lung. West et al.

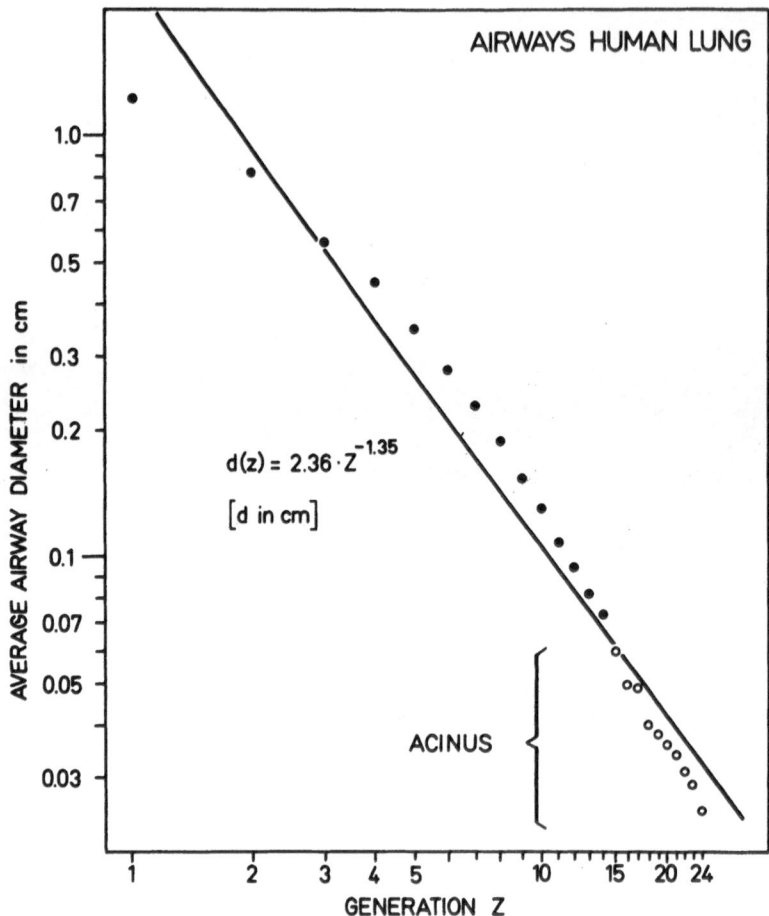

Fig. 9   Log-log plot of airway diameters measured in human lungs against the generation of branching. [From Weibel 1991]

(1986) have concluded from this kind of plot that the airway tree can be described
as a fractal tree, but that a harmonic variation is superimposed which is similar
for all species examined, except that there is a phase shift in the human lung with
respect to that of other species (Nelson et al. 1990).

   The dimensional changes in the airway tree can also be interpreted differently.
In our original analysis of the human airway tree (Weibel and Gomez 1962) we
asked the bioengineering question whether the dimensions of the successive generations were optimized for minimal work of maintaining air flow. The total work is
determined by the flow resistance which increases as the dimensions decrease and
the work required to move the mass of air which increases as dimensions increase
(Wilson 1967). It is well known in hydrodynamics that the total work of flow in

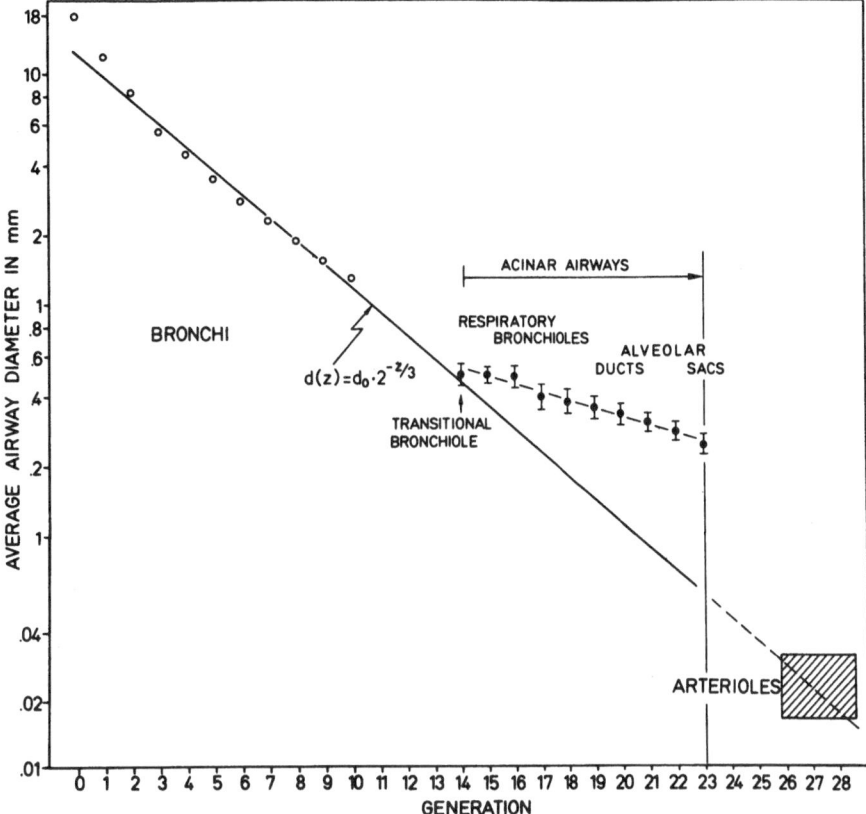

Fig. 10 Exponential plot of the same data as in Fig.9 reveals different regressions for conducting airways (open circles) and alveolated airways (closed circles) in the acinus as gas exchange unit (Fig.8b). The plot is extended to show the size of the arterial end branches which are located in generations 26–28.

a branched system of pipes is minimal if the diameter of the pipes is reduced by the cube root of one half, or $2^{-1/3}$, at each dichotomous branching. The system of pipes that distribute gas in a city are designed according to that law. D'Arcy Thompson (1942) was first to notice that this rule often applied with respect to biological vascular systems. When we plotted our measurements of airway diameter against the generation of branching we found that the first 14 generations follow this law (Fig. 10): the diameter in generation z is $d(z) = d_0 \cdot 2^{-z/3}$. The last generations do not follow the same line, but we will see that they serve their function in a different way.

Mandelbrot (1983) discusses this result in terms of fractal geometry and notes that in a space-filling tree the diameter exponent of the sequence of branches should be close to 3 and that accordingly the diameter branching ratio should be $2^{-1/3}$,

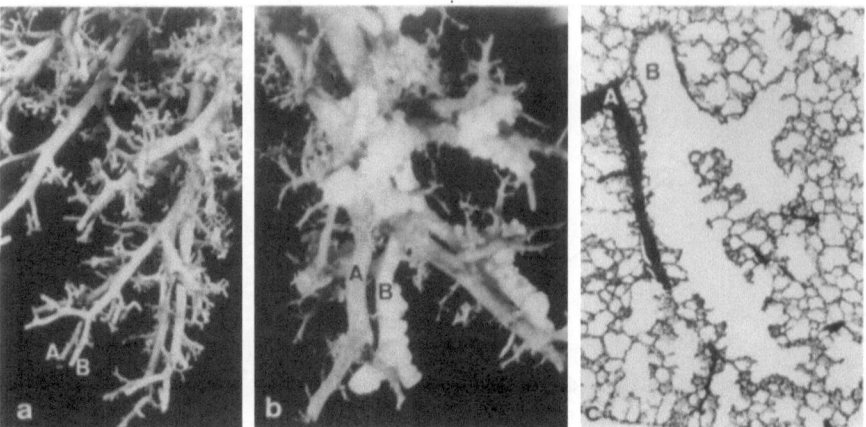

Fig. 11 The relations between pulmonary arteries (A) and bronchi (B) show that they run in parallel
and that their diameters are similar. When arteries penetrate into the acinus (compare Fig.8b)
they branch more frequently than the airways so that their diameter decreases more rapidly,
as shown in Fig.10. (a) and (b) are resin casts of human lungs magnified 3 × and 8 ×,
respectively; in the light micrograph of a rabbit lung (c), magnified 25 ×, the blood vessels
have been highlighted by black stain of the blood plasma.

as observed. Thus, even according to this analysis the human airway tree has the
properties of a fractal tree, at least as far as the airways are simple air conducting
tubes.

Beyond generation 14 the nature of the airways changes (Fig. 8b): their wall
is formed by alveoli which accommodate the large gas exchange surface around
the about eight generations of terminal branchings within the gas exchange units
called acinus (Haefeli-Bleuer and Weibel 1988). The volume of these airways
becomes greatly expanded, and even the central air ducts around which the sleeve
of alveoli is arranged takes on a larger cross-section than predicted from the $2^{-1/3}$
rule (Fig. 10). The functional explanation for this change is that, within the acinar
airways, air convection is slowed down and the oxygen transfer to the large gas
exchange surface is accomplished to a significant part by molecular diffusion of
oxygen within the air phase. This process is greatly improved if the crossection
of the diffusion front is as large as possible; the observed slower reduction of
airway diameter appears to strike an optimal balance between the conditions for
convective mass flow of air and oxygen diffusion within the air phase. Mandelbrot
(1977) argues that this feature furthermore forms good conditions for what he
calls the «alveolar inner cut-off» of a space-filling tree. Thus we conclude that the
airway system of the human lung is well designed, both in terms of solving its
physiological functions (Wilson 1967) as in terms of its morphogenetic properties,
and that both are linked to the airways' fractal nature.

Finally, a few brief remarks on the design of the pulmonary vascular trees,
arteries and veins, which develop in concert with the airways. Consider the pul-

monary arteries: they follow the course of the airways, and their diameters are very nearly matched to those of the associated airways (Fig. 11a and b). When arteries penetrate into the acinus additional branches are formed, i.e. the arteries branch over a total of about 28 generations, on average, before they reach their endbranches that discharge blood into the capillary network on the alveolar surface (Fig. 11b and c). When arterial diameters are plotted against generations, as in Fig. 10, we find that they are reduced according to the $2^{-1/3}$ rule all the way out to the endbranches (Weibel and Gomez 1962); their design thus appears governed by the requirements for optimized flow conditions. The functional importance of these design principles has recently been established by a very thorough analysis of the pulmonary arterial tree as a fractal continuum model (Krenz et al. 1992). Arteries that supply tissues such as heart muscle show a very similar design (Fig. 7b). Mandelbrot (1983, chapter 17) discusses this in terms of fractal design and notes that the diameter exponent should be somewhat smaller than that of the airways, and that this is confirmed by empirical data (Suwa and Takahashi 1971). It has indeed been shown long before the advent of fractal geometry that arteries and veins form trees along the same principles (Hess 1903).

We therefore arrive at the conclusion that airways as well as blood vessels are constructed like fractal trees. Their design principle is that of a hierarchy of self-similar tubes which develop in morphogenesis by a well balanced succession of branching and growth. The functional result of this design is that all points on the gas exchange surface of the lung are at nearly the same distance from the source of fresh air and blood: the main central airways and the heart, respectively. Furthermore, the design of such a tree allows air and blood flow through the tubes to remain laminar throughout such that the work of flow is minimal; but the «mixing» of air and blood at the gas exchange surface is just as efficient as if the two media were mixed by turbulent flow: it can be said that in the design of airway and vascular trees the turbulence has become «frozen into structure».

The fundamental question that may need to be addressed in future research is whether the apparent fractal nature of airways and vessels in the lung is the result of interactions at the physical level during morphogenesis, that is whether it is simply the expression of an adaptive malleability of structures to imposed stresses, the result of an attraction-repulsion equilibrium (Sernetz et al. 1988); or whether the fractal design is predetermined, programmed in the genetic instructions for design, including the maintenance of strict proportionality. The latter alternative has a high degree of likelihood because much of the basic morphogenesis of the lung occurs before birth, that is before the lung is subjected to the functional stresses imposed by gas exchange between the organism and the environment. It may be provocative to postulate that the organism may use «fractal genes» for efficient programming of some of its many hierarchical structures, such as airways and blood vessels.

5   The design of bio-organisms

We have looked at cells and their membranes, where most metabolic functions take place, and found that they have fractal properties, at least within certain bounds. We have looked at the lung as the vital supplier of oxygen, and have found fractal properties in the gas exchange surface as well as in the airways and vessels that supply it with air and blood. We have also noted that the blood vessel systems that supply the cells with oxygen and substrate are designed in much the same way; they are fundamental design elements of the organism as a whole, elements that connect the units, bind them into an integral system in the functional sense. Such observations provide strong support to the hypothesis that fractal geometry is a fundamental design principle for biological organisms. Although it may not be at the basis of all morphogenetic processes fractal geometry still appears to be a determinant of many of the key elements of animal design.

This, finally, leads to the consideration of whole animal design. Animals come in all sizes — from the smallest mammal, the Etruscan shrew that weighs 2 grams to the elephant of 5 tons — and, although their basic design is similar, many or most of the parameters that determine their function and design are non-linearly related to body size. This is particularly true for metabolic functions (Kleiber 1961; Schmidt-Nielsen 1984; Weibel and Taylor 1981; Weibel at al. 1992). Oxygen consumption rate per unit body mass falls with the 1/4 power of body mass so that a mouse has a six times greater mass-specific metabolic rate than a cow, and heart rate also falls with the 1/4 power of body mass. The surface area of mitochondrial inner membranes, on which oxidative energy metabolism takes place, varies in proportion to oxygen consumption and is thus non-linearly related to body size: in accordance with functional demand mitochondrial membranes are more densely packed in the cells of small animals than in those of large animals. Accordingly, the design of the vascular systems must be adjusted to supply a denser capillary network with adequate blood flow (Hoppeler and Kayar 1988). Such observations have led Sernetz et al. (1989) to postulate that metabolism is organized on fractal principles, and Spatz (1991) to speculate that this is related to the necessity to design the blood vessels as fractal trees. That this may not be as simple is seen in the fact that design is generally determined by well balanced combinations of structural and functional parameters, and that such vital structures as the pulmonary gas exchanger may be designed with considerable safety factors (Weibel et al. 1992).

6   Concluding remarks

What can we conclude from this analysis which focussed on internal surfaces and the tree structures that access them from central sources? In each case we have found several elements that suggested fractal design. How much of this is fact, and how much is fiction? Clearly, we «proved» the fractal nature of some of these design properties by log-log regressions. But is this enough?

With respect to the functional significance of these design properties we concluded (a) that the packing of surfaces into cells and organs is best achieved by using fractal structures, and (b) that to build vascular trees, airways, and even botanical trees, without fractal principles does not lead to intelligent solutions. But this was simply obtained by looking for analogies or similarities in the structure of fractal models and in real biostructures. This coincidental evidence is interesting, but it becomes heuristically productive only if we can take the analysis one step further to ask whether morphogenesis is, indeed, governed by fractal principles, be it at the genetic or epigenetic level. This proof has still to be provided.

There is a fundamental likeness between fractals and genes that determine biostructures: they are both «constructive algorithms» that create structures with precise connectivity and proportionality, while still allowing for (epigenetic) variation under external and internal influences, or imposed stresses. This, therefore, is the fundamental provocative question linking fractals to biological design: *do genes contain fractal algorithms?* Do we not have to suspect this to be true? Consider the ubiquitous occurrence of some basic structures of the Mandelbrot set, such as the logarithmic spiral: we find it in snails, molluscan shells, or the developing fern, and this fundamental structure also appears to underlie the model for airway trees. But this is no more than speculation; it becomes valid only if it is formulated into a coherent research strategy which can be tested against the facts of nature.

Acknowledgement:
This work has been supported by grants from the Swiss National Science Foundation Nos. 31-30946.91 and 31-28610.90, and by a grant from the Maurice E. Müller Foundation. I gratefully acknowledge the help received in the preparation of this paper by Karl Babl, Elsbeth Hanger, and Barbara Krieger.

References

[1] Gehr P., M. Bachofen, and E.R. Weibel (1978) The normal human lung: ultrastructure and morphometric estimation of diffusion capacity. Respir. Physiol. 32: 121–140.

[2] Haefeli-Bleuer B., and E.R. Weibel (1988) Morphometry of the human pulmonary acinus. Anat. Rec. 220: 401–414.

[3] Hess W.R. (1903) Eine mechanisch bedingte Gesetzmässigkeit im Bau des Blutgefässsystems. Arch. Entwickl. Mech. Org. 16: 632–641.

[4] Hoppeler H., and S.R. Kayar (1988) Capillarity and oxidative capacity of muscle. New Physiol. Sci. 3: 113–116.

[5] Keller Hj., H.P. Friedli, P. Gehr, M. Bachofen, and E.R. Weibel (1975) The effects of optical resolution on the estimation of stereological parameters. Proc. Fourth Internat. Congr. Stereology. Gaithersburg, 1975. National Bureau of Standards, Spec. Publ. 431: 409–410.

[6] Kleiber M. (1961) The Fire of Life. New York, Wiley.

[7] Krenz G.S., J.H. Linehan, and C.A. Dawson (1992) A fractal continuum model of the pulmonary arterial tree. J. Appl. Physiol. 72-2225-2237.

[8] Loud A.V. (1968) A quantitative stereological description of the ultrastructure of normal rat liver parenchymal cells. J. Cell Biol. 37:27.

[9] Mandelbrot B. (1977) Form, Chance, and Dimension. New York, Freeman.

[10] Mandelbrot B. (1983) The Fractal Geometry of Nature. New York, Freeman.

[11] Nelson T.R., and D.K. Manchester (1988) Modeling of lung morphogenesis using fractal geometries. IEEE Trans. Med. Imaging 7: 321–327.

[12] Nelson T.R., B.J. West, and A.L. Goldberger (1990) The fractal lung: universal and species-related scaling patterns. Experientia Basel 46: 251–254.

[13] Paumgartner D., G. Losa, and E.R. Weibel (1981) Resolution effect on the stereological estimation of surface and volume and its interpretation in terms of fractal dimensions. J. Microsc. 121: 51–63.

[14] Rigaut J.P. (1984) An empirical formulation relating boundary lengths to resolution in specimens showing «non-ideally fractal» dimensions. J. Microsc. 133: 41–54.

[15] Schmidt-Nielsen K. (1984) Scaling: Why is animal size so important? Cambridge UK, Cambridge University Press.

[16] Sernetz M., H.R. Bittner, and P. Wlczek (1988) Fraktale biologische Strukturen. Spiegel der Forschung 5: 8–11.

[17] Sernetz M., H. Willems, and H.R. Bittner (1989) Fractal organization of metabolism. In: Energy Transformations in Cells and Organisms. Eds: W. Wieser and E. Gnaiger. Stuttgart-New York, Thieme, 82–90.

[18] Spatz H.-C. (1991) Circulation, metabolic rate, and body size in mammals. J. Comp. Physiol. B 161: 231–236.

[19] Suwa N., and T. Takahashi (1971) Morphological and Morphometrical Analysis of Circulation in Hypertension and Ischemic Kidney. Munich, FRG: Urban and Schwarzenberg.

[20] Thoma R. (1901) Ueber den Verzweigungsmodus der Arterien. Arch. Entwicklungsmech. 12: 352–413.

[21] Thompson, D'A.W. (1942) On Growth and Form. Cambridge UK, Cambridge University Press.

[22] Weibel E.R. (1963) Morphometry of the Human Lung. Heidelberg, Springer-Verlag; New York, Academic Press.

[23] Weibel E.R. (1964) Morphometrics of the lung. In: The Handbook of Physiology. American Physiological Society, Respiration Section, Vol. 1, Chapter 7, 285–307.

[24] Weibel E.R., G.S. Kistler, and W.F. Scherle (1966) Practical stereological methods for morphometric cytology. J. Cell Biol. 30: 23–38.

[25] Weibel E.R. (1979) Stereological Methods. Vol. I: Practical Methods for Biological Morphometry. London-New York-Toronto, Academic Press.

[26] Weibel E.R. (1984) The Pathway for Oxygen. Cambridge MA, Harvard University Press.

[27] Weibel E.R. (1991a) Design of airways and blood vessels considered as branching trees. In: The Lung: Scientific Foundations. Eds: R. G. Crystal, J.B. West, P.J. Barnes, N.S. Cherniack and E.R. Weibel. New York, Raven, 711–720.

[28] Weibel E.R. (1991b) Fractal geometry: a design principle for living organisms. Am. J. Physiol. 261: L361–L369.

[29] Weibel E.R., and D.M. Gomez (1962) Architecture of the human lung. Science 137: 577–585.

[30] Weibel E.R., W. Stäubli, H.R. Gnägi, and F.A. Hess (1969) Correlated morphometric and biochemical studies on the liver cell. I. Morphometric model, stereologic methods and normal morphometric data for rat liver. J. Cell Biol. 42: 68–91.

[31] Weibel E.R., and C.R. Taylor (1981) Design of the mammalian respiratory system. Respir. Physiol. 44: 1–164.

[32] Weibel E.R., C.R. Taylor, and H. Hoppeler (1992) Variations in function and design: Testing symmorphosis in the respiratory system. Respir. Physiol. 87: 325–348.

[33] West B.J., V. Barghava, and A.L.Goldberger (1986) Beyond the principle of similitude: renormalization in the bronchial tree. J. Appl. Physiol. 60: 1089–1097.

[34] Wilson T.A. (1967) Design of the bronchial tree. Nature 213: 668–669.

Fractal and Non-Fractal Growth of Biological Cell Systems

Petre Tautu
Department of Mathematical Models
Research Program Bioinformatics
German Cancer Research Center
D-69009 Heidelberg

Abstract. The paper gives a general view of the microscopic growth rules and the behaviour of the spatial cell systems which generate fractal and non-fractal structures, with the purpose of finding their mathematical characteristics.

1 Introduction

The present paper finds its motivation in the following questions: **(A)** Which are those random spatial models that would be candidates for describing and *explaining* the fractal growth of (large) biological cell systems? **(B)** Under the modest assumption that fractals are not everywhere, are there random processes which generate non fractal morphologies? **(C)** Or, is there a distinct «regime» in the growth dynamics geometrically represented by a random fractal *transient* structure?

The questions above have arisen out of the analysis of a stochastic spatial growth process with local interactions as a model for carcinogenesis, originally called the Markov configuration model [57]. Numerical experiments have shown (under the caution that numerical experiments are never rigorous proofs) the formation of compact and quasi-circular patterns with fluctuating irregular shape. The conjecture made by the authors (Conjecture 3.2 in [57]) follows the well-known Richardson theorem (1972) which is set in a first-passage percolation framework. In [59], this Markovian model is associated with the original Eden models (1958, 1961) and the Williams-Bjerknes models (1972) in the family of *interacting biological cell systems*. They are characterized by an asymptotic shape which is non-fractal.

In the present paper, the term «fractal» is strictly employed only for those structures which are invariant under contraction or dilation, and possess the property of self-similarity. According to the broad definition given by B.B. Mandelbrot [40], «mathematical and natural fractals are shapes whose roughness and fragmentation *neither* tend to vanish, *nor* fluctuate up and down, but remain *essentially unchanged* as one zooms in continually and examination is refined.» Scale-invariant fractal objects are «open and diffuse», containing large empty regions («holes») on every length scale: this configuration is the origin of non-trivial scale invariance [65, p. 15] and occurs if there are long-range correlations in the pattern.

However, objects having the same fractal dimension may have different «lacunarity»; also, there are non-lacunar fractals. Recently, it is shown that lacunar DLA clusters become compact (and not self-similar) if the aggregate enlarges up to 30 million particles [41]. According to the strictly limited definition adopted in this paper, such objects are no more ordinary fractals. It appears that the fractal dimensionality — and its corresponding geometry — associated with simple aggregation models often varies continuously with some parameters used to define the model [45]. After all, fractal structures represent a geometrical expression of critical phenomena.

Then, the term «non-fractal» will be used here for any other irregular, not self-similar morphology. The reasons for this sharp distinction will be given in the next paragraph.

2 On fractals and non-fractals

2.1 A fundamental distinction

In the present paper, scale invariant and statistically self-similar structures are considered *stricto sensu* as genuine fractals. Accordingly, other irregular-looking structures are denoted on the whole as «non-fractals». This discrimination appears as unusual but avoids a certain ambiguity and its consequences.

The scaling hypothesis in statistical mechanics states that a physical system at its critical point is invariant under scale transformations, and near its critical point this invariance is broken and a «preferred» length scale — the correlation length — appears. In addition, the scaling hypothesis requires that small scale details must be ignored in order that scale invariance can be possible. For a mathematical review see [34]. The concept of non-integral (fractal) dimension is strongly connected to uniform (self-similar) scaling, so that ideal fractals are self-similar on all length scales [1]. Nevertheless, statistical self-similarity reflects the fact that a structure may look *statistically* similar while being different in detail at different scales.

Habitually, the generic term «fractal»conceals structures already defined as not identical, like self-similar and self-affine. Indeed, self-affinity describes a nonuniform scaling where shapes are statistically invariant under transformations that scale different coordinates by different amounts. The extension of fractal dimension to self-affine surfaces appears as «tricky» [63, p. 59]. Except for the singular case when the fractal dimension D, the surface «fractal» dimension D_S and the Euclidean dimension d are equal, objects cannot be «mass-fractal» and «surface-fractal» on the same length scale [54]. In addition, it is likely to be difficult to obtain a general expression of self-affine sets: this is an intractable problem [34, pp. 128–129].

If under the label «fractal» one collects geometrical objects that are self-similar, self-affine, piecewise self-affine, «fat» fractals and so on, one would come to the idea that the classical geometrical forms are only particular cases in the fractal geometry (see the note in [16, p. 179]), that is, «familiar objects such as the line, square, and cube are also fractals» [70, p. 666].

Finally, one must notice that in some cases self-similarity is «a side-effect of computational universality», since the behaviour of the computer on a subset of its input space may replicate its behaviour on the whole input space [4].

2.2 Scale invariance and morphogenesis

The actual paradigm in developmental biology postulates that morphogenesis and morphologic evolution depend upon a special set of molecular regulatory mechanisms mediated by the cell surface as it interacts at particular places in the embryo with other surfaces both cellular and noncellular [15, p. 4]. Although it was often indicated that reaction-diffusion (RD) models are inadequate to explain morphogenesis, there are some interesting RD models where diffusion is driven by nonlinear collective interactions («cooperative topobiological interactions») which are characteristic to the interacting cell biological systems. For instance, in [50] it is assumed that the morphogen diffusivity may be controlled by the concentration of a diffusible regulatory chemical that is produced at a constant rate by all cells. The desired «perfect» invariance can be obtained by adjusting the production rate as well as the rate at which the regulatory chemical leaks into the surroundings. One may presume that in morphogenesis there exist dynamical mechanisms which drive a cell system into a spatially and temporally scale invariant state. The system might show *self-organized criticality*: it describes the tendency of dissipative systems to drive themselves to a critical state with a wide range of length and time scales. In such self-organizing systems may occur simultaneously global criticality and local complexity.

Analyzing the aspects of dimensionality and the size of a cell system developing on a plane lattice, M.D. Hatlee and J.J. Kozak [29] considered that there are changes of structures determined by changes in the nature of boundary conditions («passive» to «active»). The most interesting hypothesis is that there may be two stages of morphogenesis: the early stage showing biases toward normal growth in integral dimensions, and the late stage showing «volume-preserving growth in fractal dimensions». This makes explicit an optimality principle, namely the maintenance of the area (volume) constant while increasing the edge length (surface area) in order to have the reactive sites distributed on the boundary of the system. The hypothesis in [29] might suggest a partial answer to Question **C** above and also to a question posed by H.E. Stanley [58]: «what evolutionary advantage does a fractal morphology convey?».

2.3 Irregular cell surfaces

In principle, a surface is the boundary of an object embedded in the three-dimensional Euclidean space. If $d = 2$, this boundary can be called «contour.» As a matter of fact, many biological proliferative cell systems may be studied as surface phenomena. The most interesting instance is *surface morphogenesis* [28] which presents the changes in shape of a thin sheet of cells due to intrinsic growth processes generated from inside the sheet itself. If one takes into consideration the isotropic growth of a curved surface from a flat sheet, the «Gaussian curvature» of the final structure determines whether the growth rate of the surface is sub- or superharmonic [61]. By definition, a «rough» surface is an irregular one on which there are no projections («overhanging regions») or where at least these protrusions do not dominate the scaling properties [23]. This justifies the introduction of a function $h(\mathbf{x}, t)$, $\mathbf{x} \in R^d$, $d > 1$, $t \geq 0$, which gives the «height» of the surface at position $\mathbf{x}$ on the original $d - 1$ dimensional flat surface, and at time t. For details the reader is referred to the indispensable survey [18].

Irregular surface is a subtle subject in itself, but as H.E. Stanley repeatedly pointed out [58], there are many different surfaces, depending on the considered physical process. It is evident that, in a biological context, the surface of a growing cell system must be distinguished from the surface (membrane) of a single living cell. The cytoskeleton governs the form of the cell surface which is locally and globally modulated [15]. Unicellular organisms (e.g., algae) show fractal and non-fractal structures that can be classified according to the values of the coefficients α, β, and γ of a deterministic system of differential equations [51, Eqs. 2 and 4] governing the evolution of the cell boundary.

Growing cell systems, and particularly interacting cell systems, show a variable irregular surface of their resulting compact clusters. Numerical experiments with Eden, Williams-Bjerkness or Markov configuration models make visible changing boundaries from one step to another (see, e.g., [10, pp. 20 and 49] and [57, 52]).

2.4 A note of caution

In his investigations on mosaic patterns in chimaeric rats, P.M. Iannaccone [31] showed that characteristic cell clusters («patches») formed in the mosaic liver are fractals (Fig. 2 and Table 1 in [31]). Interpreting his observations, he claimed that (i) self-similar processes are responsible for the generation of these fractal cell patterns, and that (ii) mosaic pattern «may be definable as equations with relatively few elements known as iterating functions sets». These suppositions are reproduced in [47] and sustained by computer simulations [48]. The conclusion is that «iterating self-similar division rules without focal centers of growth or cell movement were responsible for increase in parenchymal mass». Because these claims reveal a certain language confusion some brief definitions are necessary.

(i) Rigorously speaking, self-similar processes are characterized as limits *in distribution* of some stochastic processes $\{\xi(t), t \geq 0\}$ under joint rescaling of space and time. The law $\mathbf{P}$ of the new process $\{X(t), -\infty < t < \infty\}$ must verify for all t the relation

$$\mathbf{P}\{X(at) \leq x\} = \mathbf{P}\{a^H X(t) \leq x\}, \quad \text{for all real } a > 0, H \in R.$$

In distribution, a dilation by a of the time axis and a dilation by a^H of the x-axis does not change the representation of the trajectories of this process. The common example of a self-similar process is the Brownian motion with $H = 1/2$ and more generally, a stable process is self-similar with parameter $H = 1/\alpha$, $0 < \alpha \le 2$. As limits, self-similar processes have to generalize the role of stable laws, e.g., the Cauchy distribution, and processes within the limiting theory of sums of independent random variables. Birth-and-death processes converge to self-similar diffusions satisfying Eq. 5.2 in [36]. Undoubtedly, these limit processes are not the processes P.M. Iannaccone meant.

(ii) Iterated function systems were introduced in [3] as a unified way of generating a broad class of fractals. Formally, an iterated function system is a triple $\{(K, d), w, p\}$ where (K, d) is a compact metric space with distance function $d(x, y)$, $x, y \in K$, and $w = \{w_i, 1 \le i \le n\}$ is a finite collection of Lipschitz functions $w_i : K \to K$. Clearly, $p = \{p_i, 1 \le i \le n\}$ is a set of probabilities.

In a natural way, an associated Markov chain $\{\xi_n, n \in N\}$ arises by iteratively applying the maps w_i: its initial distribution is concentrated at an arbitrarily given $x \in K$ and its random transition mechanisms can be described as follows: given ξ_n, one has $\xi_{n+1} = w_i(\xi_n)$ with probability $p_i(\xi_n)$, $n \in N$, $1 \le i \le r$. The transition operator $\mathbf{U}$ associated with ξ_n is given by

$$\mathbf{U}f(x) = \sum_{i=1}^{r} p_i f[w_i(x)], \quad x \in K,$$

for any $f \in C(K)$, where $C(K)$ is the space of all bounded measurable real-valued functions defined on K [26].

Actually, the iterated function systems represent the basis of a probabilistic algorithm for the computation of images but not the description of a (biological) growth process! One must remind the term «growth of figures» introduced in 1962 by S.M. Ulam: by numerical experiments many different (fractal) patterns can be produced, i.e., figures «growing» according to certain recursive rules. Yet, we were warned: the amount of information contained in these objects is «therefore quite small, despite of their apparent complexity and unpredictability» [55]. It is not surprising that iterated function systems generate fractal forms: both K. Falconer [17, p. 113 and Th. 9.1] and G.A. Edgar [16, p. 105] begin to specify invariant sets by introducing iterated function schemes (or systems). In a beautiful dissertation on modelling marine organisms by iterative geometric constructions [32], some realistic restrictions are introduced, e.g., the exclusion of multiple occupancy of a site and the influence of the environment on the generator rules. Such restrictions will obviously disturb the self-similarity.

A particular form of iteration is the *stochastic iteration* of stable processes, introduced in [2], as a generalization of the notion of branching process. For instance, self-annihilating branching processes, continuous state space branching processes, state dependent branching processes, etc. can be constructed as stochastic iterates of a sequence of i.i.d. random walks with particular properties.

3 Spatial growth models of cell systems

3.1 Definition and examples

Let us begin with a brief definition of interacting cell systems on a regular lattice Z^d, $d \geq 1$. For technical details the reader is referred to [39]. Let S be the state space $S = \{\text{ all subsets of } Z^d\}$ and set $\Xi = S^{Z^d}$ which is the state space of the stochastic process. The elements ξ of Ξ are called configurations and the elements x of Z^d are called sites. The evolution of the interacting system $\{\xi(t), t \geq 0\}$ is determined by its jump (flip) rates $c(x, \xi)$ which possess the properties of (i) finite range, (ii) translation invariance, and (iii) attractiveness.

Definition

A family $\{\xi(t)\}$ of S-valued Markov processes is called a *stochastic growth model* if $\emptyset$ is an absorbing state and the family is a FTA system (i.e. a finite range, translation invariant, attractive system of infinitely many interacting particle system).

The interacting biological cell systems discussed in this papers are generally one-site («seed») growth processes which also are permanent (irreversible, immortal, or supercritical), i.e. $\mathbf{P}\{\tau = \infty\} > 0$, where $\tau = \min\{t : \xi_t^0 \neq \emptyset\}$ (see also [52]).

In [13], R. Durrett and D. Griffeath considered the following models as examples of stochastic spatial growth processes: (1) the Richardson model, (2) the coalescing random walks with nearest neighbour births, (3) the Williams-Bjerknes models, and (4) the basic contact process. Lately, an epidemic model [11], the oriented site percolation and a process with sexual reproduction [12] are included.

3.2 Microscopic growth rules

This paragraph will deal with the analysis of the primary differences between the above cell interacting systems on Z^d, $d \geq 1$ (and other cell systems) and the known growth models generating fractal structures. We will refer to the essential microscopic growth rules (for details see [59]).

(i) The initial configuration. Most of cell interacting systems are one-site growth models: the process starts with one cell at site $0 \in Z^d$, $d \geq 1$, or with a small group of adjacent cells (nearest neighbours). In the Markov configuration model [57] generating cells of different types, the process begins with one normal cell. On the contrary, in the DLA model, particles are added one at a time to a growing aggregate of particles via random walk (Brownian) trajectories, but these particles start outside of the region occupied by the initial aggregate. Actually, they are launched from a randomly selected point on a circle that just enclose the cluster [45]. However, DLA-like bacterial colonies *(B. subtilis)* can be obtained if the bacteries are inoculated at the centre of a Petri dish and the peptone concentration is up to 1g/l [42, 22]. In other experiments [66, 43], the «seed» is a line: the inoculation of some microorganisms *(E.coli, B. subtilis, Aspergillus oryzae)* is made along a straight line («strip geometry»).

(ii) The multiplication process. Biological growth is characterized by cell division. At each site on the lattice a birth process (or a birth-and-death process) of a particular type takes place (remind that the jump rate is dependent on the configuration ξ). In [57] the cell cycle is explicitly introduced. As a consequence of binary fission, a daughter cell must look for a vacant site. This search is governed by the following

(iii) Occupation rules. (1^0) Multiple occupancy is excluded. (2^0) Occupation of a nearest neighbour vacant site *(short range)*: the choice can be made with (a) equal or (b) unequal probabilities. If the configuration is compact, the cell proliferation is restricted at the boundary. The most familiar example of an asymmetrical cell growth system on Z^2 was given by M. Eden where the probability p_1 of adjoining a cell at a vertical edge can be an integral multiple of the probability p_2 of adjoining to a horizontal edge. One can assume, for instance, that an anisotropy in the nutrient medium of a tissue culture may determine configurations with elliptic shape. It is known that lattice anisotropy has an important effect on the growth of DLA clusters, and that this effect is very much smaller for hexagonal and triangular lattices than for the square lattice [45]. (3^0) Competitive occupation of a nearest neighbour site (if it is occupied). This kind of short-range interaction is assumed in [57]; in this model a newly born cell can choose (with equal or unequal probabilities) an already occupied site and eliminate the «old» cell. In an interacting cell system this competition of the «fittest» may be interpreted as the general dynamics of «cell loss». Another kind of local competition is assumed in the Williams-Bjerknes model, where the malignant cell possess a «carcinogenic advantage»; in the theory of interacting particle systems, the infinite Williams-Bjerknes cell system is known as the «biased voter model» [39, 10]. (4^0) Occupation of a distant vacant site *(long range)*. A new cell must go on to search for a vacant site until it finds one (spending no time at the intervening occupied sites). The resulting migration process is called *long range exclusion process* because cells can migrate very long distances in short times when a large configuration exists [38]. In addition, the rate at which a cell at lattice site x attempts to move to site y may depend on the configuration at other sites («speed change»).

Except for the random walk, DLA is perhaps the simplest process *(Laplacian growth model* [1]) that generates a fractal structure. The more interesting is then the «internal» variant of DLA that generates a non-fractal structure [37]. In this model, particles are repeatedly dropped at the origin $\mathbf{0} \in Z^d$, $d \geq 2$; one by one, each particle performs an independent simple symmetric d-dimensional random walk until it «sticks» at the first site not previously occupied. The asymptotic shape of an internal DLA configuration is an Euclidean ball [37, Th. 1].

The modest conclusion is that the above microscopic biological growth rules suggest fine distinctions between fractal and non-fractal growth processes; however, two supplementary conditions, namely (a) the size of the observed configuration and (b) the external growth circumstances, must also be taken into consideration.

(a) For example, numerical experiments have shown that the shape of a DLA configuration evolves from a (more or less) circular shape, characteristic of very small clusters, to a cross-like shape of a cluster containing more than 10^6 occupied sites, with an intermediate passage to a diamond-like shape if the configuration consists of about 10^5 sites [45]. Also, the simulated Eden cluster shows the size dependence of its shape: even for $d = 2$ very large clusters are not exactly circular.

(b) Bacteriological experiments show that the morphology of growing colonies depends on the nutrient concentration, e.g. DLA-like structures at small concentrations of peptone (1g/l), round and smooth forms, «similar to petals which are fused together into one», at intermediate concentrations (4g/l), and round colonies at high nutrient concentrations (64g/l). Obviously, the structure of the support is also important: at a low concentration of agar, the bacterial colonies have a «dense branching morphology» [22], and, simulated on a continuum, the Eden cluster is no more compact. The compactness of the morphology is a function of the viscosity ratio η in the general *gradient governed growth* (GGG) model: for $\eta = 0.0001$ the realized structure is fractal, for $\eta = 0.01$ the structure is irregular but not fractal, and one obtains an Eden-like structure if $\eta = 10.0$ [49].

3.3 The asymptotic shape (I)

Strictly speaking, the shape of a geometrical object represents those geometrical attributes that remain invariant under any translation, rotation and scaling. The reader is referred to [7] and [62] for applications in morphometrics (= measurement of shapes, their variation and changes). Particularly interesting in [62, Ch. 3] is the application of Dirichlet principle which may supply a quantitative definition of the term «smooth». (See the definitions of *smoothness* and *roughness* in [7] with the analytic approximation.)

This paragraph is motivated by the following simple question: If the random walk is the simplest example of a random fractal, what about its shape? An answer is given in [53] where the results were obtained in the form of a power series expansion, called *1/d expansion*. A different, probabilistic approach will be presented in the sequel.

Let $\{X_i, 1 \le i \le n\}$ be a sequence of independent, identically distributed, d-dimensional random variables $(d > 1)$ with a common distribution F and, as usual $S_n = \sum_{i=1}^{n} X_i$. We say that S_n is the position at instant n of a particle performing a d-dimensional random walk. In order to obtain asymptotic estimates for the probability that S_n is contained in a cube (or a ball) centered at the origin, another random variable X with the same distribution as X_1 is introduced. If X satisfies two special geometrical conditions, namely the direction and the cone conditions, and if it is, in addition, genuinely d-dimensional (i.e. its distribution is not supported on a $(d - 1)$-dimensional hyperplane), then there exist positive constants c_1, c_2 and λ_0 such that for all $\lambda \geq \lambda_0$ and all n,

$$c_1\left(\frac{\lambda}{a_n} \wedge 1\right)^d \leq \mathbf{P}\{S_n \in C(0,\lambda)\} \leq c_2\left(\frac{\lambda}{a_n} \wedge 1\right)^d, \tag{1}$$

where $C(0,\lambda)$ is the cube of side length 2λ centered at the origin [25, Th. 2.1]. This is a consequence of a local limit theorem in the case that the additional random variable X is in the domain of attraction of a stable law.

According to W. Feller [21], the variables X_i are said to belong to the domain of attraction of a nondegenerate distribution G if there exist real constants $a_n > 0$ and b_n such that $\mathbf{P}\{(S_n - b_n)/a_n \leq x\} \to G(x)$, as n grows to infinity. It is important to point out that the localization of the variable X in the domain of attraction of a stable law with index α, $0 < \alpha \leq 2$, means that there is a centering sequence $\{b_n\}$ such that $\{(S_n - b_n)/a_n\}$ is stochastically compact [21], that is, it is a tight sequence and all limit laws are nondegenerate. Typical local limit theorems depend on compactness. The sequence $\{a_n\}$ is used, when suitably chosen, to normalize $\{S_n\}$ for weak convergence.

Two additional remarks must be made: (a) the problem is dependent on dimensionality because in $d \geq 2$ the spread of $\{S_n\}$ may vary greatly in different directions and the centering task becomes more complicated, and (b) the theorem (1) does not apply to all variables in the domain of attraction of a stable law [25].

Remind that the normal distribution ($\alpha = 2$) is the only member of the class of stable laws that has a finite second moment: for all other members of the class, the calculations using characteristic functions, show infinite variance (see, e.g., [24]). The characteristic exponent α measures the «thinness» of the tails of the distribution and also indicates the existence and nature of the moments. A small value of α will mean the setting of considerable probability mass in the tails of the distribution and thus, a long-tailed distribution implies a distribution with infinite variance. It is known that if $\alpha \downarrow 0$, the Lévy distribution has the lognormal distribution as limiting form.

It is very important to point out that special random walks which generate self-similar clusters [30] possess an infinite mean-squared displacement per jump and their distribution is long-tailed. This might be the mathematical characterization of typical random fractal growth processes.

A similar problem can be formulated for the Galton-Watson branching process $\{Z_n, n \in N\}$. If the process is supercritical and the basic distributions $\{p_i\}, i \in N$, are normally attracted to a particular stable law, then there exist constants a_n, b_n, c_n, for which the limit

$$\lim_{n \to \infty} \mathbf{P}\left\{\frac{Z_n - b_n}{a_n} \leq x \mid Z_0 = c_n\right\} = G(x) \tag{2}$$

holds with G as the corresponding stable law [35, Coroll. 3].

More adequate in a biological context is the question about the asymptotic shape of a branching random walk. If $I^{(n)}$ is the set of positions of the n-th generation of individuals generated by a supercritical branching random walk on R^d, scaled by $1/n$, then $I^{(n)}$ looks like a convex set I for large n, provided that the process survives. Then I is the asymptotic shape of this branching random walk, and according to Theorem A in [6], it is a smooth rounded shape. Moreover, J.D. Biggins proved that the upper bound C on the asymptotic shape Γ of Mollison's contact birth process equals this shape, $\Gamma = C$. This leads to the conclusion that the propagation of a monotone continuous time Markov contact process in a particular direction is bounded: the limits for individual direction can be «stuck together» into a convex body Γ such that the convex hull H_t of inhabited points converges in shape and velocity (Eq. 3.29 in [46]). This result can be correlated with the observation of M. Eden that in his spatial model for embryogenesis configurations with many short branches are more probable than those with a few long branches.

3.4 The asymptotic shape (II)

This paragraph is devoted to the presentation of the asymptotic geometrical behaviour of known interacting cell systems by introducing the essential arguments derived from the theory of subadditive stochastic processes. Indeed, the main idea which dominates this approach is that the spread times (occupation of a site, infection, etc.) are subadditive. The interesting result is that the set of occupied sites grows linearly with t (possibly at an infinite rate) and has an asymptotic shape which is not random [34] (see also [10, 11c]). It must be pointed out that the argument behind the proof of the asymptotic shape theorem is the evidence of the linear growth in a fixed direction; this implies the right growth rate in any finite number of directions simultaneously, and therefore the requirement of some uniform estimate on different growth rates in two directions which are close together (Lemmas 3.5 and 3.6 in [33]). The problem of self-affinity for such interacting systems is then solved.

The asymptotic shape of all interacting cell systems is investigated in the same framework, e.g., the Richardson model [14], the Williams-Bjerknes model [8], and the Markov configuration model [57]: they may be considered as asymptotically *nonfractal cell growth models*. Actually, D. Richardson was the first author who compared the Eden model (his Example 9) with the Williams-Bjerknes model with infinite carcinogenic advantage (his Example 6). In [56] K. Schürger introduced two generalized Eden models and demonstrated a similar asymptotic geometric behaviour. This generalized Eden model is a discrete time interacting cell system with state space Ξ the set of all configurations; Ξ_0 is the set of all configurations $\eta \in \Xi$ having a finite non-void support. Because this model was frequently investigated especially for its nontrivial surface (see a collection of papers in [19]), we include here the probabilistic approach of its asymptotic geometric behaviour. The Eden growth process $\eta_n^{(k)} \in \Xi_0$ chooses at random one of the sites (say y) on its k-boundary $\partial_k(\eta_n^{(k)})$. This choice is equiprobable. The process makes a jump from state $\eta_n^{(k)}$ to state $\eta_{n+1}^{(k)} = \eta_n^{(k)} \cup \{y\}$. In order to formulate the shape theorem, one must take into account the monotonicity condition, the formal definition of a norm, and the hypothesis of a finite span of the process. One obtains the following

Theorem

(Schürger, 1981). Let $k \geq 1$. Then there exists a norm $N^*(\cdot)$ on R^d such that, for all $\xi \in \Xi_0$ and $0 < \epsilon < 1$, we have for the generalized Eden process $\{\eta_n^{(k)}\}$ for which $\eta_0^{(k)} = \xi$, that

$$\{x | x \in Z^d, N^*(x) \leq (1 - \epsilon) n^{1/d}\} \subset \eta_n^{(k)} \subset \{x | x \in Z^d, N^*(x) \leq (1 + \epsilon) n^{1/d}\}$$
$$(3)$$

almost surely for all sufficiently large n.

Such results might be thought of as a *strong law of large configurations* because for all $\xi \in \Xi_0$, the set $(1/t)\{x | \tau(x) \leq t\}$ (pointwise scalar multiplication) converges in some sense — which is common in measure theory — a.s. ($\mathbf{P}_\xi$) to the set $\{x | N(x) \leq 1\}$ as $t \to \infty$. Obviously, $\tau(x)$ is the first instant at which site x is occupied [56].

The result above formalizes the original observation made by M. Eden that «the colony is essentially circular in outline». The biological analogue of this model can be, for example, the two-dimensional growth of a cell clone: «wherever there is a nutrient medium we can be quite sure that the colony will have the largely circular morphology exhibited by the model».

This is just the moment for two important remarks:

(1) In Richardson's paper, the probability that a site will be occupied is denoted by p. His first remark is that as p varies from 1 to 0, the unit ball of the associated norm N varies from a diamond to a circle («however, no proof of this is available», he added). It is now possible to prove this conjecture in terms of F, the distribution of the passage time. One assumes that F is exponential, i.e. $F(x) = 1 - e^{-x}$, if $x \geq 0$. Let $\lambda(F) = \inf\{x : F(x) > 0\}$ be the left endpoint of the support of F. Also, denote by $B_0 \subset R^d$ a nonrandom convex set having nonempty interior. According to a proposition given in [33, Prop. 6.15], if $\lambda(F) > 0$ and F is not concentrated on λ, then this convex set is strictly contained in the «diamond» $\{x \in R^d : |x| = 1/\lambda\}$. Moreover, for high dimensions and exponential F, it is not a ball [33, Coroll. 8.4].

The reader should peer at Durrett's simulation of an epidemic model (or «forest fire») in order to realize the structure changes (black burnt sites): a diamond form (with a «flat edge» as in [14]) if $p = 0.75$, a tendency to circularity at $p = 0.60$, and a clear fractal structure at $p = 0.50$ [10, pp. 239–241].

(2) D. Richardson pointed out an essential fact: as p decreases (p=0.1), the roughness of the boundary seems to increase. In consequence, the natural norm imposed by the lattice breaks down «and the Euclidean norm appears to assert itself».

3.5 Crinkliness

Re-analyzing the Williams-Bjerknes model (believed by its authors to be a fractal model), D. Mollison introduced a measure of surface irregularity

$$C(\xi) = \frac{1}{4\sqrt{|\xi|}} \#\{(x, y) | x \in \xi, y \notin \xi, x \text{ and } y \text{ adjacent}\}, \quad d = 2, \quad (4)$$

called *crinkliness*. The formula (4) implies that $C(\xi) = 1$ for the square arrays, all other structures having $C(\xi) > 1$. It appears that for $|\xi| = n$ small, the crinkliness has a great variability, while for increasing n, $C(\xi)$ tends to a limit which depends upon κ, the carcinogenic advantage. It is conjectured in [57, Conj. 3.1] that $\lim_{t\to\infty} C(\xi_t) = \alpha$, almost everywhere (P_ξ), in the case of the Markov configuration model, where $\alpha > 0$ depends on cell cycle and cell differentiation intensities. According to the numerical experiments [57], $C(\xi)$ varies form 1.38 to 1.82 (the total length of the boundary increases and the growth quotient t^2/n slowly decreases). There is a close connection between crinkliness and spread velocity [46].

4   The diffusion-injection fractal process

Based on his data on agricultural uniformity trials, H. Fairfield Smith suggested in 1938 that the covariance $\Gamma(r)$ of yield at points a distance r apart falls off as a power of r at large distance. The same behaviour was reported by flood height, roughness of roads and runways, yarn diameter, and response from population samples. These data stimulated P. Whittle [67] to invent a diffusion model with particular characteristics of its moments (variance, autocovariance). Lately, the author called it a *diffusion-injection model* [68] which actually is a d-dimensional anisotropic random field (see also [64]).

Let us consider a random variable $\xi(x, t)$, $x = (x^1, \ldots, x^d)$, $d > 1$, which may be interpreted in the agricultural example as the concentration of a nutrient in a field (or the height of the runway, the nutrient concentration in a bacterial culture, or the density gradient of a growth factor in a tissue culture, etc.). $\xi(x, t)$ obeys the stochastic differential equation

$$\frac{\partial \xi}{\partial t} = \nabla^2 \xi - \alpha^2 \xi + \epsilon. \quad (5)$$

This is a diffusion equation driven by a Wiener process $\epsilon(x, t)$ which is explained as follows: variation is «injected» continuously by the noise process and is spread out in space by the diffusion term $\nabla^2 \xi$. The strength of the field tends to be damped by the term $-\alpha^2 \xi$. In the original paper $\xi(x, t)$ is called a «fertility» function and $\nabla^2 \xi = \sum_j \left[\frac{\partial}{\partial x_j}\right]^2 \xi$, is the physical diffusion of nutrient laterally in the soil, tending to equalize concentration. $\{\xi(x, t)\}$ is a d-dimensional random field.

The following characteristics of the model are of importance:

(i) The process $\{\xi(x,t)\}$, $x \in R^d$, $d \geq 1$, $t \geq 0$, is stationary with respect to both x and t; the input process $\{\epsilon(x,t)\}$ is stationary in its arguments, too.

(ii) If one defines the spatial autocovariance function by

$$\text{Cov}[\xi(x,t), \xi(x+y,t)] = \Gamma(r) = \frac{e^{-r\sqrt{2\alpha}}}{4\pi r}, \quad x,y \in R^d, d \geq 1, |y-x| = r, \quad (6)$$

then, if $d > 2$ and $\alpha = 0$,

$$\Gamma(r) = c r^{-d+2}, \quad c = \text{const}, \quad (7)$$

[67, Eq. 7], where r is called the lag.

(iii) The spectral density of the process $\{\xi(x,t)\}$, i.e. the Fourier transform of the autocorrelation $A(r)$, can be related with the covariance $\Gamma(r)$. Let ω be the frequency and ν the wave-number vector, $\nu = (\nu^1, \ldots, \nu^d)$, $d \geq 1$, these being complementary to t and x, respectively. If $\{\xi(x,t)\}$ has spectral density function $f(\nu,\omega)$, then its autocovariance becomes

$$\Gamma(r) = \frac{1}{(2\pi)^d} \int e^{i\nu r} f(\nu) d\nu. \quad (8)$$

Clearly, the covariance is dimension sensitive: if $d \geq 2$, the integral does not converge absolutely, and if $d < 2$, it will diverge as $\alpha \downarrow 0$ due to the singularity of $f(\nu)$ for small ν. This divergence corresponds to genuine non-stationarity of the process for $d < 3$ (Th. 20.3.1 in [68]).

One might say that the driven diffusion process possesses a covariance $\Gamma(r)$ but not a variance: actually, the variance is infinite due to the fact that the diffusion mechanism does not smooth ϵ sufficiently powerfully. However, a mild degree of spatial correlation in the input will also ensure that $\xi(x,t)$ has finite variance [67].

More generally, a spectral density $f(\nu) \propto |\nu|^{-2}$ corresponds to an autocovariance

$$\Gamma(r) \propto |r|^{2-d}, \quad d > 2 \quad (9)$$

[68, Eq. 20.4.3]. In this case, one demonstrates the existence of a self-similar power law $\Gamma(r) \sim c r^{-\lambda}$ with $1/2 \leq \lambda \leq 3/2$, with strong peaking in the frequency at $\lambda = 1$, and lesser peakings at the extreme values $\lambda c(1/2, 3/2)$. Indeed, the data give $D = 1.6$–1.8, estimated from block variance [9].

The above scrutiny of Whittle's model pinpoints the intrinsic conditions a stochastic process which generates fractal structures must satisfy. A noteworthy remark is that Whittle's process is a zero-mean generalized Gaussian process and, in the terminology of D. Williams [69], a X *harness process*. Accordingly, it is claimed that P. Whittle arrived at his process by using a *serial harness* approach [27, Sect. 7].

Recently, the diffusion-injection model was mentioned as an example of *spatial long memory* [5]. Formally, a stationary process is said to exhibit long-range dependence if the spectral density is

$$f(\nu) \sim L_1(\nu)|\nu|^{1-2H}, \quad H \in (1/2, 1), \tag{10}$$

where $L_1(\cdot)$ is slowly varying for $|\nu| \to 0$.

Finally, one must mention the diffusion model reported by P. Meakin [44]: it is assumed that the nutrient is supplied by the boundary conditions which maintain a fixed concentration x of nutrient on a circle of lattice sites surrounding the growing cluster. The growth probability p is proportional to some power λ of x, so that if $p \sim x^2$, the model becomes equivalent to the dielectric breakdown model, known as Laplacian growth model [1].

References

[1] Aharony, A.: *Fractal growth*. In: Fractals and Disordered Systems (A. Bunde, S. Havlin eds), pp. 151–173. Berlin: Springer (1991).

[2] Athreya, K.B.: *Stochastic iteration of stable processes*. In: Stochastic Processes and Related Topics (M.L. Puri ed.), Vol. **1**, pp. 239–247. New York: Academic Press (1975).

[3] Barnsley, M.F., Demko, S.: *Iterated function systems and the global construction of fractals*, Proc. Roy. Soc. (London) A **399**, 243–275 (1985).

[4] Bennett, C.H.: *Dissipation, information, computational complexity and the definition of organization*. In: Emerging Syntheses in Sciences (D. Pines ed.), pp. 215–233. Redwood: Addison-Wesley (1988).

[5] Beran, J.: *Statistical methods for data with long-range dependence*, Statist. Sci. **7**, 404–427 (1992).

[6] Biggins, J.D.: *The asymptotic shape of the branching random walk*, Adv. Appl. Probab. **10**, 62–84 (1978).

[7] Bookstein, F.L.: *The Measurement of Biological Shape and Shape Change*, Lecture Notes Biomath. **24**. Berlin: Springer (1978).

[8] Bramson, M., Griffeath, D.: *On the Williams-Bjerknes tumour growth model, I, II*, Ann. Probab. **9**, 173–185 (1981); Math. Proc. Camb. Phil. Soc. **88**, 339–357 (1980).

[9] Burrough, P.A.: *Fractal dimensions of landscapes and other environmental data*, Nature **294**, 240–242 (1981).

[10] Durrett, R.: *Lecture Notes on Particle Systems and Percolation*. Pacific Grove (CA): Wadsworth & Brooks/Cole (1988).

[11] Durrett, R.: *Stochastic growth models: Recent results and open problems*. In: Mathematical Approaches to Problems in Resource Management and Epidemiology (C. Castillo-Chaves et al. eds) [Lecture Notes Biomath. **81**], pp. 308–312. Berlin: Springer (1989).

[12] Durrett, R.: *Stochastic growth models: Bounds on critical values*, J. Appl. Probab. **29**, 11–20 (1992).

[13] Durrett, R., Griffeath, D.: *Contact processes in several dimensions*, Z. Wahrscheinlichkeitstheorie verw. Gebiete **59**, 535–552 (1982).

[14] Durrett, R, Liggett, T.M.: *The shape of the limit set in Richardson's growth model*, Ann. Probab. **9**, 186–193 (1981).

[15] Edelman, G.M.: *Topobiology. An Introduction to Molecular Embryology*. New York: Basic Books (1988).

[16] Edgar, G.A.: *Measure, Topology, and Fractal Geometry*. New York: Springer (1990).

[17] Falconer, K.: *Fractal Geometry. Mathematical Foundations and Applications*. Chichester: Wiley (1990).

[18] Family, F.: *Dynamic scaling and phase transitions in interface growth*, Physica **A 168**, 561–580 (1990).

[19] Family, F., Vicsek, T. (eds): *Dynamics of Fractal Surfaces*. Singapore: World Scientific (1991).

[20] Feller, W.: *The asymptotic distribution of the range of independent random variables*, Ann. Math. Statist. **22**, 427–432 (1951).

[21] Feller, W.: *On regular variation and local limit theorems*, Proc. 5th Berkeley Symp. Math. Statist. Probab., Vol. II, Part 1, pp. 373–388. Berkeley: Univ. California Press (1967).

[22] Fujikawa, H., Matsushita, M.: *Bacterial fractal growth in the concentration field of nutrient*, J. Phys. Soc. Japan **60**, 88–94 (1991).

[23] Gouyet, J.-F., Rosso, M., Sapoval, B.: *Fractal surfaces and interfaces*. In: Fractals and Disordered Systems (A. Bunde, S. Havlin eds), pp. 229–261. Berlin: Springer (1991).

[24] Granger, C.W.J., Orr, D.: *«Infinite variance» and research strategy in time series analysis*, J. Amer. Statist. Assoc. **67**, 275–285 (1972).

[25] Griffin, P.S.: *An integral test for the rate of escape of d-dimensional random walk*, Ann. Probab. **11**, 953–961 (1983).

[26] Grigorescu, S., Popescu, G.: *Random systems with complete connections as a framework for fractals*, Stud. Cerc. Mat. **41**, 481–489 (1989).

[27] Hammersley, J.M.: *Harnesses,* Proc. 5th Berkeley Symp. Math. Statist. Probab., Vol. III, pp. 89–117. Berkeley: Univ. California Press (1967). (Reprinted in [38])

[28] Hart, T.N., Trainor, L.E.H.: *Geometrical aspects of surface morphogenesis,* J. Theor. Biol. **138**, 271–296 (1989).

[29] Hatlee, M.D., Kozak, J.J.: *Stochastic flows in integral and fractal dimensions and morphogenesis,* Proc. Natl. Acad. Sci. USA **78**, 972–975 (1981).

[30] Hughes, B.D., Shlesinger, M.F., Montroll, E.W.: *Random walks with self-similar clusters,* Proc. Natl. Acad. Sci. USA **78**, 3287–3291 (1981).

[31] Iannacone, P.M.: *Fractal geometry in mosaic organs: a new interpretation of mosaic pattern,* FASEB J. **4**, 1508–1512 (1990).

[32] Kaandorp, J.A.: *Modelling growth forms of biological objects using fractals,* Ph. D. Thesis, Univ. of Amsterdam (1992).

[33] Kesten, H.: *Aspects of first passage percolation.* In: École d'Été de Probabilités de Saint-Flour XIV-1984 [Lecture Notes Math. **1180**], pp. 125–264. Berlin: Springer (1986).

[34] Kesten H.: *Percolation theory and first-passage percolation,* Ann. Probab. **15**, 1231–1271 (1987).

[35] Lamperti, J.: *Limiting distributions for branching processes,* Proc. 5th Berkeley Symp. Math. Statist. Probab., Vol. II, Part 2, pp. 225–241, Berkeley: Univ. California Press (1967).

[36] Lamperti, J.: *Semi-stable Markov processes,* Z. Wahrscheinlichkeitstheorie verw. Gebiete **22**, 205–225 (1972).

[37] Lawler, G.F., Bramson, M., Griffeath, D.: *Internal diffusion limited aggregation,* Ann. Probab. **20**, 2117–2140 (1992).

[38] Liggett, T.M.: *Long range exclusion processes,* Ann. Probab. **8**, 861–889 (1980).

[39] Liggett, T.M.: *Interacting Particle Systems.* New York: Springer (1985).

[40] Mandelbrot, B.B.: *Fractal geometry: what is it, and what does it do?,* Proc. Roy. Soc. (London) **A 423**, 3–16 (1989).

[41] Mandelbrot, B.B.: *Plane DLA is not self-similar; is it a fractal that becomes increasingly compact as it grows?,* Physica **A 191**, 95–107 (1992).

[42] Matsushita, M., Fujikawa, H.: *Diffusion-limited growth in bacterial colony formation,* Physica **A 168**, 489–506 (1990).

[43] Matsuura, S., Miyazima, S.: *Self-affine fractal growth of Aspergillus oryzae,* Physica **A 191**, 30–34 (1992).

[44] Meakin, P.: *A new model for biological pattern formation,* J. Theor. Biol. **118**, 101–113 (1986).

[45] Meakin, P.: *Models for colloidal aggregation*, Ann. Rev. Phys. Chem. **39**, 237–267 (1988).

[46] Mollison, D.: *Spatial contact models for ecological and epidemic spread*, J. Roy. Statist. Soc. **B 39**, 283–326 (1977).

[47] Ng, Y.-K., Iannaccone, P.M.: *Fractal geometry of mosaic pattern demonstrates liver regeneration is a self-similar process*, Develop. Biol. **151**, 419–430 (1992).

[48] Ng, Y.-K., Iannaccone, P.M.: *Experimental chimeras: Current concepts and controversies in normal development and pathogenesis*, Current Topics Develop. Biol. **27**, 235–274 (1992)

[49] Nittmann, J., Daccord, G., Stanley, H.E.: *When are viscous fingers fractal?* In: Fractals in Physics (L. Pietronero, E. Tosatti eds), pp. 193–202. Amsterdam: North Holland (1986).

[50] Othmer, H.G., Pate, E.: *Scale-invariance in reaction-diffusion models of spatial pattern formation*, Proc. Natl. Acad. Sci USA **77**, 4180–4184 (1980).

[51] Pelce, P., Sun, J.: *Geometrical models for the growth of unicellular algae*, J. Theor. Biol. **160**, 375–386 (1993).

[52] Röthinger, B., Tautu, P.: *On the genealogy of large cell populations*. In: Stochastic Modelling in Biology (P. Tautu ed.), pp. 166–235. Singapore: World Scientific (1990).

[53] Rudnick, J., Gaspari, G.: *The shapes of random walks*, Science **237**, 384–389 (1987).

[54] Schaefer, D.W., Bunker, B.C., Wilcoxon, J.P.: *Fractals and phase separation*. Proc. Roy. Soc. (London) **A 423**, 35–53 (1989).

[55] Schrandt, R.G., Ulam, S.M.: *On recursively defined geometrical objects and patterns of growth*. In: Essays on Cellular Automata (A.W. Burks ed.), pp. 232–243. Urbana: Univ. Illinois Press (1970).

[56] Schürger, K.: *On a class of branching processes on a lattice with interactions*, Adv. Appl. Piobab. **13**, 14–39 (1981).

[57] Schürger, K., Tautu, P.: *A Markov configuration model for carcinogenesis*. In: Mathematical Models in Medicine (J. Berger et al. eds) [Lecture Notes in Biomath. **11**], pp. 92–108. Berlin: Springer (1976).

[58] Stanley, H.E.: *Fractals and multifractals: The interplay of physics and geometry*. In: Fractals and Disordered Systems (A. Bunde, S. Havlin eds), pp. 1–49. Berlin: Springer (1991).

[59] Tautu, P.: *On the qualitative behaviour of interacting biological cell systems*. In: Stochastic Processes in Physics and Engineering (S. Albeverio et al. eds), pp. 381–402. Dordrecht: Reidel (1988).

[60] Tautu, P.: *Interacting biological cell systems*. In: Stochastic Modelling in Biology (P. Tautu ed.), pp. 50–90. Singapore: World Scientific (1990).

[61] Todd, P.H.: *Gaussian curvature as a parameter of biological surface growth*, J. Theor. Biol. **113**, 63–68 (1985).

[62] Todd, P.H.: *Intrinsic Geometry of Biological Surface Growth*. Lecture Notes Biomath. **67**. Berlin: Springer (1986).

[63] Tsonis, A.A.: *Chaos. From Theory to Applications*. New York: Plenum Press (1992).

[64] Vecchia, A.V.: *A general class of models for stationary two-dimensional random processes*, Biometrika **72**, 281–291 (1985).

[65] Vicsek, T.: *Fractal Growth Phenomena*. Singapore: World Scientific (1989).

[66] Vicsek, T., Cserzö, M., Horváth, V.K.: *Self-affine growth of bacterial colonies*, Physica **A 167**, 315–321 (1990).

[67] Whittle, P.: *Topographic correlation, power-law covariance functions, and diffusion*, Biometrika **49**, 305–314 (1962).

[68] Whittle, P.: *Systems in Stochastic Equilibrium*. Chichester: Wiley (1986).

[69] Williams, D.: *Some basic theorems on harnesses*. In: Stochastic Analysis (D.G. Kendall, E.F. Harding eds), pp. 349–363. London: Wiley (1973).

[70] Yates, F.E.: *Fractal applications in biology: Scaling time in biochemical networks*. In: Numerical Computer Methods (L. Brand and M.L. Johnson eds) [Methods in Enzymology **210**], pp. 636–675. San Diego: Academic Press (1992).

Evolutionary Meaning, Functions and Morphogenesis of Branching Structures in Biology

Giuseppe Damiani
I.D.V.G.A.-C.N.R.
c/o Dipartimento di Genetica e Microbiologia
Universitá di Pavia
via Abbiategrasso 207
27100 Pavia, Italy

Abstract. A causal explanation of the regularity and invariance of biological forms must relate to morphogenetic generative laws, to allometric constraints of growth processes, to biunique relations of an organism with environment and other organisms, to the pattern of evolutionary processes, and to physical causes as the forces present in the environment and the mechanical properties of the available building materials. The evaluation of the relative importance of these different causes is not always easy. Are evolutionary trends of organic forms the results of functional adaptation or of random processes? A general model to explain the ontogenetic and phylogenetic upward trend of complexity of biological branching structures is proposed. The function of these structures in living organisms is to distribute or to gather biological material and physical entities. Optimization processes produced by natural selection led to minimization of the networks length (the Steiner problem) and of the amount of information needed for the construction of these structures. What is the optimization method chosen by nature? Experimental data and computer simulations suggest a simple way to generate minimized network: sensitive entities spread or concentrate in a surface or in a volume responding to morphogenetic gradients of physical forces or of chemical substances. Recursive local rules of repulsion or attraction leads to the formation of fractal, branching, vascular and global networks. Slight changes of morphogenetic fields and of responsiveness of sensitive entities influence shapes of produced patterns. Natural selection allows survival of the biological structures more suitable for their physiological functions: to distribute or to gather something in an economic (cost functions minimization), uniform (space-filling) and size-independent (self-similar scaling) way. Therefore biological branching structures are the results of functional adaptation processes. Computer simulations by means of genetic algorithms based on the same optimization method chosen by nature may be applied successfully to solve a wide variety of scientific and engineering problems.

1 Introduction

The aim of scientists is to discover and to explain regularities in observations of natural phenomena. These regularities can be described by natural laws which are expressed in mathematical terms. Euclidean geometry and analytical calculus, typically in the form of linear differential equations, are the standard way of modelling the laws of nature. These mathematical models allowed the formulation of laws describing simple physical mechanic systems, but complex systems, as the biological ones, often remained unexplained. Moreover many patterns and shapes of natural objects and phenomena cannot be represented by Euclidean geometry. Recently, with the advent of computers, a new kind of mathematics was developed which deals with discrete entities rather than continuos sets of points and with approximate numerical solutions of both linear and non linear equations rather than exact analytical solutions of linear equations. In particular B.B. Mandelbrot developed the fractal geometry which describes shapes with similar properties at different scales (Mandelbrot, 1975). The mathematical community debates the value of these new discrete models which are mainly deduced through computer simulations instead of a series of theorems and proofs. Some traditional mathematicians believe fractals are just pretty pictures. Mandelbrot replies that «pretty pictures in the appropriate minds lead to pretty problems and entire new fields». Undoubtedly computer simulation and fractal geometry are very useful tools to understand structures and dynamics of complex systems generated by simple iterative rules. A wide range of natural structures have been quantitatively characterised using the idea of a fractal dimension (Mandelbrot, 1982). Our universe is a fractal (Damiani, 1984; Maddox, 1987) with fractals everywhere (Barnsley, 1988). In particular many biological structures have fractal, branching and vascular morphologies: trees, venation of leaves, lungs and in general most circulatory and transport systems in living organisms, suture lines of ammonoids, aggregator structures produced by myxobacteria and myxomyceta, cromatophores of some animals, nervous cells and many others. As written recently, «the mathematical concept of fractal scaling brings an elegant new logic to the irregular structure, growth and function of complex biological forms» (West & Goldberger, 1987).

2   Structure and function

The formation of structures and shapes in living organisms has long been considered a mysterious and supernatural phenomenon. When the problem was examined by Darwin (1859) in the context of the evolutionary theory, the forms of the living organisms has been explained largely in terms of adaptation produced by natural selection. A natural or artificial selection process requires the following steps: (1) reproduction of discrete units; (2) occurrence of heritable differences among these units; (3) different rate of reproduction of the different units. Individuals with the best fitness to a certain environment have the greatest probability to generate progeny. Biotic and physical factors present in a certain environments are responsible for selective reproduction of living organisms. Selection is cumulative: each generation is build on the successful adaptations of the generation before. Therefore a structural and functional optimization of biological structures is achieved by evolutionary processes. The strong correspondence between form and function supports the neo-Darwinian adaptationist explanation for the biological shapes. However, forms are not only moulded by functional selection, but are subjected to genetic, developmental and physical constraints (Russel, 1916). If functionalism presupposes the primacy of function over form, on the other hand, structuralism emphasises the primacy of form over function (Rieppel, 1988). Internal factors might be more important than external selection. Only few morphological transformation are permitted by the available genetic information and developmental pathway. Only few types of genomic reorganization are viable and economic solutions. Moreover relationships between different structures of an organism tend to vary during its growth. These genetic and structural constraints limit the number of the possible morphogenetic changes which are the substrate for adaptive selection. A compromise solution between the functionalistic and the structuralistic viewpoints was proposed: environment may create form, but only within the limits of the types created by ontogeny (Alberch, 1980).

3   Functional adaptation or random choices?

Some researchers reject the adaptationist program of neo-Darwinian biology (Gould & Lewontin, 1979). For Lewontin evolution is a very complex phenomenon and adaptation is insufficient to explain most of the patterns of change: «Genes, organisms, and environments are in reciprocal interaction with each other in such a way that each is both cause and effect in a quite complex way» (Lewontin, 1983). For Stephen Jay Gould non-adaptive processes, acting by differential origin and extinction of species, direct evolutionary trends within clades (macroevolution), while natural selection, acting by differential birth and death of individual, directs evolutionary change only within populations (microevolution): «Two classical trends are the complexification of suture lines between chambers in ammonite shells and increasing symmetry of the cup in Palaeozoic crinoids. If we explore the speculative explanations offered to account for them we find that these explanations presuppose a direct advantage for some particular morphology under a phyletic regime of natural selection. The complex sutures strengthen the ammonite shell, for example, or perfection of radial symmetry permit sessile organisms like crinoids to collect food from all directions. We also find that such explanations have been notoriously unsuccessful, a plethora of contradictory speculations that never brings resolution. Perhaps the problem lies deeper than our failure to devise a good adaptive story. The geometry of punctuated equilibrium means that we cannot extrapolate natural selection within populations to produce evolutionary trends because trends are a product of the differential fate of species considered as stable entities. Trends may occur simply because some kinds of species speciate more often than others, not because the morphologies so produced have any advantages under natural selection (indeed, such a trend will occur even if extinction is completely random).» (Gould, 1982). Raup and Gould (1974) showed with stochastic simulation of evolutionary processes that many morphological patterns of apparent order arise at unexpectedly high frequency in random models. Are macroevolutionary trends simply extrapolation of microevolutionary events? Are internal factors more important than external selection? Are symmetry and regularity of natural forms and patterns the results of adaptive processes or of random fluctuations? Are there simple universal laws that allow us to comprehend them? It is difficult to answer these general questions but in particular cases it is possible to search a solution. One of these cases is the evolution of biological branching structures.

4   The optimal form

D'Arcy Thompson (1942) tried to identify the most important mathematical and physical forces which moulds the form of organisms so as to assure their functional adaptation: «Few had asked whether all the patterns might be reduced to a single system of generating forces, and few seemed to sense what significance such a proof of unity might possess for the science of organic form.» If some general laws of organic form exist we must search for them in universal patterns which are evolved independently in different groups of organisms. Many structures constructed by growing organisms converge repeatedly on a limited number of architectural designs (Hildebrant & Tromba, 1985). For example the analysis of the organizational properties of animal skeletons in a theoretical morphospace suggest that «organic structure must necessarily approach recurrent elements of design.» (Thomas and Reif, 1993). Undoubtedly the fractal, branching and vascular morphologies are very common biological structures. Many researchers have studied the geometrical and topological properties of these structures which were usually described by centrifugal or centripetal schemes grouping the different bifurcation into different orders (Weibel, 1963; MacDonald, 1983). Quantitative analysis of several parameters, as branch length, branch diameter, and bifurcation angle, demonstrates the presence of general laws and regularities. For example in human lung the diameter of a branch and the length to the branch from the base of the trunk are correlated according to an approximate power law (Weibel, 1963). Some of these laws indicate the geometrical self-similarity of the biological branching structures. Three main processes need to explain regularity and invariance of biological structures: (1) compatibility with organism growth, (2) physical optimization in relation to their physiological functions, and (3) minimization of the energetic cost for their construction and maintenance.

5   Scale invariance during growth

Galileo was one of the first to describe the allometric problem. The change of scale during growth is one of the most important structural problems of pluricellular organisms (McMahon, 1973). Many physiological processes are influenced by the size. For example the mass and the nutritional requirements of an organism increases as the cube of length but surface area only as the square. The problem is that exchanges of matter and energy between a living organism and its environments depend on its surface area. Evolution solved this scaling problems by means of fractal geometry. A fractal structure is endowed with scale-invariance: it reveals the same pattern, more or less ordered, at different scales when examined with magnifying lenses of different strength. The exponent of the allometric equations of Tessier (1931) and Huxley (1932) is the fractal dimension D of Mandelbrot (1975) and describes the self-similarity at different scales of the biological structures and processes. The dimensional analysis explains the meaning of the value of this exponent. In biological branching and vascular structures this value has a spatial meaning. It is between 1 and 3 indicating how a fractal pattern fills a plane (1<D<2) or a space (2<D<3). For example the fractal dimension of human lung is approximately 2,2. An extensive review of the studies on the relationships between size and shape in biology was written by McMahon and Bonner (1983).

6   Physiology of transport systems

The function of branching structures in living organisms is to distribute or to gather biological material (as lymph, blood, etc.) or physical entities (as light, air, bending stresses, etc.). Therefore optimization processes produced by natural selection led to minimization of their length and of their resistance to the flow of the transported entity and to maximizatation of their resistance to mechanical stress. Gutfraind and Sapoval (1993) studying the Laplacian potential around irregular object show that «the fractal geometry can be the most efficient for a membrane or electrode that has to work under very variable conditions». Generally the problem is to transport something from a plane (D = 2) or a volume (D = 3) to a point (D = 0) or viceversa. What is the shortest network of line segment interconnecting an arbitrary set of points? This mathematical problem, known as the Steiner problem, became popular in 1941, when Courant and Robbins included it in their book «What Is Mathematics?». The solution to this problem «has eluded the fastest computers and the sharpest mathematical minds» (Bern and Graham, 1989). An application of the Steiner problem is the construction of minimum-spanning-tree representing the evolutionary relationships. Another variants of the Steiner problem is the construction of a rooted tree to connect a fixed point to a set of points uniformly spread in a limited plane or volume. The biological branching structures are good solutions of this problem. Lungs and plants are two paradigms of the biological transport systems showing their physiological meaning. As Leonardo da Vinci observed: «The total amount of air that enters the trachea is equal to that in the number of stages generated from its branches, like a plant in which each year the total estimated size of its branches, when added together, equals the size of the trunk.» Weibel (1963) described the morphology of the lung airways as a fractal binary tree using a centrifugal scheme. McMahon (1975) and McMahon and Kronauer (1976) have highlight the mechanical principles underlying the branching patterns of trees. Moreover the evolution of land plants has been computer-simulated considering some characteristics of branching patterns and several parameters for quantifying the physiological advantages of the different morphologies (Niklas, 1986). Three characteristics define a universe of possible branching patterns: probability of branching, branching angle and rotation angle. The selective parameters considered are: the light gathering ability, the resistance to mechanical stresses and the capacity to shade neighbouring plants or part of itself. The simulated evolutionary trends are in good agreement with trends found in the fossil record (Niklas, 1986).

7 The ammonoid shell

When a special geometry is required to solve particular physiological functions it is difficult to understand the relationship between function and form. For example the meaning of the evolution of ammonoid septa and their associated suture lines is not obvious. The ammonoid phragmocone was a rigid floating structure functioning like a small submarine. The functional problem is to have maximum volume with minimum weight. The analysis of the mechanical principles involved in the static of ammonoid shell (Seilacher, 1975; Damiani, 1986; Hewitt & Westermann, 1986; Hewitt & Westermann, 1987) leads to the conclusion that continuous increase in septum complexity serves to increase shell resistance to hydrostatic pressure applied via the external shell wall (Westerman, 1975) and the body chamber (Pfaff, 1911). The problem is to gather the bending stress on a limited surface of the external shell ($D = 2$) and to transmit these stresses to the perpendicular surface of the internal septum ($2<D<3$). Moreover the septum is generally subjected to different perpendicular pressure on the two sides. The ontogenetic and phylogenetic development of septum and suture lines has improved a balanced distribution of the deformation energy in every element of the shell. The septum is an equipotential fractal surface. The junction of the external shell with the internal septum is named suture line ($1<D<2$) and it has a fractal geometry. Everybody can observe the morphological analogy between the septum and the plots of equipotential lines around a fractal cluster (Mandelbrot et al., 1990) and between complex suture lines and the tree-like fractals structures (Damiani, 1989). In particular, the ontogenetic and phylogenetic development of Mesozoic ammonoid suture lines is analogous to the development of a line discovered in the 1904 by the mathematician Koch (Damiani, 1987; Damiani, 1989).

8 Minimization of energetic cost

Structures construction and storage of genetic information for the morphogenetic processes have an energetic cost. Therefore optimization processes produced by natural selection led to minimization of morphogenetic changes and of the information amount needed for construction of the biological structures. Iteration of a simple rule can generate very complex system. The somatic development of an organism requires the following steps: (1) reproduction of discrete units; (2) occurrence of differences among these units; (3) different rate of differentiation and reproduction of the different units (Eigen and Schuster, 1979). These rules are similar to those required for an evolutionary process: evolution works at individual level and development at cellular level. Even if ontogenesis is often a recapitulation of phylogenesis, clearly the spatial and temporal self-organisation of a pluricellular organism is a process more deterministic than the evolutionary dynamics of a population. Undoubtedly computer simulations are very useful tools to understand structures and dynamics of complex systems generated by simple iterative rules. A branching structures is produced by the repetition of a set of simple rules called generator: every branch splits in two other branches in each generation. The number of the branches is duplicated in each generation cycle. Other generators producing fractal structure have some characteristics in common with the generators producing the tree-like branching structures: the new tips are located at the free extremity of the old tip in the tree-like lines, at an asymmetric position in the Mandelbrot line and at the middle of the old tip in the Cesaro line (and also in most of the other Koch-like lines). A great variety of different structures is produced by slight changes of generation rules. It is not surprising that recursive branching is a good metaphor for organisms evolution and for embryonic development. Dawkins (1986) uses simple branching rules to simulate the spectacular evolution and development of an unlimited number of Biomorphs in the computer world.

9   Morphogenetic rules

The description of a biological branching structure as a tree has some limitations. Asymmetrical branching and cross-linking of branches, technically known as anastomosis, are not considered in these schemes. Moreover the structural analogy between biological branching forms and Koch-like lines do not explain the mechanism of morphogenetic processes. The question of how cells estimate their location within the body has preoccupied embryologists for the past century. The concept of gradients of unknown diffusible substances, first propounded by Morgan in the 1905, have been invoked as determinants of polarity in a wide variety of organisms. In the 1952 Turing realised that a particular level of the postulated gradients might determine the developmental patterns. At any given position cells are able to feel the local concentration of a morphogenetic gradient and they may move and/or differentiate in reply to it. In some cases cells are sensitive to physical stresses and therefore the ontogenetic development is determined by interaction between genetic and environmental elements. This is the case of ammonoid septum construction: successive septa of the same shell vary their patterns in response to the corrugations of the shell wall which produce different distribution of the bending stresses (Seilacher, 1975). Oster et al. (1980) suggest that mechanical deformation influencing both the viscosity and the elasticity of the cellular cytogel are the major factor determining morphogenetic transformation. The morphogenetic mechanisms suggested by Turing can explain the formation of a wide range of patterns (Gierer and Meinhardt, 1972).

10 Cellular automata and viscous fingers

Simple computer simulations of the Turing reaction-diffusion model are based on the so called «cellular automata». A cellular automaton is a set of interacting entities that can replicate themselves (von Neumann, 1966). The evolution of a cellular automaton depends on the initial configuration of the cells and on the rules for the calculation of the next state of each cell in each generation (Hayes, 1984; Wolfran, 1984). An important programme of this type is the diffusion limited aggregation (DLA) model developed by Witten and Sander in the 1981 to simulate the dendritic growth (Sander, 1986; Sander, 1987). A programme similar to the DLA is the dielectric breakdown model (DBM) by Niemeyer, Pietronero and Wiesmann (1984) in which the tip splitting and growth are regulated by different parameters. These models produces branched patterns similar to the «viscous finger» formed by a fluid when it is forced under pressure into another immiscible fluid of higher viscosity (Nittmann, Daccord & Stanley, 1985; Nittmann, & Stanley, 1986). The experimental study of these phenomena involves an apparatus called the Hele-Shaw cell which consists of viscous fluid confined between two parallel plates. If a less viscous fluid is injected into the middle of the cell, it breaks up into many branched fractal structures displacing the higher viscous fluid. Another simple experimental apparatus for studying the viscous finger patterns is obtained by squeezing a drop of a viscous substance between two glasses which are then pulled off slowly (Damiani, 1984; Damiani, 1986). The attractive force among the molecules of the viscous substance and the tendency to go toward the interface between the substance and the air produce convergent vascular structures.

11 How evolution solved the Steiner problem

I have developed and studied many cellular automata producing different kinds of divergent or contracting many-branched, fractal and vascular structures (Damiani, 1984). A general characteristic of these programmes is that when uniformly distributed entities are diffused or concentrated by a repulsive or attractive force, they respectively produce divergent or contracting fractal and vascular structures. For example the DBM simulate a diffusion process with tip splitting and DLA describe the fractal growth by means of random aggregation. A very important aspect in these simulated processes is that recursion of few rules produces complex multidimensional lattices beginning with simple monodimensional entities or viceversa. Similar processes and rules produce a variety of inorganic and organic branched patterns. What are the recursive rules chosen by nature to generate minimal networks? Experimental data and computer simulations suggest two complementary ways to construct Steiner networks: punctiform entities are spread or concentrate in a surface or in a volume responding to gradients of physical forces (as bending stresses) or of chemical morphogenetic elements. In the diffusion model the sensitive entities go away according to a repulsive force and when one entity is too isolate it splits into two parts. In the contraction model uniformly distributed entities converge according to an attractive force and when two entities are very close they stick together. These simple recursive local rules leads to the formation of fractal, branching, vascular and global networks. Slight changes of morphogenetic gradients and of responsiveness of sensitive entities influence the shapes of produced patterns. Natural selection allow survival of the biological structures more suitable for their physiological functions. Computer simulations by means of genetic algorithms (Bound, 1987) reconstruct morphological changes found in the fossil record with reasonable accuracy.

12 Conclusion

The results of computer and mechanical simulations strongly support the hypothesis that some biological forms, as the branching structures, are the most functional design in relation to the material properties of the universe. These forms «are topological attractors that evolution cannot avoid.» (Thomas and Reif, 1993). A paradigm of this functional design is the complex structure of ammonoid septum which increases the shell resistance to mechanical stresses. The adaptionist program of neo-Darwinian biology is often sufficient to explain evolutionary trends. When it fails perhaps the problem is our failure to devise a good adaptive story. The optimized structures and procedures developed by natural selection may be applied successfully to solve a wide variety of scientific and engineering problems (Otto, 1982).

References

[1] Alberch P., Ontogenesis and morphological diversification, Amer. Zool., 20, 653–667 (1980)

[2] Barnsley M., Fractals everywhere (Academic Press, Inc.) (1988)

[3] Bern and Graham R.L., The shortest-network problem, Scientific American, 60, 66–71 (1989)

[4] Bound D.G., New optimization methods from physics and biology, Nature, 329, 215–219 (1987)

[5] Courant R. and H. Robbins, What is mathematics?, (Oxford University Press), (1941)

[6] Damiani G., Il gioco della vita, (E.I.A.ed., Roma) (1984)

[7] Damiani G., Significato funzionale dell'evoluzione dei sette e delle linee di sutura dei naulilordi e degli ammonoidi, In: Fossili, Evoluzione, Ambiente (Pergola, Italy) 1, 123–130 (1984)

[8] Damiani G., Simulations of some ammonoid suture lines, In: Fossili, Evoluzione, Ambiente (Pergola, Italy) 2, 221–228 (1987)

[9] Damiani G., I frattali e le linee suturali delle ammoniti, Le Scienze, 245, 60–68 (1989)

[10] Darwin C., On the origin of species (Murray J. ed., London) (1859)

[11] Dawkins R., The blind Watchmaker (Longman) (1986)

[12] Eigen M. & Schuster P., The Hypercycle, (Springer Verlag) (1979)

[13] Gierer A. & Meinhardt H., A theory of biological pattern formation, Kybernetik, 12, 30–39 (1972).

[14] Gould S.J., The meaning of punctuated equilibrium and its role in validating a hierarchical approch to macroevolution. In: Prespectives on evolution, (Milkman R. ed., Sinauer) 83–104 (1982)

[15] Gould S.J. & Lewontin R.C., The spandrels of San Marco and the Panglossian paradigm: a critique to the adaptationist programme, Proc. Roy. Soc. Lond. Ser. B., 205, 581–598 (1979)

[16] Gutfraind R. and Sapoval B., Active surface and adaptability of fractal membranes and electrodes, submitted to Journal de Physique I (1993)

[17] Hayes B., The cellular automaton offers a model of the world and a world unto itself, Scient. Am., 250, 10–16 (1984)

[18] Hewitt, R.A. & Westermann G.E.G., Function of complexly fluted septa in ammonoid shells. I. Mechanical principle and functional models, N. Jb. Geol. Palaont., Abh., 172, 47–69 (1986).

[19] Hewitt, R.A. & Westermann G.E.G., Function of complexly fluted septa in ammonoid shells. II. Septal evolution and conclusions, N. Jb. Geol. Palaont., Abh., 174, 135–163 (1987)

[20] Hildebrant S. & Tromba A.J., Mathematics and optimal form (Freeman, New York) (1985)

[21] Huxley, J.S., Problems in Relative Growth (Methuen, London) (1932)

[22] Lewontin R.C., Gene, organism and environment, In: Evolution from molecules to men (Bendall D.S. ed., Cambridge University Press) (1983)

[23] Koch H. von, Sur une courbe continue sans tangente, obtenue par une construction geometrique elementaire, Arkiv for Matematik, Astronomioch Fysik, 1, 681–704 (1904)

[24] MacDonald N., Trees and networks in biological models, (Wiley J. and sons ed.) (1983)

[25] McMahon T.A., Size and shape in biology, Science, 173, 1201–1204 (1973)

[26] McMahon T.A., The mechanical design of trees, Scientific American, 233, 92–102 (1975)

[27] McMahon T.A. & Kronauer R.E., Tree structures: deducing principles of mechanical design, J. theor. Biol., 59, 443–466 (1976)

[28] McMahon T.A. & Bonner J.T., On size and life (Freeman, New York) (1983)

[29] Maddox J. The universe as a fractal structure, Nature, 329, 195 (1987).

[30] Mandelbrot B., Les objets fractals: forme, hasard et dimension, Flammarion, Paris (1975)

[31] Mandelbrot B., The Fractal Geometry of Nature (Freeman, San Francisco) (1982)

[32] Mandelbrot B. & Evertsz C.J.G., The potential distribution around growing fractal clusters, Nature, 348, 143–145 (1980)

[33] Morgan T.H., The pysical basis of heredity (Lippincott Co., Philadelphia) (1919)

[34] Neumann von J., Theory of self-reproducing automata, (University of Illinois Press, Urbana) (1966)

[35] Niemeyer, Pietronero & Wiesmann, Fractal dimension of dielectric break-down, Phys. Rev. Lett., 52, 1033–1036 (1984)

[36] Nittmann J., Daccord G. & Stanley H.E., Fractal growth of viscous fingers: quantitative characterization of a fluid instability phenomenon, Nature, 314, 141–144 (1985)

[37] Nittmann J. & Stanley H.E., Tip splitting without interfacial tension and dentritic growth patterns arising from molecular anisotropy, Nature, 321, 663–668 (1986)

[38] Niklas K.J., Computer-simulated plant evolution, Scientific American, 254, 68–75 (1986)

[39] Oster G.F., Odell G.M., Alberch P. & Burnside B., The mechanical basis of morphogenesis, Devel. Biol., 85, 446–462 (1981)

[40] Otto F., Natürliche Konstruktionen, (Deutsche Verlags, Stuttgart) (1982)

[41] Pfaff, E., Über Form und Bau der Ammonitensepten und ihre Beziehungen zur Suturlinie, Iber. nieders. geol. Vers., 4, 207–233 (1911)

[42] Raup D.M. & Gould. S.J., Stochastic simulation and evolution of morphology — towards a nomothetic paleontology, Systematic Zoology, 23, 305–322 (1974)

[43] Rippel O.C., Fundamentals of comparative biology (Birkhäuser Verlag, Basel) (1988)

[44] Russel E.S., Form and function (Murray J, ed., London) (1916)

[45] Sander L.M., Fractal growth processes, Nature, 322, 789–793 (1986)

[46] Sander L.M., Fractal growth, Scientific American, 256, 82–89 (1987)

[47] Seilacher A., Mechanische Simulation und funktionelle Evolution des Ammonitenseptums, Paleont. Z., 49, 268–286 (1975)

[48] Tessier G., Recherches morphologiques et physiologiques sur la croissance des insectes, Travaux de la station biologique de Roscoff, 9, 27–238 (1931)

[49] Thomas R.D.K. & Reif W.E., The skeleton space: a finite set of organic designs, Evolution, 47, 341–359 Thompson D'Arcy W., On Growth and Form, (Cambridge University Press) (1917)

[50] Turing A.M., The chemical basis of morphogenesis, Philosophical Transactions of the Royal Society, B237 (1952)

[51] West B.J. & Goldberger A.L., Physiology in fractal dimensions, American Scientist, 75, 354–365 (1987)

[52] Weibel E.R. (1963) Morphometry of the human lung (Springer, Berlin) (1963)

[53] Wetermann G.E.G., Model for origin, function and fabrication of fluted cephalopod septa, Paleont. Z., 49, 235–253 (1975)

[54] Witten T.A. & Sander L.M., Diffusion limited aggregation, a kinetic critical phenomenon, Phys. Rev. Lett., 47, 1400–1403 (1981)

[55] Wolfran S., Cellular automata as models of complexity, Nature, 311, 419–424 (1984)

Relationship Between the Branching Pattern of Airways and the Spatial Arrangement of Pulmonary Acini — A Re-Examination from a Fractal Point of View

Hiroko Kitaoka[1] and Tohru Takahashi[2]

1) Department of Internal Medicine, Kitaoka Hospital, Kurayoshi 682, Japan
2) Department of Pathology, Institute of Development, Aging and Cancer, Tohoku University, Sendai 980, Japan

Abstract. The upper lobe of the left lung, surgically removed from an adult for a small carcinoma, was subjected to serial slicing and computer assisted 3-D reconstruction of airways and acini. A fractal dimension of the airways of about 1.74 was obtained applying a 3-D box-counting to the graphics data. At the same time, volumetry was performed of the acinus, the structural and functional unit of the lung, using the same software. The volumes obtained from a total of 130 acini proved to be comparatively uniform, showing a normal type distribution with a mean of 173 ± 38 mm^3. The volume of acini had no correlation with their location in the lung or with the generation number of the terminal bronchioles supplying them, and this was considered to be reflecting a homogeneous 3-D arrangement of acini in the organ. Furthermore, a simple 3-D model of lung was introduced to analyze the relationship between the branching pattern of airways and the volume of acini supplied by them. It was shown that the self-similarity of the spatial distribution of airways is consistent with a volumetrically homogenized space division of the lung.

1   Introduction

The structure of the human lung can be simplified as a finely branching tree, with the principal bronchus corresponding to its stem. Two parts are discriminated in this tree: the airways and the air spaces. The airways are thick segments composing the conducting part of the tree, starting at the principal bronchus and gradually thinning as the segments divide several times over the bronchi and bronchioli until they end at the terminal bronchioles (TBs). Division further continues, but the area of lung tissue belonging to one TB is called an acinus (Fletcher et al. 1957). The acinus forms a respiratory part of the lung and serves as a structural and functional unit. The lung consists of a vast number of acini that are packed in the space according to some rule. Thus, there are two aspects in the structure of the lung: a tree, and at the same time, an aggregation of unitary bodies, both closely correlated in the organ formation.

The concept of the diameter exponent was proposed about one century ago by Thoma (1901), and recently referred to by Mandelbrot (1982) as a fractal property. In a branching ductal structure, this allows one to estimate the flow Q through a branch of d in diameter by

$$Q = Cd^n \tag{1}$$

where n is diameter exponent and C is a constant. When the branching pattern is assumed to be a symmetric regular dichotomy: a parent branch divides into two daughter branches, each having the same length and diameter, this equation is equivalent to the following:

$$d_z = d_0 2^{-z/n} \tag{2}$$

where d_z is the diameter of the branch in generation z. Weibel (1963, 1991) stated that the airway diameter was accordant with the above equation in which n was 3, in the range of z from 2 to 10, with the trachea being the 0th generation.

The diameter exponent of human systemic arteries n was estimated at 2.6–2.7 by Suwa and Takahashi (1971). Suwa (1981) attempted to induce this value of n from a polyhedral space division point of view. Here he assumes that all the supracellular structures of organs are subordinate to a particular space division which he designated as «equilibrium space division», implying that it minimizes the potential energy of the constituent units in a field of mechanical force. Aside from his statements, the equation concerning the diameter exponent is likely to express at least the equivolumic space division by the branching structure. It may be reasonable to assume that when terminal branches have an equal diameter, the flow of air or blood is equal and, therefore, also the volume of tissues supplied by the branches must be equal.

In this article, we present an attempt at correlating the branching pattern of human airways with the volume of acini, by performing 3-D reconstruction from serial slices of a human lung. With an additional analysis of 3-D airway models, discussion will be extended to the relationship between the branching pattern and the space division of lung.

2 The spatial distribution of the airways

2.1 Preparation of a lung specimen

Our investigation of the 3-D structure of lung was greatly facilitated by introducing Heitzman's fixation of lung (Heitzman, 1973). This technique allows a fixed lung specimen to retain an inflated state without being collapsed even in dry conditions. The lung specimen thus fixed was cut into serial thin slices using a microslicer, a devise for slicing soft tissues at a constant thickness of several hundred microns.

A left upper lobe was obtained from a female, aged 51, in whom lobectomy was performed for small lung cancer of the apicoposterior segment. Besides the cancer, there were morphologically no abnormalities in airways or air spaces. The resected lobe was fixed by Heitzman's method we modified, using a fixative containing polyethylene glycol (Kitaoka and Itoh 1991, 1992). After fixing for two days at a constant intrabronchial pressure of 25 cm H_2O, a block, about 5.5×4.5 cm, was taken from the anterior segment to study the 3-D properties of airways and acini contained. This block was sliced serially at a thickness of 0.5 mm with a microslicer and 48 serial slices were obtained (Figure 1).

The other part of the lobe was sliced serially at a thickness of 2 mm, in order to reproduce the airway tree of the lobe. All the slices were submitted to radiography. The radiographs presented the internal structure of the slices as in Figure 2. The entire airway tree from the left upper bronchus down to TBs was followed in the slices, and the generation numbers were determined for all the TBs, according to Weibel's definition (Weibel 1963). In the next place, the contours of airways in each slice were inputted into a microcomputer (PC-9801; NEC, Tokyo) by digitization, using a software designed for 3-D visualization and morphometry. Three-D images of airways were then integrated in a computer display as in Figure 3.

2.2 Measurement of the fractal dimension for the conductive airways

The software was also designed so as to apply the 3-D box-counting method to the data of airways thus inputted. The size of cubic box for sampling was variable, and had a multiple of 0.5 mm, the thickness of a single slice. The software allowed a computer to count the number of boxes containing the reconstructed airways. Three places in the specimen were selected for the measurement of fractal dimension (Figure 4). At one of them, a cube with an edge of 24 mm was set (Space A), and at each of the other two, a cube with an edge of 18 mm (Spaces B and C). In Space A, most branches proved to rise from a bronchus of sixth generation. Space B contained a tree branching from a bronchus of 9th generation. Space C had two airway trees.

Figure 5 shows log-log plots where a, the edge length of the sampling box in the abscissa, is related with $N(a)$, the number of boxes containing the airways shown in the ordinate. There are close linear correlations in the range of edge length from 0.5 to 6 mm. The slopes were almost the same in the three samplings, being 1.74, 1.73 and 1.74 for Spaces A, B and C, respectively. The branching generations of the airways contained in Space A ranged from 6 to 14. Those in Space B from 9 to 16, larger by about three generations than in Space A. Spaces B and C, having the same volumes (18^3 mm^3), significantly differed in the volume of the airways contained: the volume in Space C was about three-fourths of that in Space B. This is reflected in the different y — intercept in Figure 5. However, the slopes of the regression lines for the three spaces were almost the same. Thus, despite such local differences as revealed with conventional metric quantities like the number of branching generation or the volume density, the airways have a constant fractal dimension, which gives an integrated description for the spatial distribution of airways.

It has been reported several times that the cluster-cluster aggregation in the 3-D space gives rise to having a fractal dimension of about 1.75. Meakin (1987) estimated the fractal dimension by computer simulation. According to Weitz and Oliviera (1984), who experimentally estimated the fractal dimension of the aggregating gold colloids in the 3-D space by transmission electron microscopy, the

Fig. 1   Forty-eight serial slices from a fixed lung specimen. Thickness of each slice is 0.5 mm.

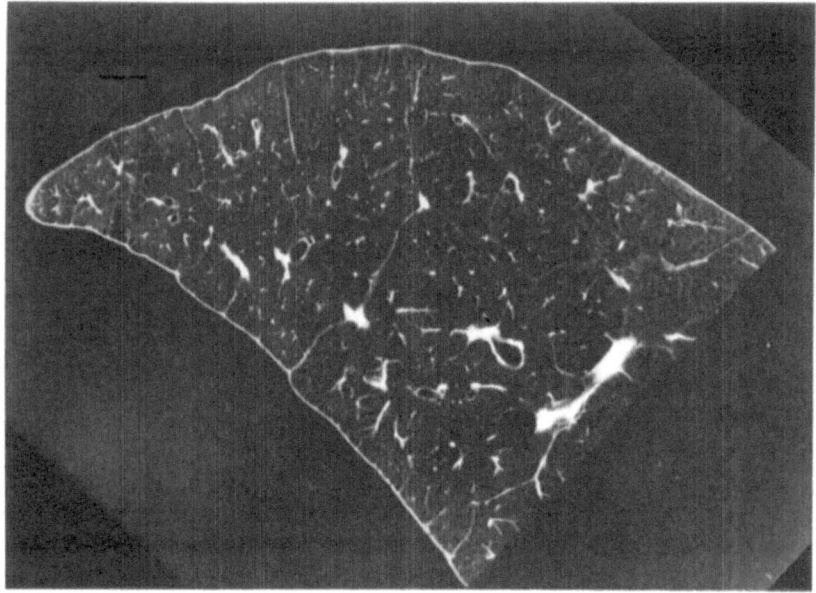

Fig. 2   Radiograph of a slice. Bar represents 5 mm.

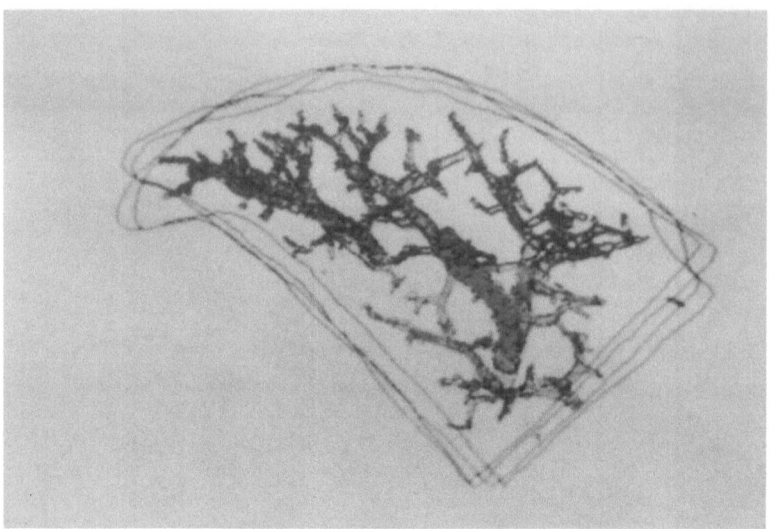

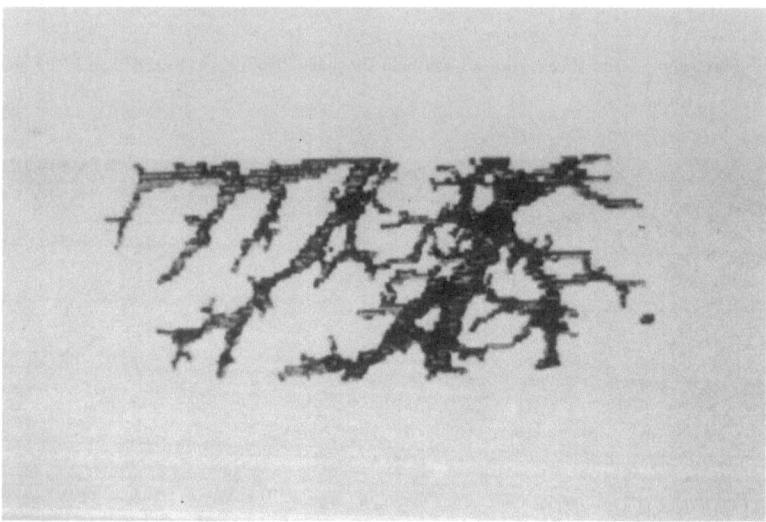

Fig. 3   Three-D reconstruction of the airways in the computer display. The upper figure presents an upward view from the bottom. The lower one presents a frontal view.

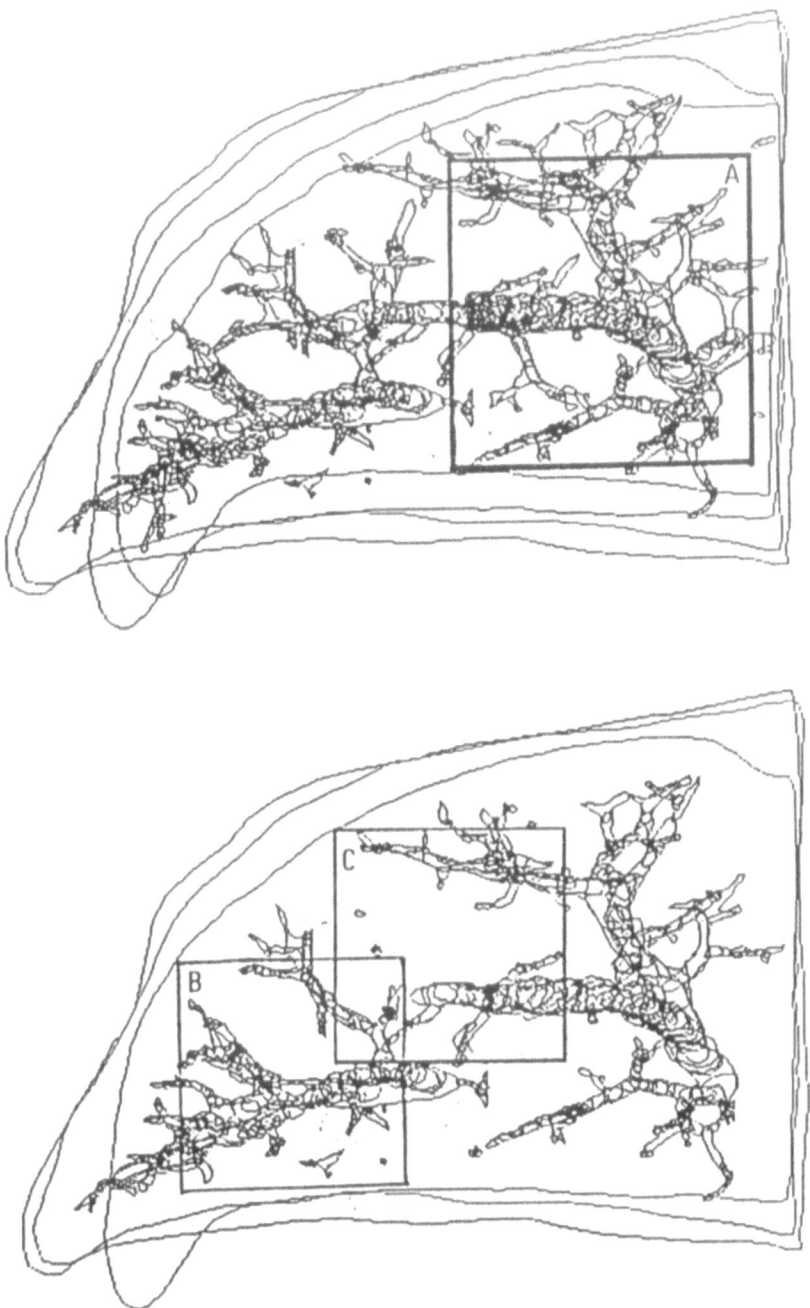

Fig. 4 Spaces for measuring of fractal dimensions. Space A is a cube with an edge of 24 mm, and Spaces B and C are cubes of with edges of 18 mm.

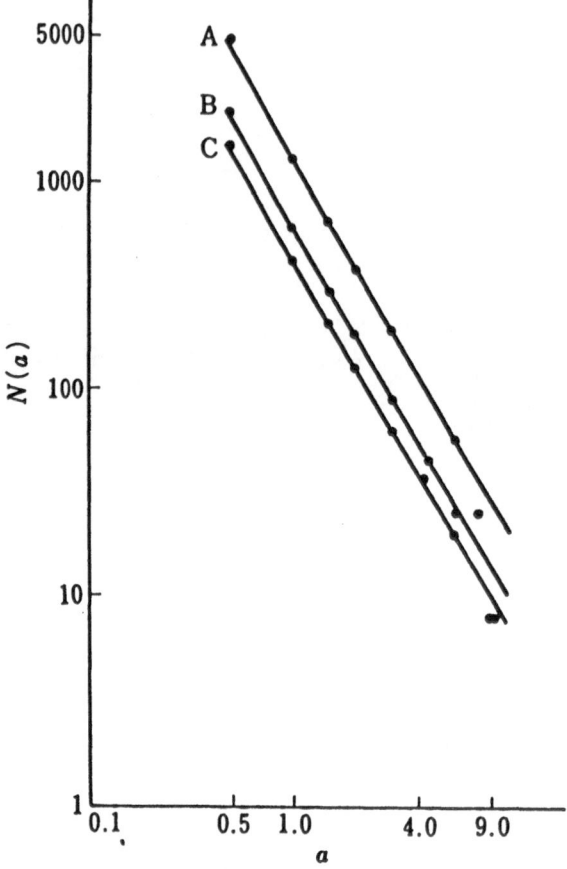

Fig. 5   Log-log plots of the edge length of the box a, versus the number of boxes containing the airways $N(a)$.

fractal dimension also proved to be about 1.75. The proximity of the airway fractal dimension to these suggests that there is a principle common to living and non-living matters.

3   Volumetry of the pulmonary acini

It was confirmed in a 3-D scanning through the 48 serial slices, that there were altogether 171 TBs. Of these, 130 had their acini wholly contained in the volume. For all these 130 acini, the contour was defined on the slices and inputted into the computer using the 3-D reconstructionsoftware. The volume of individual acini was calculated on Cavalieri principle.

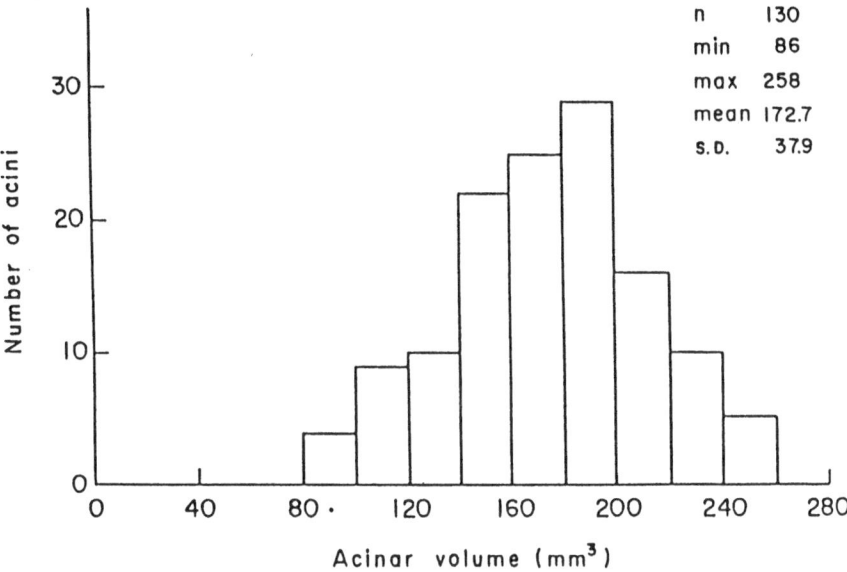

Fig. 6   The histogram of acinar volumes estimated.

The estimated volumes of the 130 acini showed an approximately normal distribution with a mean of 173 ± 38 mm³ (Figure 6). There was no significant correlation between the volumes of acini and the generation numbers of the TBs supplying them (Figure 7). Furthermore, in another investigation, it was proved that the location of an acinus had nothing to do with its volume (Kitaoka and Itoh 1992). The acini were classified into three groups according to their location. Group A comprised those facing the costal surface, Group B those facing the mediastinal surface and Group C those having no pleural surface. As shown in Table 1, Group A proved to differ significantly from the other groups in the generation number of the supplying TB. The mean in Group A was larger than the other groups by about 2 generations. This difference suggests that the TBs in the outer zone of the lung like Group A were more distant from the central airways than in the inner zone like Groups B and C. In the acinar volume, however, there was no significant difference among the three groups.

These results suggest that the airway tree is designed so as to assign a constant volume to the acini they supply. On the other hand, the branching pattern itself has been proved to be statistically self-similar. It appears that the self-similar branching of the airways serves, somehow, to realize the equivolumic division of the air spaces.

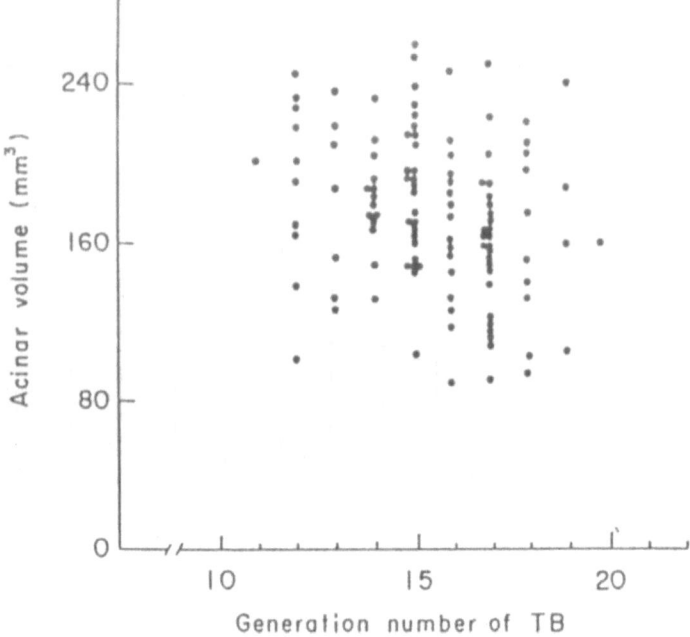

Fig. 7 Correlation between the number of generation of TBs and the acinar volume. The correlation coefficient is −0.24, which is not significant.

Group	n	Volume of acini (mm³)	Generation number of TBs
A	43	174 ± 35	16.7 ± 1.2*
B	23	176 ± 38	14.8 ± 1.6
C	61	170 ± 40	14.6 ± 1.7

Group A, acini facing the costal surface; Group B, those facing the mediastinal surface; Group C, those not facing any pleural surfaces.
* $p < 0.01$ between A and B and also between A and C.

Table 1 Correlation among the location of acini, the volume of acini and the generation number of TBs.

4   Three-D models of the branching tree and the space division

In order to give a ground for this presumption, 3-D models of airways are to be introduced. Model 1, as in Figure 8a, has a symmetric dichotomous branching. The figure presents a view from a Z direction. Suppose that the whole space is a cube with an edge of 24 mm and divided into 64 small cubic cells, each with an

Generation	Diameter (mm)	Length (mm)
0	18	120
1	12.2	47.6
2	8.3	19.0
3	5.6	17.6
4	4.5	12.7
5	3.5	10.7
6	2.8	9.0
7	2.3	7.6
8	1.86	6.4
9	1.54	5.4
10	1.30	4.6
11	1.09	3.9
12	0.95	3.3
13	0.82	2.7
14	0.74	2.3
15	0.66	2.0
16	0.60	1.65

Table 2 Human airway Model A (Weibel 1963).

Generation number	Diameter (mm)	Length (mm)
5 6 7	4	12
8 9 10	2	6
11 12 13	1	3

Table 3 Diameters and lengths in the airway model.

edge of 6 mm. Let us assume that these small cubic cells correspond to pulmonary acini. The end points of the tree are located at the centers of the cubes. Referring to the data of the specimen used in this study and those of Model A of Weibel (1963, Table 2), the generation numbers, diameters and lengths of the branches were determined as in Table 3. Both the diameter and length of branches were assumed to decrease to half after branching three times. This is nearly synonymous to say that the diameter exponent of the airway is 3. Strictly speaking, the volumes of cubic acini differs to a certain extent because they contain different volume of airways. The volume of the largest acinus was calculated at 212 mm^3, that of the smallest at 172 mm^3, with a mean of 198 mm^3. However, the difference is so little as to be disregarded. Figure 8b shows the frontal view of the space. Now, we imagine that we cut the whole space into 48 serial slices at a thickness of 0.5 mm and follow the same steps we did in measuring the fractal dimension in the actual lung specimen. However, the edge length of the sampling box was limited to three levels of 0.5 mm, 1.5 mm and 3.0 mm, because all airways in this model were running along the edges of acini or penetrating toward their centers. As shown in Figure 8c, the log-log plot shows a linear relation with a slope of 1.74, the same as the fractal dimension in the actual lung.

Model 2 was constructed so that the volumes of the acini might not be equal. Figure 9a shows a symmetric dichotomous branching pattern except for the pairs of terminal branches whose lengths are not equal, with the larger one three times longer than the other. Correspondingly, different volumes were assigned to the acini at a ratio of 1:3. In this model, log-log plot was not linear (Figure 9b). Model 3 was proposed as an asymmetric branching model (Figure 10a). In this model, each acinus has the same volume as in Model 1, but the last three branches were modified as in Figure 10b. The generation numbers of the terminal branches were counted from 11 to 15. The branching pattern of airways in the actual lung is more similar to this model than to Model 1. The result of 3-D box-counting in Model 3 was almost the same as in Model 1, giving a fractal dimension of 1.74.

In addition, another model (Model 1′) was proposed, changing the diameters of branches in Model 1 (Fig. 11a). All the branches were given the same diameter of 1 mm so that the diameter exponent was ∞. In this model, log-log plot was not linear (Fig. 11b). $N(a)$ calculated in the four models are presented in Table 4.

These results seem to be suggesting that self-similarity of a branching tree is realized when the following two conditions are satisfied: the equivolumic space division and a diameter exponent of near by 3. The results obtained in Models 1 and 3 indicates that whether the branching routes to the end points are symmetric or not has nothing to do with the statistical self-similarity if the two conditions are met. In Model 3, the generation numbers of the terminal branches ranges between 11 and 15 because of asymmetric branching, though their diameters are the same. In an asymmetric branching system, which is commonly seen in living organisms, the generation number seems to bear little significance. This may explain why relationship between the generation number and the diameter of airways defined by Weibel (1963) does not apply to the range beyond the 11th generation.

Of course, the acinus in the actual lung is not cubic, and the end points of TBs are not always located at their centers. However, these models indicate at least the relationship between the branching pattern and the space division in a simplified way.

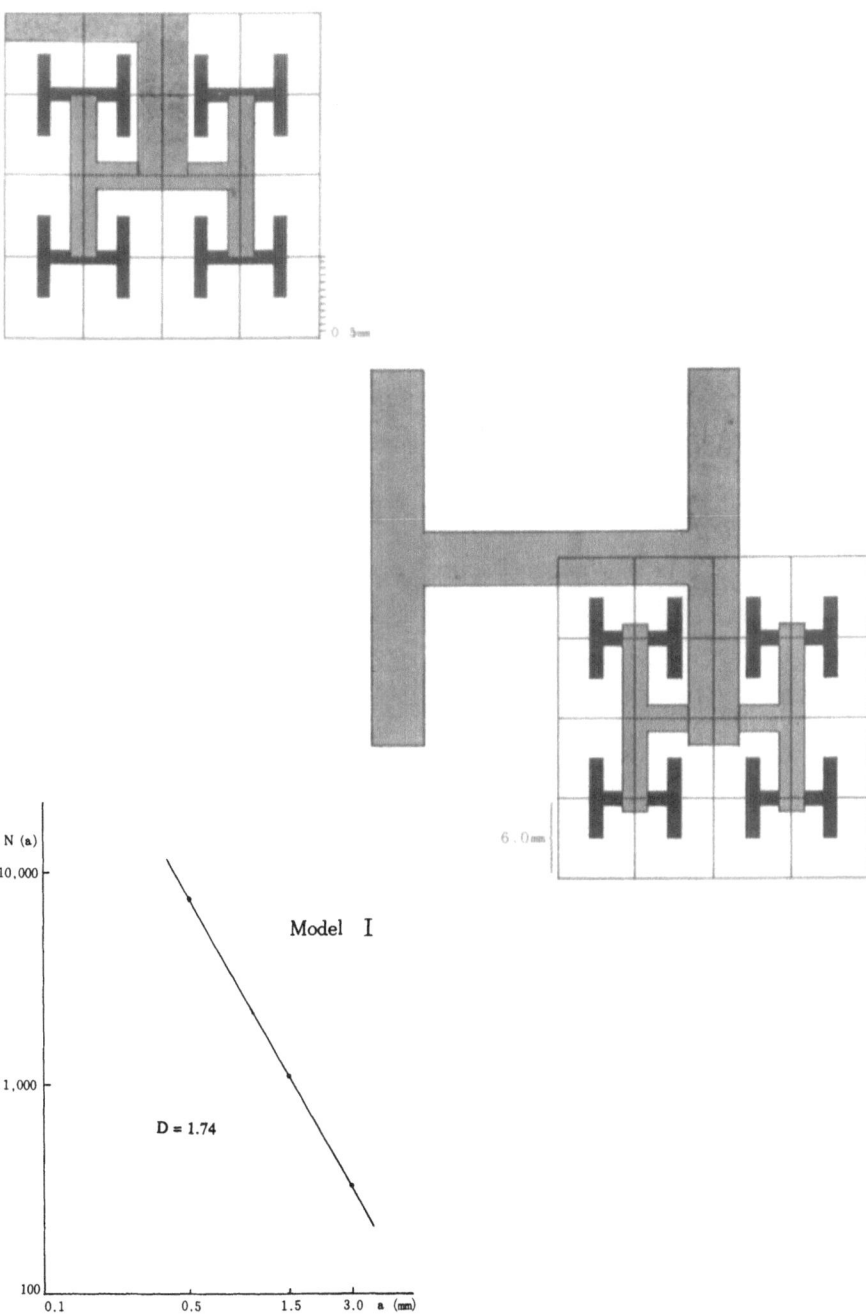

Fig. 8 A symmetric dichotomous branching (Model 1). A view from a Z direction (a) and a frontal view (b). The whole space, a cube with an edge of 24 mm, is cut into 48 serial slices at a thickness of 0.5 mm. The fractal dimension is estimated at 1.74 (c).

	N(0.5)	N(1.5)	N(3.0)	D
Model 1	7359	1048	325	1.74
2	6563	993	397	×
3	7363	1030	328	1.74
1'	3339	921	325	×

Table 4 Calculated N(a) and fractal dimension D in the four airway tree models.

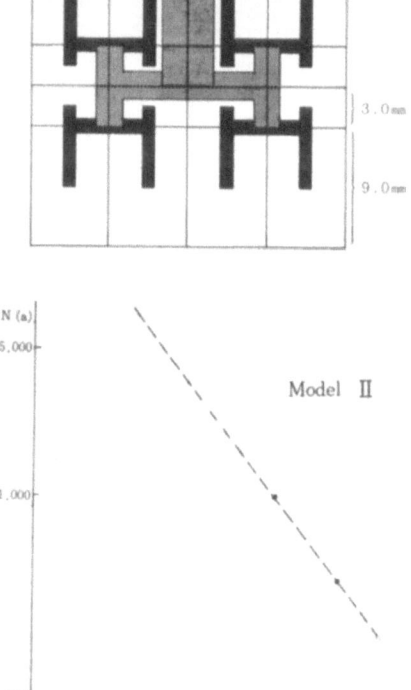

Fig. 9 Model 2 is constructed so that the volumes of acini might not be equal. The pairs of terminal branches are different in length (a). Log-log plot is not linear (b).

5   Conclusion

Though as yet unable to give a coherent explanation, we may assume that there is a principle which closely relates the self-similarity of airways with the space division of lung tissue into unitary acini of an equal volume. Room is left for the

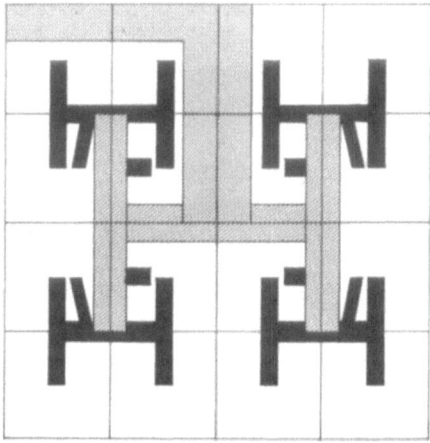

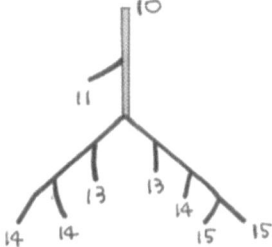

Fig. 10 Model 3 is an asymmetric branching in which acini have the same volumes (a). The branching
pattern of the last three branches is modified (b).

concept «equilibrium space division» by Suwa to be re-examined from a statistical
physics point of view. However, this concept seems to provide an important key
to disclose a fractal-related design principle for living organisms.

References

[1] Heitzman, E.R. (1973) The lung. Radiologic pathologic correlation. Mosby,
    St. Louis.

[2] Kitaoka, H. and Itoh, H. (1991) Spatial distribution of the peripheral airways
    — application of fractal geometry. Forma 6, 181–191.

[3] Kitaoka, H. and Itoh, H. (1992) Computer-assisted three-dimensional volume-
    try of the human pulmonary acini. Tohoku J. Exp. Med. 167, 1–12.

[4] Mandelbrot, B.B.(1982) The fractal geometry of nature. Freeman, San Fran-
    cisco.

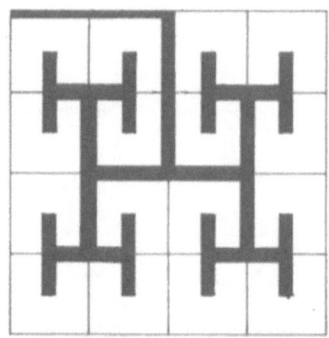

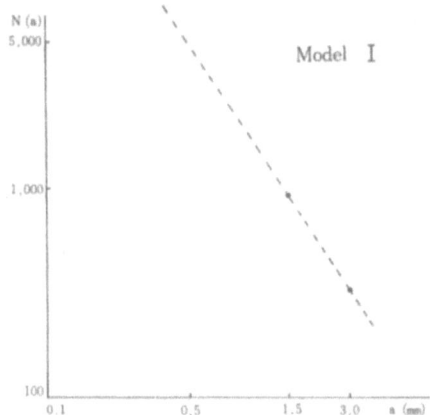

Fig. 11 Model 1′ consists of branches whose diameters are all equal (a). The branching pattern is the same as in Model 1. Log-log plot is not linear (b).

[5] Meakin, P. (1987) Diffusion-limited cluster-cluster aggregation. In: Domb, C. and Lebowitz, J.L. eds. Phase Transitions and Critical Phenomena, Academic Press, New York, pp. 432–439.

[6] Suwa, N. and Takahashi, T. (1971) Morphological and morphometrical analysis of circulation in hypertension and ischemic kidney. Urban & Schwarzenberg, München-Berlin-Wien.

[7] Suwa, N. (1981) Supracellular structural principle and geometry of blood vessels. Virchows Archiv A 390, 161–179.

[8] Thoma, R. (1901) Über den Verzweigungsmodus der Arterien. Arch. Entwicklungsmechanik, 12, 352–414.

[9] Weibel, E.R. (1963) Morphometry of the human lung. Springer, Heidelberg.

[10] Weibel, E.R. (1991) Fractal geometry: a design principle for living organisms. Am. J. Physiol. 261, L361–369.

[11] Weitz, D.A. and Oliveria, M. (1984) Fractal structures formed by kinetic aggregation of gold colloids. Phys. Rev. Lett. 52, 1433–1436.

Multivariate Characterization of Blood Vessel Morphogenesis in the Avian Chorioallantoic Membrane (CAM): Cell Proliferation, Length Density and Fractal Dimension

Haymo Kurz, Jörg Wilting and Bodo Christ
Anatomisches Institut II der Universität
D-79001 Freiburg

Abstract. The development of blood vessels in the CAM of chicks between 6 and 19 days of development was characterized as a time series of their fractal dimension D. The growth rate of endothelial and other cells that constitute the vessel walls, was estimated in terms of numerical density N_A, and the arterial vessel length density L_A was also assessed. Image analysis and stereology were used for this quantitative study. D changed between day 10 and 14 from the initial 1.3 to 1.5, whereas N_A dropped from 2500 mm^{-2} to 250 mm^{-2}. L_A showed only an increase from 4 mm^{-1} to 6 mm^{-1}, and the total vessel length was estimated to expand from 2.5 m to 50 m. The effect of Vascular Endothelial Growth Factor on the bifurcation frequency was documented by an increased D of 1.8. Multiple scatter plots were used for visualization of the data, which suggest mainly «intussusceptive» (i.e. by mesh formation without sprouting) capillary growth. This explains the increased density of hemodynamic sources and sinks in the capillary layer, which is the anatomical correlate of a postulated functional shunt. The very rich vascularization of the older CAM rather is a side-effect of angiogenic factors acting in the embryo than optimal metabolic adaptation.

1 Introduction

Living structures may be described as being in self-organizing fluctuating steady-state far from equilibrium [1]. Fractal geometry, on the other hand, has been applied for physical phenomena far from equilibrium, like turbulence (cf. chpt. 10 in [2]). With the cardio-vascular system, which amounts to approximately 50% the weight of early developmental stages of birds, the first fully functioning system is organized, and it is just this system, which establishes steady-state and far-from-equilibrium conditions over several orders of magnitude. It therefore is not too astonishing that vessel systems have been characterized with fractal dimensions (cf. chpts. 15–17 in [2], [3]), and that they have been modeled as fractals with branching algorithms [4]. While the significance of «fractal organization» is fairly accepted with respect to the limiting physical conditions of blood circulation [5,6,7], this concept seems rather speculative in view of the genetic and developmental realization in the living organism. We therefore attempted for the first time to describe the pattern formation of a real vessel system by determining its fractal dimension D. Since D is a number without physical dimension, additional information is needed to more precisely describe the actual biological generator and the physiology of the system, and to combine this knowledge with existing morphological and physiological findings. Hence, the density of proliferating

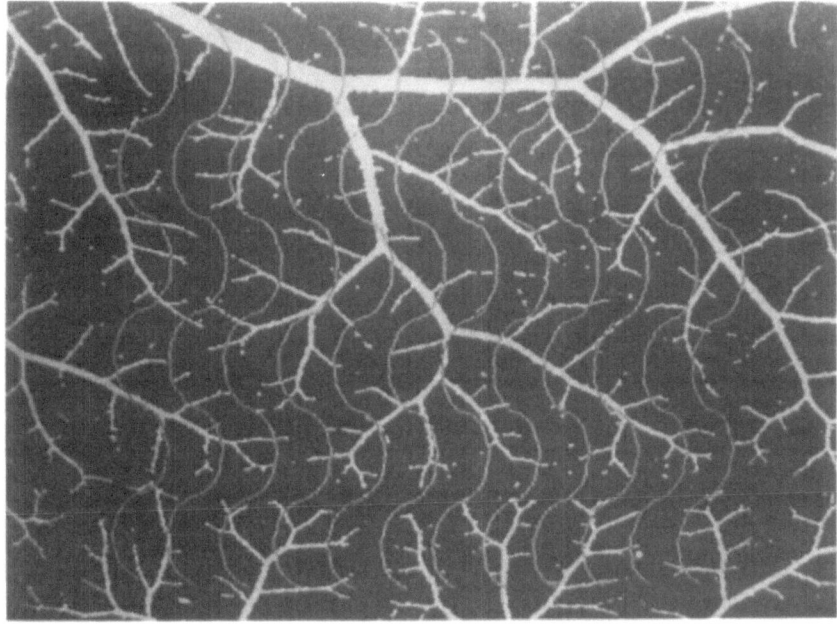

Fig. 1   Binary image of arterial vessels in the 14-day CAM. The cycloidal waves are superimposed with the XOR function to show the counted intersections in gray.

cells that contribute to the developing vessel pattern, N_A, and the observed vessel length density L_A were determined with stereological methods, and combined by explorative multivariate statistics.

The CAM was chosen for methodological and practical reasons: Being the largest extra-embryonic and the major respiratory organ of the embryo, it grows in intimate neighborhood to the egg shell and is easily accessible. It displays a comparatively simple, layered structure, and the vessel «tree» hence grows in a — topologically two-dimensional — tissue sheet. Its fractal dimension consequently lies between 1.0 and 2.0 and thus may be treated with standard image analysis. Moreover, the CAM assay has become a standard for the evaluation of angiogenetic substances [8], and for the grafting of e.g. tumors [9], and the physiology of avian gas exchange is known since long [10].

Two major questions were to answer with the work presented here: Does the quantitative analysis of blood vessel formation give evidence for a possible mechanism of angiogenesis [11] in the CAM? Can the concept of an arterio-venous shunt, which was derived from respiratory data [12,13], be supported by the structural data?

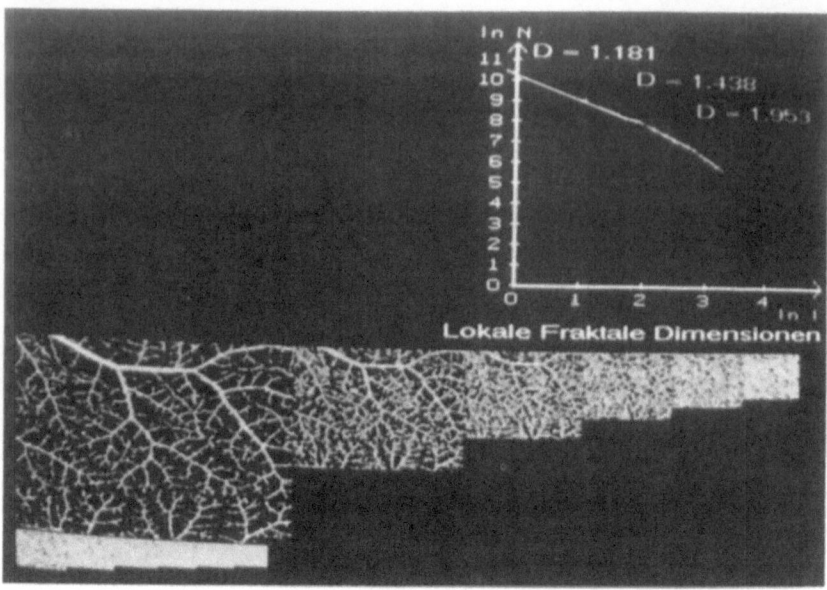

Fig. 2 Box counting method with the SIS system. The reduced pattern indicates the scaling. Three regions of different slopes are distinguished in the log-log plot.

2   Materials and Methods

After removal of ca. 1 ml albumen, White Leghorn chicken eggs were fenestrated at day 3 or 4 of incubation at 310 K and 80% humidity. CAMs were obtained at days 6 through 19 ($n = 10$ each); after fixation in situ, rectangular pieces of about 1 cm^2 were transferred to a petri dish with fixative on an illuminated box. The native contrast of the erythrocyte-filled arteries in these whole mounts was viewed via a B/W CCD with Macro objective 1:2, with one randomly oriented specimen as one sampling unit. The SIS (Münster) system was used for 8 bit gray value, 512^2 pixel image processing at a spatial resolution of 15 μm/pixel. The images were treated with automatic shading correction, gray value standardization and binarization at a fixed threshold; isolated pixels were eliminated with the connectivity operation. For the dimensional analysis with the box counting method (BCM, cf. [2,14]), D was taken as the negative slope of the log-log plot of box edge length (1,3,5,... 25 pixels) vs. number of boxes hitting the border of the binary vessel image. In a separate series of preparations [15], vascular endothelial growth factor (VEGF) was locally applied to the CAM.

The binary image was also used for stereological estimates of $L_A = \pi(I_x + I_y)/4c$ by automatically counting the result of an AND operation with a cycloidal grid, with I_x: intersection count in a random orientation x, I_y: count at right angles to x, c: length of test lines in one sampling frame (modified from [16]).

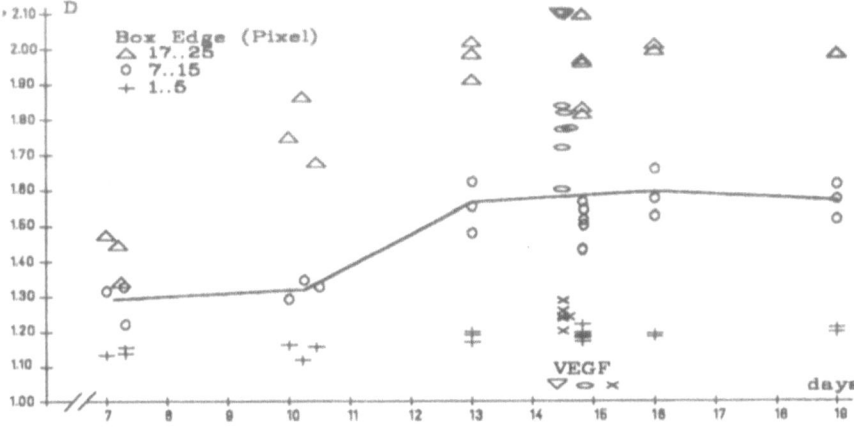

Fig. 3 Dependence of D from incubation time and box size. The application of VEGF leads to an elevated D mostly for medium sized, but also small boxes.

For the estimation of N_A, 0.3 ml of 40 mM bromo-deoxyuridine (BrdU, Sigma) was applied to the CAM for 45 minutes before fixation. The incorporation of BrdU into the nuclei of proliferating cells was detected with monoclonal anti-BrdU antibody (Dakopatts) after acid hydrolysis. The diamino-benzidine (DAB) reaction indicated the labelled nuclei with brown precipitate. With whole mounts embedded in glycerol on a clean slide, nuclei in the capillary layer of the CAM were counted at $16 \times$ microscopic magnification in stripes of 0.04 mm^2 in a systematic random sample.

Whole mounts were also stained in DAB without pretreatment to obtain strong positive contrast of the arterial and capillary vessels, filled with erythrocytes. This allowed for microscopic registration of the vessel «endpoints» in the capillary layer at $16 \times$ magnification. Tessellations around these were generated by mathematical morphology (skeletonization, dilation), and arterio-venous distances (AVD) measured in two-dimensional random samples.

3   Results

The projected image of a CAM arterial tree is shown in Fig. 1. Statistical self-similarity appears to be present in the branching pattern. The dimensional analysis of Fig. 1 with the BCM is shown in Fig. 2. The results for the first time-series of experiments are combined in Fig. 3. As this natural «fractal» has upper and lower limits, the local dimensions D of the branching vessels were determined for three regions: for box lengths below 90 μm, $D = 1.1$ varied only slightly with development, and for box lengths over 230 μm, D approached 2.0. For intermediate box lengths, a considerable rise of D with development was measured between day 7 (1.3) and 13 (1.5), with no further increase later. The application of the

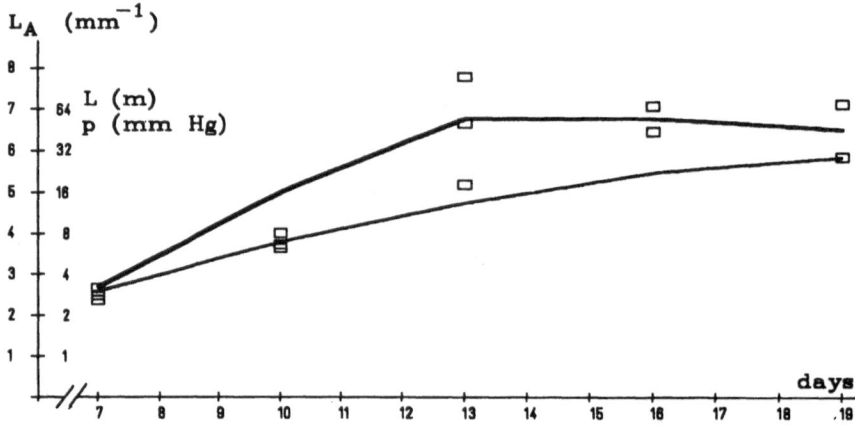

Fig. 4   Change of L_A (squares), L (thick line) and p (thin line, from [17]) with development. The logarithmic ordinate applies to L and p.

potent angiogenetic substance VEGF [15] led to a $D = 1.8$ (day 13..16) under the carrier, as compared to the $D = 1.5$ outside (Wilcoxon $p < 0.05$).

The length density of arteries with apparent diameters larger than 15 μm, together with the calculated total arterial vessel length and the mean blood pressure [17] is plotted against incubation time in Fig. 4. L_A shows approximately linear growth with only two-fold increase between day 7 and 13; as the CAM itself continues to expand during this period, L, however, grew exponentially from 2.5 m to about 50 m.

Multiple scatter plots combine the three numerical descriptors of vascular morphogenesis in Fig. 5. N_A declined from about 2500 mm^{-2} (day 8) to 1200 mm^{-2} (day 14) and then dropped to 250 mm^{-2} from day 16 on. Again, the sudden jump of D and the gradual increase of L_A is found as before.

A microscopic image of the terminal arterial and venous vessels, together with their pattern of hemodynamic sources and sinks is shown in Fig. 6.

The histogram of AVDs is given in Fig. 7 for the 10 and 16 day CAM. The large variability is obvious in both, but the occurrence of AVDs shorter than 100 μm is typical only for the older CAM.

4   Discussion

The CAM arterial vascular pattern has been portrayed as growing in a planar, tree-like fashion. The thickness of the CAM is known to be about 100–150 μm, with all the larger vessels remaining in a plane parallel to and underneath the capillary layer, i.e. the chorionic surface. Only the immediate pre- and post-capillary vessels can deviate considerably from this plane, but their diameters are below 15 μm and

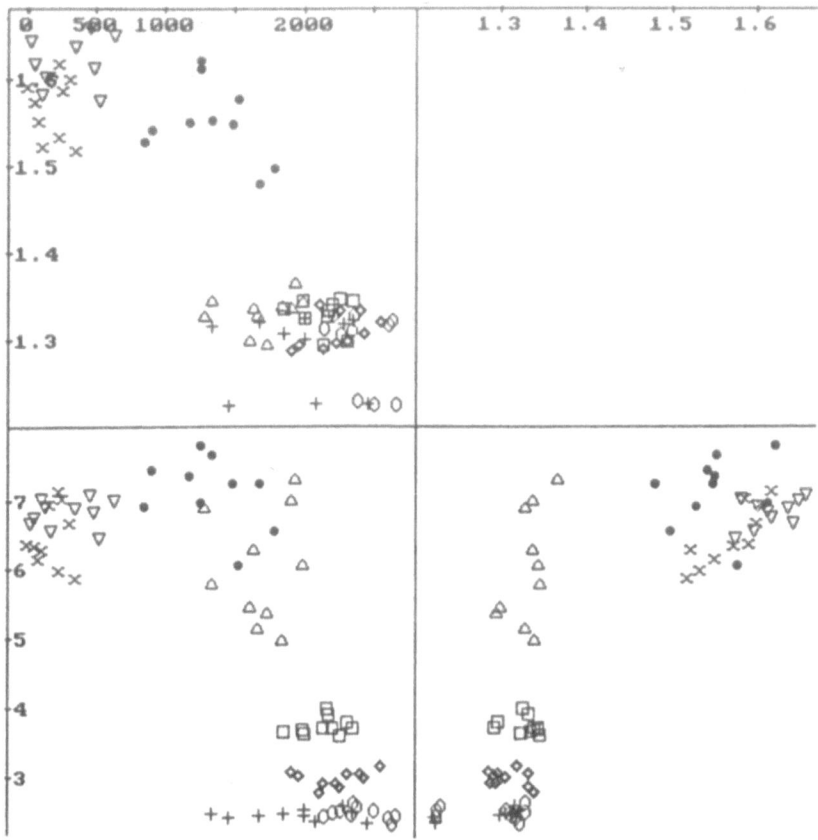

Fig. 5   Scatter plots of N_A vs. D (top), N_A vs. L_A (bottom left), and D vs. L_A (bottom right). Age in days (d) is indicated by symbols: 6d +; 7d 0; 8d ◇; 10d □; 12d △; 14d ●; 16d ▽; 18d ×. Note the well separated 12d and 14d clusters.

they remain invisible to the CCD with macro objective, which guarantees that no detail is lost by evaluating the projection. A certain amount of «roughness» of the apparent vessel border is lost, however, by eliminating isolated pixels, which were mostly due to noise. This may account for the rather low value of $D = 1.1$ for small boxes. The projection through a random fractal, though, should not reduce its dimension (p. 378 in [2]), if D is smaller than the topological dimension, which is equal to two in our case.

Vessel «trees» are characterized by two different parameters, D and the diameter exponent Δ [2,7]; in our case, self-similarity and thus $D = \Delta$ can be assumed [2]. The estimation of fractal dimensions also depends on the chosen method [14]. As the Sierpinski triangle test with our BCM shows a difference in D of only 0.01, we are confident that D is determined precisely in view of the biological variability. The BCM is found to be a reproducible measure for box edge lengths

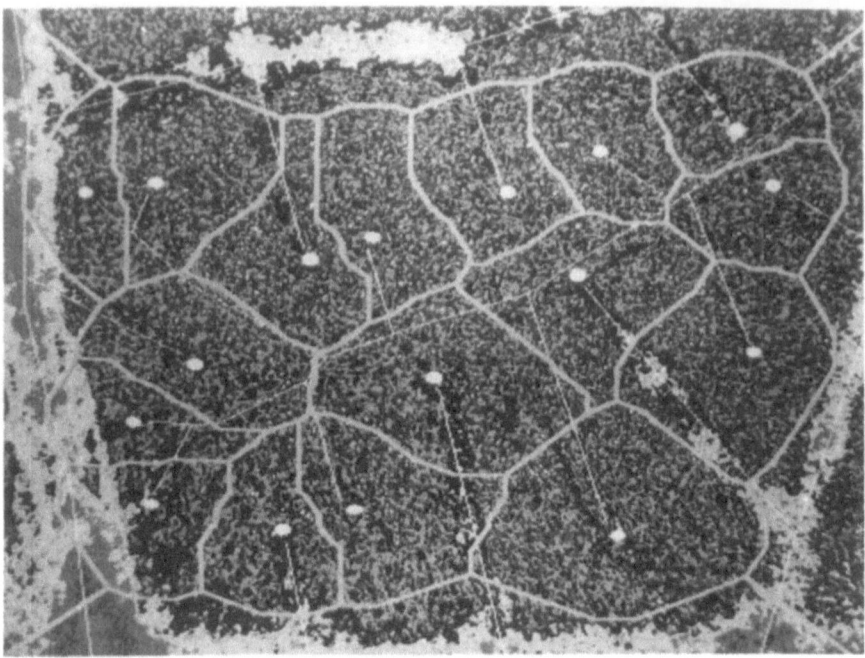

Fig. 6   Patterning of the CAM capillary bed (granulated grey) by arterial and venous vessel endpoints
(dots). The sources are connected to the arteries (lines along and branching off the edges),
the sinks drain to the central vein (oblique middle line).

between 5 and 19 pixels. The best linearity, however, and the most striking differences are seen in the region 7 . . . 15 pixels. We use this «local» D as a measure of
complexity, one interpretation being that vessel bifurcation intensity is enhanced
during the CAM growth period, just when the CAM covers the entire shell from
inside after day 12. The elevated D after application of VEGF corresponds to the
even more enhanced bifurcation frequency and to the many irregular vessels found
in histological preparations [15].

The stereological estimation of L_A by counting intersections of the vessel
image with reference lines has the advantage of working around the problems of
defining length in the quadratic pixel raster. It thus is rather independent from the
BCM, which counts pixels directly. The cycloidal lines were found to be very
effective in that the variation coefficient of L_A between different orientations on
one pattern always was below 2%. This regularity also suggests that the vessels
in any single CAM represent a stationary and isotropic fibre process, although the
variability between the CAMs of one age group becomes large after day 10 (cf.
Fig. 5).

From the triple scatter plot it is obvious that the proliferation rate in the
capillary layer is highest, as long as larger vessels grow, while D stays at 1.3 during

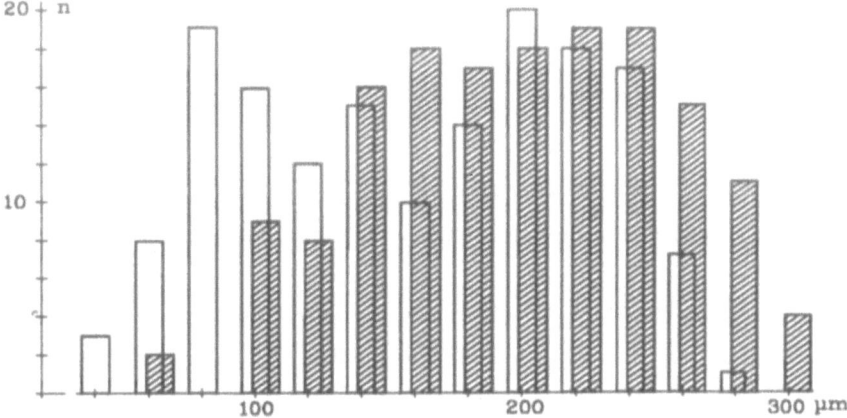

Fig. 7   Distribution of AVDs from random samples of 10-day CAM (hatched, $n = 160$), and 16-day CAM (white, $n = 169$), in which the proportion of AVDs larger than 250 μm is reduced in favor of rather short AVDs below 100 μm.

this period. It jumps to values above 1.5, when L_A has reached its maximum, and N_A has dropped to half the earlier value. This suggests a sequence of first purely intussusceptive capillary growth, which later appears as «sprouting» of precapillary vessels [9,11,15]. With D, the conductivity also increases [7], which can compensate the still low blood pressure.

The measurement of arterio-venous endpoint distances, and their distribution in the capillary layer may be easily related to the great variability of blood flow seen in the living CAM, which appears as a non-stationary temporal pattern. The small AVDs appear as a side-effect of angiogenic factors needed for the embryo proper, and explain the functional shunt, so that the structure of the older CAM seems to be not optimally adapted to metabolic needs. For a transient organ, which is needed for less than 14 days, this seems acceptable in view of e.g. increasing hematocrit etc. [12].

In summary, our introductory questions could be answered: Explorative data analysis of cell proliferation, vessel length, and fractal dimension supports the hypothesis of intussusceptive growth in the capillary plexus, which in turn forms the major conducting vessels. The functional shunt is most likely realized by the diminished AVDs, and is caused by the conducting vessels forming more and more connections to the capillary bed.

The strength of the fractal concept of dimension lies in the description of complexity, which is generated from scarce information. Structure may thus be related to both molecular and macroscopic function, and biological and synthetical pattern formation may be compared. Following this notion, one can speculate, whether fractal geometry can be useful in extending the «hypercycle» [1] concept.

Acknowledgement:
This work was supported by DFG grant Ch 44/9-2.

References

[1] M. Eigen, Die Naturwissenschaften 58/10, 465–523 (1971).

[2] B.B. Mandelbrot, The Fractal Geometry of Nature, Freeman, San Francisco (1983).

[3] T. Matsuo, R. Okeda, M. Takahashi, and M. Funata, Forma 5, 19–27 (1990).

[4] H.R. Bittner, in: H.O. Peitgen et al. (Eds.), FRACTAL 90, Elsevier, Amsterdam (1990).

[5] M. Sernetz, B. Gelléri, and J. Hofmann, J. Theor. Biol. 117, 209–230 (1985).

[6] J.B. Bassingthwaighte, R.B. King, and S.A. Roger, Circ. Res. 65, 578–590 (1989).

[7] H.C. Spatz, J. Comp. Physiol. B 161, 231–236 (1991).

[8] J. Wilting, B. Christ, and M. Bokeloh, Anat. Embryol. 183, 259–271 (1991).

[9] D.H. Ausprunk, D.R. Knighton, and J. Folkman, Devel. Biol. 38, 237–248 (1974).

[10] H. Rahn, C.V. Paganelli, and A. Ar, Resp. Physiol. 22, 297–309 (1974).

[11] S. Patan, B. Haenni, and P.H. Burri, Anat. Embryol. 187, 121–130 (1993).

[12] H. Tazawa, Am. Zool. 20, 395–404 (1980).

[13] D. Wangensteen, and E.R. Weibel, Resp. Physiol. 47, 1–20 (1982).

[14] D. Stoyan, and H. Stoyan, Fraktale — Formen — Punktfelder, Akademie Verlag, Berlin (1992).

[15] J. Wilting, B. Christ, M. Bokeloh, and H.A. Weich, Cell Tiss. Res. (1993).

[16] E.R. Weibel, Stereological Methods I, Academic Press, New York (1979).

[17] L.H.S. Van Mierop, and C.J. Bertuch, Am. J. Physiol. 212, 43–48 (1967).

Phyllotaxis or Self-Similarity in Plant Morphogenesis

François Rothen
Université de Lausanne, BSP-Dorigny
CH-1015 Lausanne, Switzerland

1   Phyllotaxis or an exotic form of two-dimensional crystallography

Phyllotaxis refers to the geometry governing the arrangement of inner florets of a sunflower (Fig. 1), of the scales of a pineapple (Fig. 2), of the leaves around a stem and so on. The florets align with spiral whorls in the case of spiral *phyllotaxis* (daisy) or with helices in the case of *cylindrical phyllotaxis* (pineapple or fir-cone).

In Fig. 2 showing the picture of a pineapple with its hexagonal scales, the eye is at once attracted by the helices on which the neighbouring scales are aligned. A given hexagonal (rectangular) scale belongs to three (two) different helices, called *parastichies*; they bind each scale with its six (four) different neighbours.

Parastichies are grouped into families, a *parastichy family* being the set of all parastichies parallel to each other (see Fig. 3). Each family consists of a number k of parastichies which completely characterizes the family. Depending on the number of parastichies an arbitrary scale belonging to a given phyllotaxis will then be specified by a pair (k,l) or a triplet (k,l,m) characteristic of the two or three families of parastichies through the pattern of scales (this is also true for plane phyllotaxis whose parastichies are spirals rather than helices).

It was soon recognized that k,l,m are very often (but not always) successive members of the *Fibonacci sequence* [1,2]

$$\{f_n\} = 1, 1, 2, 3, 5, 8, 13, 21, 34, \ldots \tag{1}$$

defined by the recurrence law

$$f_{k+2} = f_k + f_{k+1} \qquad (f_1 = f_2 = 1). \tag{2}$$

The Fibonacci sequence is related to the *golden section* τ:

$$\lim_{k \to \infty} \frac{f_k + 1}{f_k} = \tau =: \frac{1 + \sqrt{5}}{2} = \tau^{-1} + 1 \tag{3}$$

The golden section is an irrational number whose remarkable properties cannot all be listed here; one of them however is worth mentioning. Every irrational x between 0 and 1 can be expressed in a unique way as an infinite sequence of

Fig. 1 Sunflower: a hidden self-similarity.

stacking fractions:

$$x = \cfrac{1}{a_1 + \cfrac{1}{a_2 + \cfrac{1}{a_3 + \dots}}} =: [a_1, a_2, a_3, \dots] \tag{4}$$

Fig. 2 Pineapple: a further hidden self-similarity.

Fig. 3 Looking at the pineapple, one immediately realizes that the scales are regularly disposed
along spirals, called parastichies. The three ones going through the hatched scale have been
marked here A, B, C. Parallel spirals are grouped in a given family: so are A and A'.

where a_i ($i = 1, 2, 3, \ldots$) is a positive integer [3]. The semi-infinite sequence is
called a *continued fraction*. Now $\tau^{-1} = \tau - 1$ belongs to the interval [0,1] and
its continued fraction writes

$$\tau^{-1} = [1, 1, 1, \ldots]. \tag{5}$$

Moreover all numbers whose continued fraction exhibits only ones from some
stage are called *noble numbers* [4]. For instance

$$\frac{1}{2 + \tau} = [3, 1, 1, 1, \ldots] \tag{6}$$

is called *Lucas' number*. Noble numbers share with $\tau - 1$, «the noblest of all noble
numbers», some interesting properties we shall discuss below.

Throughout history, there is an impressive number of people associated with phyllotaxis (see for instance [1] or [5]), among whom one could quote, somewhat arbitrarily: D'Arcy Thompson, the famous author of «On Growth and Form» [6], H.S.M. Coxeter, who devoted one chapter of his «Introduction to Geometry» to phyllotaxis [7] and recently Levitov who made an outstanding analysis of an equivalent phenomenon in physics [8].

2 Self-similarity in phyllotaxis

Self-similarity arises in an obvious way in the case of a nautilus (Fig. 4) as it merely reflects a growth process without any change of relative size (the proportions remain constant during the process, there are no superimposed constraints). The case of a phyllotactic pattern is however more complicated. Scales and florets have to expand on a *tissue* whose proper growth rate and shape influence not only the final geometry of the pattern but also its composing units (scales, florets).

We shall see in the next section that, according to the value of the parastichy numbers, some hidden but important self-similarity may be present in the pattern.

On the other hand, the parastichy numbers can change in a given pattern during the growth, giving rise inside the same pattern to different domains, each one characterized by a particular pair[1]) of parastichy numbers. This transition is called a bifurcation. During such a process, one pair can be replaced by some other pair (sect. 5). As a consequence, all possible phyllotactic patterns can be classified by a bifurcation tree. Interestingly enough, the tree itself shows characteristic self-similarity properties.

In the next section, we will introduce two important new geometrical parameters, i.e. divergence and plastochrone ratio, in order to discuss these various expressions of self-similarity. Sect. 4 and 5 will be devoted to the discussion of self-similarity itself.

3 Divergence and plastochrone ratio

It is useful to replace the parastichy numbers by two other parameters, the divergence and the plastochrone ratio, which allow a more precise description and better describe a phyllotactic pattern.

Let us consider Fig. 5 in which the scales of a pineapple are numbered according to the order of their birth, the older ones having the lowest numbers. The *divergence* x is the angular distance between two successive scales, i.e. between scales k and $k + 1$. Notice that x is expressed as a fraction of whole turns: $0 \leq x < 1$.

---

1)   In the following, we shall forget the fact that parastichy numbers can also occur as a triplet.

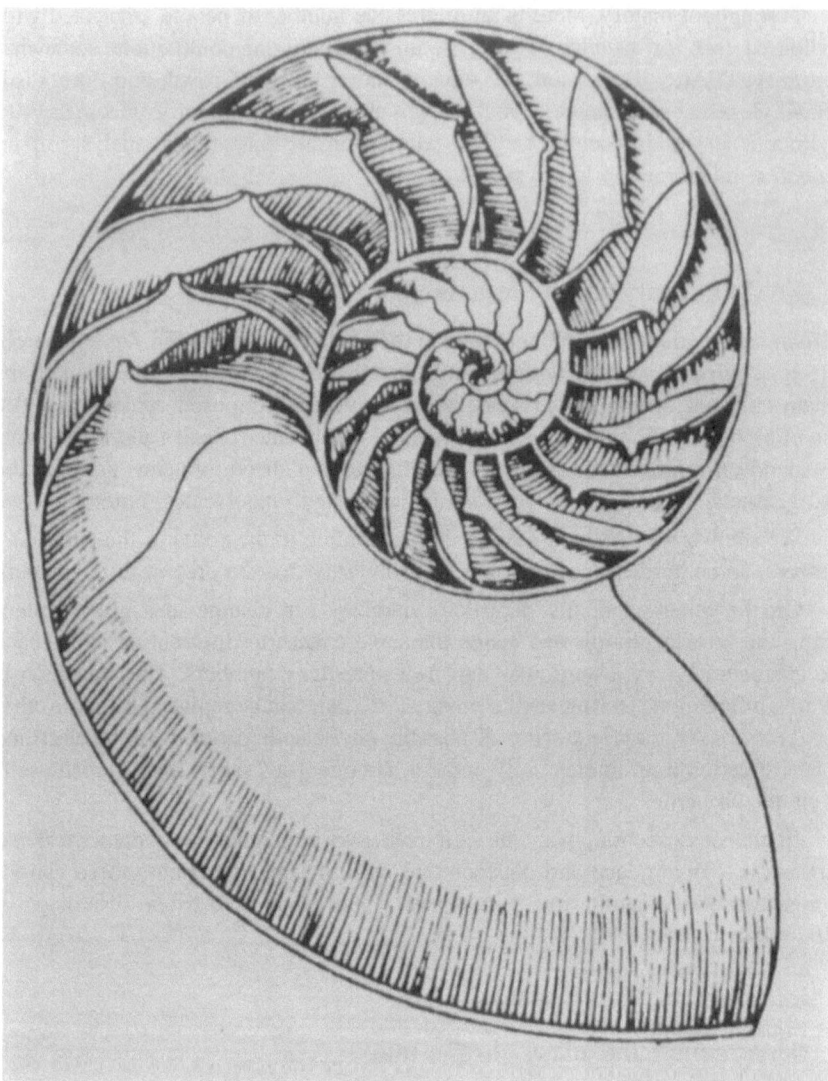

Fig. 4 Nautilus: an obvious self-similarity.

The pineapple can be schematically represented by replacing each scale by a point on a cylinder with circular cross-section and a circumference of unit length. One could even unwind the cylinder on a plane (Fig. 6a, 6c and 6d) producing a periodical two-dimensional lattice. In the case of plane phyllotaxis (sunflower), the successive florets can be represented by points aligned on a spiral (Fig. 6b) with a divergence equal to $1/2\pi$ times the angular distance between two successive

Fig. 5 The scales of pineapple have been numbered from the bottom upward. Notice that the numbers increase regularly along parastichies: 5 to 5 along parastichies belonging to the same familiy as A, which is therefore called a 5-family. Notice also the presence of an 8-family (B) as well as a 13-family (C). A, B and C refer to Fig. 3.

points. Hereafter, we shall speak of a *periodical lattice (spiral lattice)* to describe the pattern of Fig. 5d (5b).

To determine completely the geometry of both lattices, we have to add a further parameter, namely the *plastochrone ratio*[2]) z. In the case of the periodical lattice, if one chooses the origin at a point of the lattice (Fig. 6d), the plastochrone ratio is equal to the positive ordinate z_p of the point with abscissa x. In other words, $(0,0)$ and (x, z_p) are the coordinates of the points corresponding to two successive scales. In the case of the spiral lattice, the plastochrone ratio is defined as the ratio $z_s > 1$ of the distances of two successive points to the origin. This

---

2) Divergence as well as plastochrone ratio are terms coined by botanists.

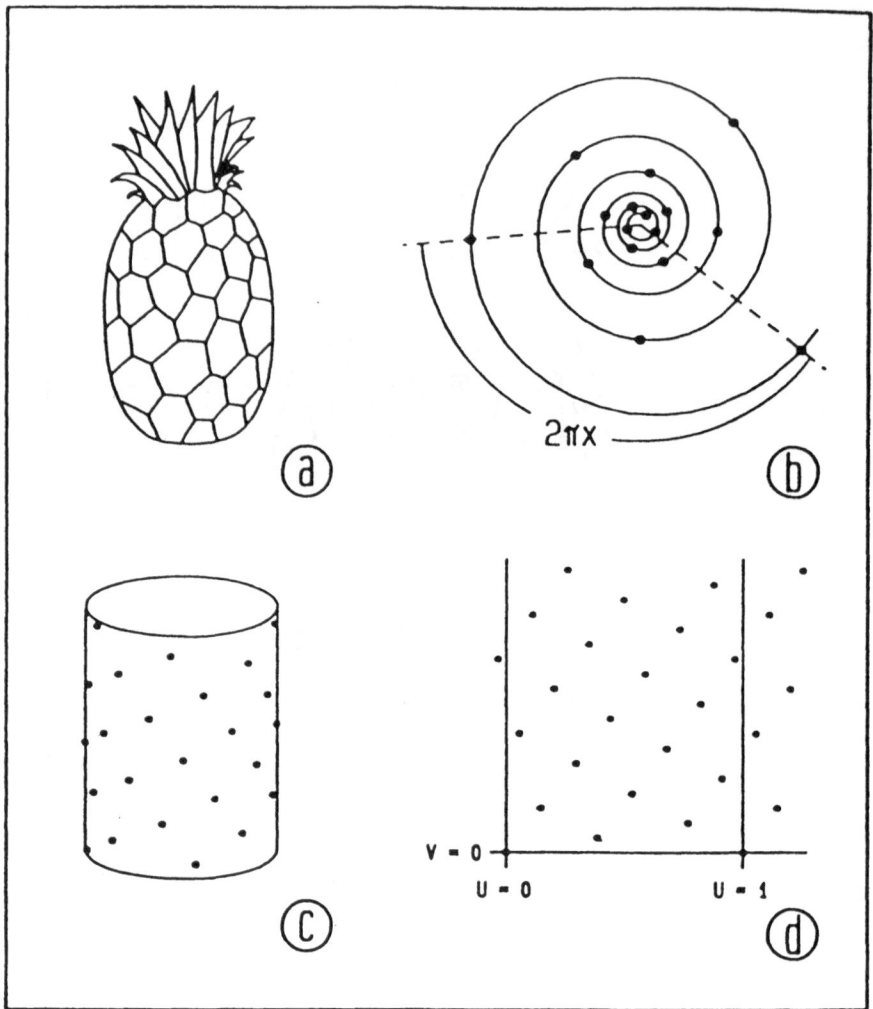

Fig. 6 The pineapple (a) is idealized as a cylinder (c). On the corresponding cylindrical lattice, each point represents the center of a scale. Unrolling the cylinder on a plane, one gets the periodical lattice (d). The spiral lattice (b) is an idealization of the pattern of a sunflower or a daisy. In the spiral lattice, the angular distance of two successive points aligning with the spiral is equal to $2\pi x$, x being the divergence of the pattern. In the periodical lattice, the point having the lowest positive vertical coordinate is specified by the pair (x, z_p) where z_p is the plastochrone ratio. The divergence is still given by x.

definition implies that the spiral is equiangular[3]) (or logarithmic), the origin of the coordinates coinciding with the point to which the spiral converges inwards. In polar coordinates, two successive points coincide with (r, q) and $(z_s r, \theta + 2\pi x)$.

4   Symmetry properties of a periodical lattice under compression

Assume for a while that the pineapple can be identified with a cylinder with circular cross-section and that its rate of growth is a constant so that the interval between the birth of two successive scales is always the same. Then the pineapple can be compared in a realistic way with the above mentioned periodic lattice.

We shall now put forward a hidden symmetry of the periodic lattice which is highest when the divergence is equal to the golden number τ; we then speak of *golden divergence*.

In the vegetable kingdom, the chirality[4]) of most patterns seems to be arbitrary (there are as many patterns turning right than left). As a consequence, for x between 0 and 1, both divergences x and $1 - x$ can be considered equivalent. For a golden divergence, this means that τ^{-1} and $\tau^{-2} = 1 - \tau^{-1}$ are equivalent to τ. Botanists prefer to use degrees to measure divergences and to avoid angles greater than 180°, so that botanists fix the value of the golden divergence to $\tau^{-2} \cdot 360° = 137°.507 \ldots$

Arithmetic gives a well-known recipe to approximate irrational numbers by rational fractions. The best approximations (to be specified [3]) are given by the successive troncations of the continued fraction corresponding to a given irrational

$$x \Leftrightarrow \left\{ [a_1] \div \frac{1}{a_1}, [a_1, a_2], [a_1, a_2, a_3], \ldots \right\} \tag{7}$$

τ^{-1} and τ^{-2} are respectively approximated by the following sequences:

$$\tau^{-1} \Leftrightarrow \left\{ \frac{1}{1}, \frac{1}{2}, \frac{2}{3}, \frac{3}{5}, \frac{5}{8}, \frac{8}{13}, \ldots, \frac{f_k}{f_{k+1}}, \ldots \right\} \tag{8}$$

$$\tau^{-2} \Leftrightarrow \left\{ \frac{0}{1}, \frac{1}{2}, \frac{1}{3}, \frac{2}{5}, \frac{3}{8}, \frac{5}{13}, \ldots, \frac{f_{k-1}}{f_{k+1}}, \ldots \right\} \tag{9}$$

These rational fractions involve only members of the Fibonacci sequence (1).

As a consequence, if the divergence of a pattern almost equals $5/13$, this means that approximately 13 points align with the spiral while it describes 5 turns (this is the case of the spiral of Fig. 6b). Florets number k, $k + 13$, $k + 26$, ... almost align with a straight line (actually a very steep spiral) issued from the

---

3)   As will be clear in the next section, this choice of a particular shape of the spiral does not restrict the range of validity of the model.

4)   In the case of an enantiomorphic object such as a cork-screw, the chirality can be identified with the direction of winding.

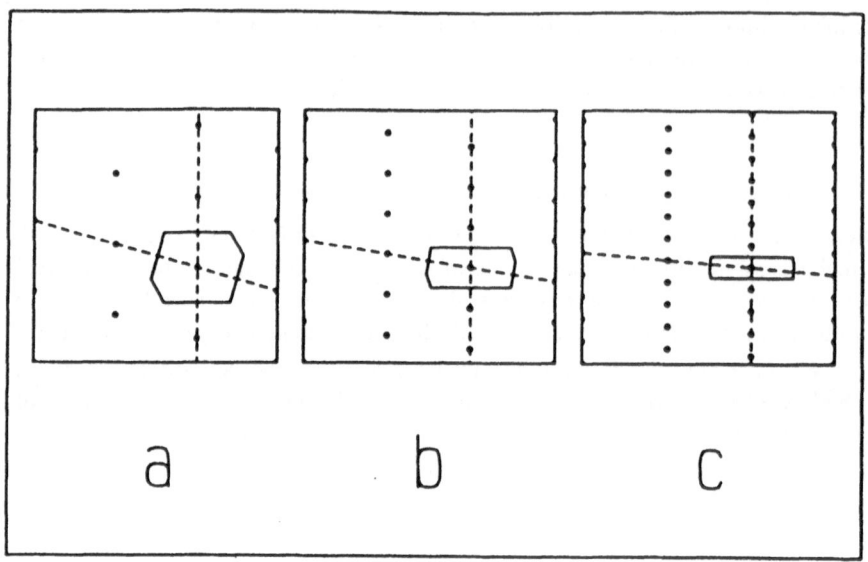

Fig. 7   If the divergence of the periodical lattice is rational (here $x = 1/3$), the points align with vertical lines. The polygons shown here in each case are the Voronoi polygons (VP), i.e. the closed curves which surround the set of all points of the plane which are nearer to a given element of the lattice than to any other element. As can be seen, there is no shape invariance of the VP when the plastochrone ratio z_p is reduced, i.e. when the lattice is compressed along the vertical direction. Dashed lines indicate two of the most visible parastichies.

origin. This remark is valid for $k = 0, 1, 2, \ldots, 12$; the steep spiral is a parastichy belonging to a family of just 13 members. Notice that 13 is the value of the denominator of 5/13. The same argument is also valid for 3/8 and 2/5 which are (less good) rational approximations of τ^{-2}. It follows that 5, 8 and 13 are the parastichy numbers corresponding to the pattern of Fig. 6b. Using the same argument, one can determine the parastichy numbers of the pineapple of Fig. 5 and they happen to be the same. Now to the central point of this section. Notice that we shall restrict ourselves to the case of the periodic lattice (recall for the present discussion that the periodic lattice is equivalent to the cylindrical lattice[5])) but the case of the spiral lattice is perfectly analogous.

Obviously, the pineapple is neither a cylinder nor is its growth rate a constant. The question therefore arises: what are the consequences of the important shape gap between botanical objects and their geometrical idealization?

Observation tells us that some hidden mechanism leads to an almost constant divergence (we shall come to this point later). The plastochrone ratio, however, very frequently changes throughout a phyllotactic pattern. First of all the rate of

---

5)   A helix with constant pitch winding around the cylinder corresponds to the straight line connecting (0,0) and (x, z_p) in the periodic lattice.

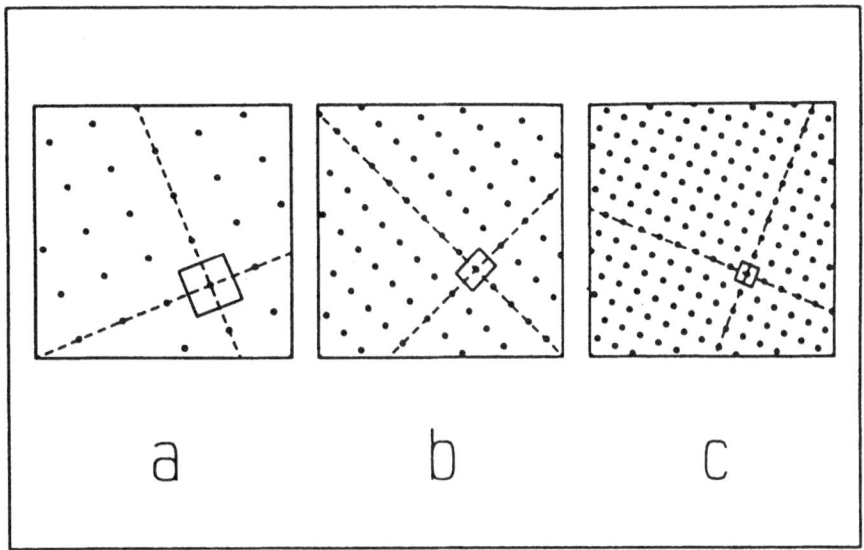

Fig. 8 If x is irrational, parastichy transitions occur during the compression (here $x = 2^{1/2} - 1$). (a) Parastichies belong to the 2- and 5-families. (b) Due to vertical compression of the lattice, a transition has occured. Parastichies now belong to the 5- and 7-families and the Voronoi polygon has been distorted and rotated. (c) Further compressing of the lattice leads to a new transition: 5- and 12-families become visible and the Voronoi polygon recovers the shape it had in (a).

growth is not always constant. On the other hand, putting some helicoidal lattice onto a surface differing markedly from a cylinder is more or less equivalent to change the plastochrone ratio of a cylindrical lattice from point to point.

The consequence of the fact that x can be considered as constant while z can vary throughout the pattern is easy to express through the use of the parastichy numbers (we consider here only parastichies occuring as pairs rather than triplets): there will be separate regions of the pattern, each one being characterized by a different pair. However, it follows from the constancy of x that each pair must correspond to the sequence of the rational approximants of x. More precisely, both members of the pair will generally be denominators of two successive approximants of x [5].

There is now a very interesting point. Let us investigate the shape of the Voronoi-polygons[6]) (VP) defined around each lattice site. In principle, their shape depends on x and z_p (or only on z_p if x is constant). VP tile the space occupied by the lattice as well as the scales of the pineapple do. One therefore expects the

---

6) The Voronoi-polygon (or Wigner-Seitz cell, or Dirichlet domain) is the set of all points (here of the cylinder) which are nearer to a given element of a discrete point lattice than to any other element.

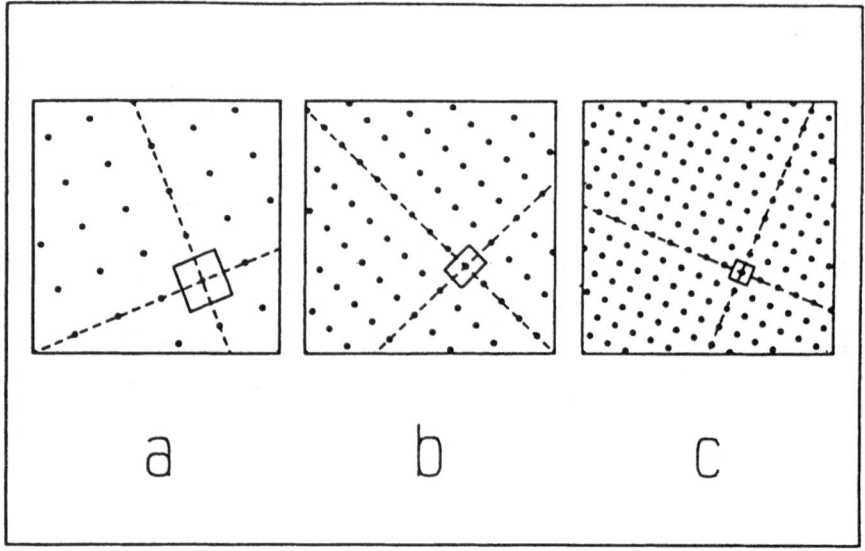

Fig. 9   If x is equal to the golden mean, compressing the lattice along the vertical axis leads to normal transitions which do not distort the shape of the Voronoi polygon. The parastichy numbers are always successive members of the Fibonacci sequence (1): one successively notices (a) (2,3); (b) (3,5) and (c) (5,8).

shape of the botanic units to be modified in the same way as the Voronoi-polygons as soon as the value of z_p changes from region to region.

In order to understand what could then happen, consider a periodical lattice with a fixed value of the divergence x but with successive values $z_1 > z_2 > z_3$ of z_p. Decreasing z_p from z_1 to z_2, then to z_3 is equivalent to a compression of the lattice along a direction orthogonal to the x-axis. The possible results of this compression can be classified according to the arithmetical properties of the divergence:

- If x is a rational number the VP get thinner and thinner as z_p decreases (see Fig. 7). The shape of a scale would greatly change throughout the pattern

- When x takes some irrational value, a new phenomenon occurs during a compression, namely the appearance of *parastichy transitions*: for special values of z, the pair of parastichy numbers jumps to another value[7]) (Fig. 8). In general, during such a transition, the shape and orientation of the VP abruptly change.

However, after a finite number of transitions, the VP always recovers its original shape. In the case of the golden divergence (Fig. 9), the shape remains

---

7)   If x is rational, parastichy transitions only occur at an early stage of the compression.

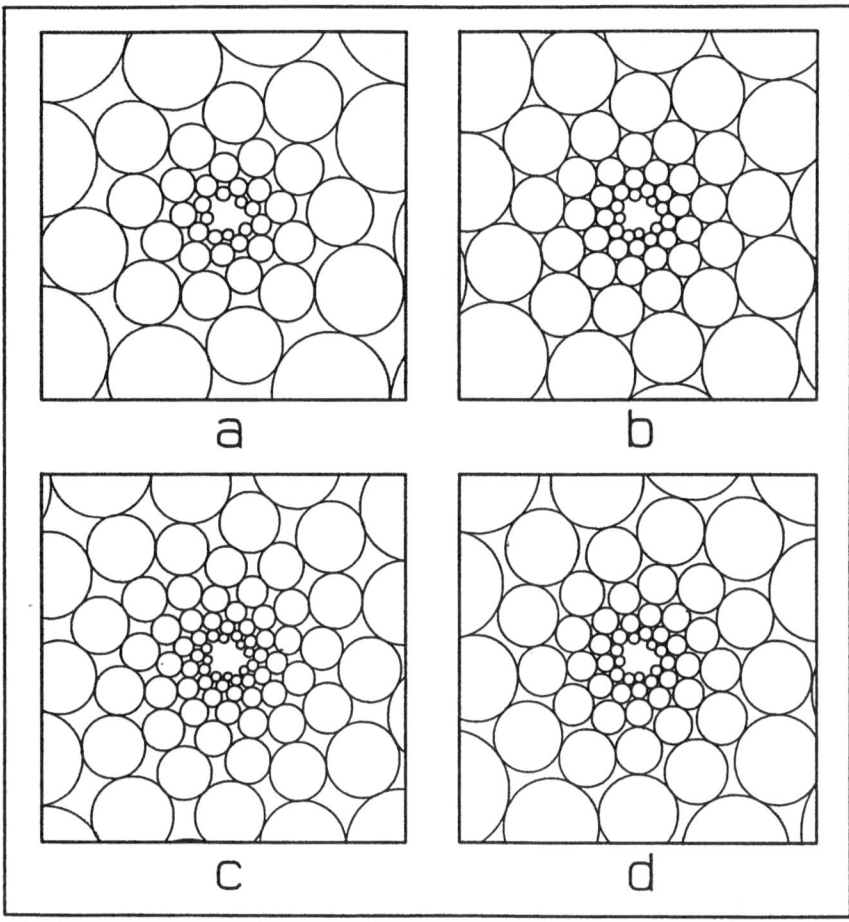

Fig. 10 Starting from a given configuration (a) with parastichy numbers (k, l) (here $k = 5, l = 8$), the plastochrone ratio z_s is slowly reduced. The lattice becomes more and more compact, up to close-packing (b) where the three parastichy families $(k, l, k+l) = (5, 8, 13)$ are apparent. Further decrease of z_s can lead either to $(l, k+l) = (8, 13)$ through a normal transition (c) or to $(k, k+l) = (5, 13)$ through a singular transition (d).

invariant during a transition. The whole succession of shape changes of the VP is directly linked to the development of x as a continued fraction [5].

The behaviour of the periodical lattice under compression has an important consequence for a real botanical pattern. If the plastochrone ratio is non uniform throughout the lattice, separate domains appear corresponding to different pairs of parastichy numbers. If the divergence is equal to τ (or to a noble number [5]), the shape of the scales will be the same in all domains. However, in the case of a rational divergence, or if x is very «far» from a noble number, either the botanical units cannot fill equally well each domain or they have to change their shape from domain to domain. Now, observation rules out these latter cases: one notices only noble divergences. Interestingly enough, this shape invariance of scales is the consequence of a seemingly pure arithmetic property, namely the occurence of an infinite sequence of ones in their continued fraction!

5   Self-similarity of the bifurcation-tree

What are the contraints ruling the parastichy transitions introduced in the preceeding section? In order to answer this question, it is necessary to add something to the point lattices considered so far, either by replacing points by non overlapping, tangent circles [9,10], or by introducing well-defined forces between the points [8,11], i.e. by transforming the system into an interacting point system.

Consider the pattern consisting of points aligned with an equiangular spiral so that the divergence and the plastochrone ratio are constant everywhere (see sect. 3). Assume that we number the points along the spiral, successive points corresponding to successive numbers (the numbers are assumed to increase outward). Number 0 is given to an arbitrary fixed point of the pattern. Assume further that the distance of 0 to the center Ω of the spiral is equal to 1. The distance between point k and Ω is equal to $z_s^k (-\infty < k < +\infty)$. If we draw a circle of radius $\rho_l = a z_s^k$ around k, we can adjust [9,10] the (positive) number a so that (see Fig. 10a):

- any circle Co is tangent with exactly 4 other circles, say C1, C1', C2, C2';
- two tangent circles, say C1 and C1', belong to the same parastichy going through Co; the second parastichy going through Co involves C2 and C2';
- there is no overlapping;
- except for scaling and rotation, any circle of the pattern is fully equivalent to any other (there is complete self-similarity of the pattern).

Now the whole system depends on a continuous way of z_s: if we slightly change z_s, we have to let x vary in order that the tangent condition remains fulfilled: x must be equal to some function $x_t(z_s)$ of z_s for the circles to remain tangent.

Assume now that, starting from some value $z_0 > 1$ of z_s, we steadily reduce z_s. For some z_1 such that $z_0 > z_1 > 1$, the lattice of circles will be *close-packed*: each circle has now two new neighbours C3 and C3' which are also tangent to Co and define a third parastichy going through it (Fig. 10b). If we further reduce z_s by preserving the whole set of tangency rules, we have to let either (C1, C1') or (C2, C2') move away. One can then prove [9,10] that, if k and l are relative primes ($k < l$) and if (k,l) represent the parastichy numbers for $z_0 > z_s > z_1$, then the parastichy transition corresponds to a *bifurcation*:

$$(k,l) \longrightarrow \begin{cases} (l, k+l) & \text{normal transition} \\ (k, k+l) & \text{singular transition} \end{cases} \qquad (10)$$

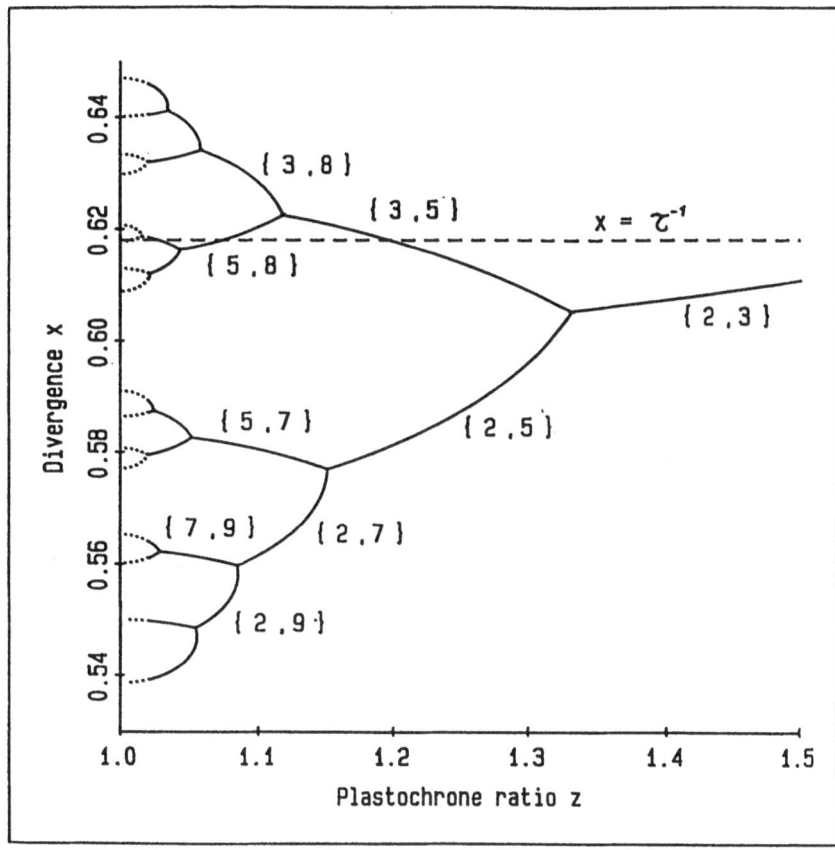

Fig. 11 The curves put forward the values of (x, z_s) for which the construction of a lattice of tangent circles is possible. Starting with a system characterized by parastichy numbers $(k, l) = (2, 3)$, a reduction of z_s leads to close-packing of circles. A further decrease of z_s may then produce two different situations: either the k-parastichies disappear (normal transition $(k, l) \rightarrow (l, k + l)$) or the l-parastichies vanish (singular transition $(k, l) \rightarrow (k, k + l)$). Further reduction of z_c leads to new transitions.

Further details can be found in reference [10]. Anyhow notice that starting from a pair (f_r, f_{r+1}) of consecutive Fibonacci numbers, a sequence of n normal transitions leads to (f_{r+n}, f_{r+n+1}) and the divergence remains a golden one.

On the other hand, it is easy to show that a normal transition always leads to a sign change of the curve $x = x_t(z_p)$. After a cascade of normal transitions, the divergence oscillates around an asymptotic value. Starting from any pair (k, l), such a cascade leads to new pairs $(k(n), l(n))$ such that $\lim_{k \to \infty} k(n)/l(n)$ is equal to some noble number. Among others, this fact leads to the self-similarity of the bifurcation tree of Fig. 11. Self-similarity is actually a consequence of the form of (9) which controls all bifurcations [11].

Levitov has recently investigated the compression along the z_p-axis of a periodical lattice of point particles [8]. The particles were interacting through an isotropic but otherwise arbitrary potential (no kinetic energy was assumed). He found a tree analogous to the bifurcation tree of tangent circles, except for an important point: all bifurcations were replaced by *quasi-bifurcations*. To perform a singular transition, the system had to jump across a gap in the (x, z_p) space while the normal bifurcations were still continuous. As a consequence, from a loose state where all interparticle distances are large, there is a unique connected way leading to a dense lattice; this way leads, through an infinite sequence of normal bifurcations, to a lattice with golden divergence! An experiment involving ferrofluid drops leads to a similar conclusion [12].

Now it is too early to say that the mechanism hidden behind the quasi-universal emergence of the golden mean has been solved. First of all, following an idea originated by Thornley [13], another investigation [14] was based on a very simple model of hormones reacting around a circle[8]). It also led to a tree of quasi-bifurcations, although the plastochrone ratio was replaced by quite a different parameter, namely the degradation and diffusion rate λ of the hormone. The self-similarity of the tree is baffling in this case. Moreover, using the same model, Guerreiro [15] was able to show that the unique connected way in the (x, λ) tree does not universally lead to a golden divergence (the asymptotic value of the divergence depends on rather unimportant details of the model).

Acknowledgement:
Knowledge about self-similarity in phyllotaxis has been gathered by many people: it is impossible to acknowledge all contributors. I would like to quote those people with whom I had special contacts, and among them G. Bernasconi, J. Boéchat, D. Bonnaz, Y. Couder, S. Douady, A. Joyet, J. Guerreiro, A.-J. Koch, M. Kunz, L.S. Levitov, Piotr Pieranski and N. Rivier. R. Huguenin was kind enough to read the text; I wish to thank M. Ede, F. Pretzsch and C. Rossier for technical help. Last but not least, I am very indebted to the organizers of the symposium «Fractals in Biology and Physics», namely G. Losa, D. Merlini and T. Nonnenmacher and to the Swiss National Foundation for Scientific Research for its material support.

References

[1] Adler I., J. Theor. Biol. 45 (1974), 1.

[2] Jean R.V., Phytomathématique, Presses de l'Université du Québec, Montréal (1978).

[3] Hardy G.H., Wright E.M., An Introduction to the Theory of Numbers, Fifth Edition, Clarendon Press, Oxford (1979).

8)   In the neighborhood of the tip of a growing stem, new sprouts are born along a ring-shaped region (the meristem); this region surrounds the apex.

[4] Schroeder M.R., Number Theory in Science and Communication, 2nd Edition, Springer Verlag, Berlin (1986).

[5] Rothen F., Koch A.-J., J. Phys. France 50 (1989), 633.

[6] Thompson D'Arcy W., On Growth and Form, 2nd Edition, Cambridge University Press, Cambridge (1942).

[7] Coxeter H.S.M., Introduction to Geometry, 2nd Edition, Wiley and Sons, New-York (1969).

[8] Levitov L.S., Phys. Rev. Lett. 66 (1991), 224; Europhys. Lett. 14 (1991), 533.

[9] van Iterson G., Mathematische und Mikroskopische Anatomische Studien über Blattstellungen, Gustav Fischer Verlag, Jena (1907).

[10] Rothen F., Koch A.-J., J. Phys. France 50 (1989), 1603.

[11] Rivier N., Koch A.-J., Rothen F., in Biology Inspired Physics, L. Peliti Ed., Plenum Press, New-York and London (1991).

[12] Douady S., Couder Y., Phys. Rev. Lett. 68 (1992) 2098.

[13] Thornley J.H.M., Ann. Bot. Fenn. 39 (1975), 493.

[14] Koch A.-J., Guerreiro J., Bernasconi G., Sadik J., submitted to J. Phys. I, France.

[15] Guerreiro J., Thesis, Université de Lausanne (1993).

Fractals in Molecular
and Cell Biology

Evolutionary Interplay Between Spontaneous Mutation and Selection: Aleatoric Contributions of Molecular Reaction Mechanisms

Werner Arber
Abt. Mikrobiologie
Biozentrum, Universität Basel
Klingelbergstr. 70
CH-4056 Basel, Switzerland

The concept of evolutionary change does not assume biological evolution to progress into a specific direction. Rather, single steps of change are at least in part aleatoric. In this view evolutionary change may be considered to be of fractal nature. However, in living systems, only a small fraction of mutants, i.e. of individuals having suffered a change in their genetic information, are maintained and propagated. Many other mutants can not compete with the successful ones and with their parents and are thus eliminated. This phenomenon is called selection.

Molecular genetics has revealed that spontaneous mutagenesis is the result of many different mechanisms, all affecting in parallel the genetic stability of living beings. This has most successfully been studied with microorganisms such as bacteria and their viruses. Theses studies have revealed that in many cases enzymes are involved in the formation of mutations. Nevertheless, mutagenic reactions usually do not reproducibly yield predictable products. Rather, the implied enzymes act as variation generators and the resulting mutations show a certain degree of randomness both with regard to their location on genomes and with regard to the time of their appearance. Probably, the intrinsic structural flexibility of biological molecules also helps to give to mutagenesis an aleatoric nature.

Let us recall that genetic information is contained in filamentous DNA molecules, in which it is encoded by the linear sequences of nucleotide pairs. Four different nucleotide pairs are available as letters to compose the linear script of the genetic information. For convenience and simplicity, we will call here a mutation any change occurring to the inherited nucleotide sequence, independent of whether such change results in an alteration of the properties, i.e. the phenotype, of the organism. Bacteria are unicellular organisms and they carry about 5×10^6 nucleotide pairs in their genome which is defined as the total genetic information of an organism. Mutational changes can affect one or more nucleotide pairs at once and can be substitutions, deletions, additions or structural rearrangements.

Upon propagation of cells, DNA of course also reproduces. In bacteria, the DNA replication rate is quite high, and the replication machinery progresses with a speed of about 1000 nucleotide pairs per second. Once in a while replication lacks fidelity, most likely due to the structural flexibility of the involved molecules. This can for example result in the substitution of one nucleotide by another. Although some sites on DNA molecules can be affected by this type of mutation somewhat more often than others, in general practically any location on a genome can be hit by a mutation due to replication infidelity. This statement applies also to a number of other mutagenesis mechanisms, so that in the course of long periods of time many different sites may suffer a mutation.

Mutations are quite often lethal, i.e. the organism is not able to propagate any longer. Other mutations can give the organism a selective disadvantage. Positive selection, in contrast, occasionally provides to a mutated organism an advantage as compared to its parent. The mutant may then eventually overgrow other forms of life. This kind of selection has been shown generally not to exert a direct influence on the mutagenesis per se. Rather it acts by the preference given to the particular mutant form of spontaneous origin. Finally, many alterations in the nucleotide sequences remain without noticeable influence to the organism. These are called neutral mutations. Some of them may have a longterm implication on genetic variation.

From these statements we can see that mutagenesis is essential for a steady progress of biological evolution. However, mutation rates should be lower than one new mutation per genome and generation. Otherwise the stability of a species would be endangered, particularly since many randomly occurring mutations are lethal or partly detrimental. Interestingly, haploid microorganisms with different genome sizes show often a different mutation rate per nucleotide pair, but have their mutation probability limited to in between 1% and 1‰ new mutations per genome and generation. This is usually the result of intrinsic antimutagenic factors which reduce the genetic instability of the cell.

A number of different enzyme-mediated processes provide means for rearrangements of the nucleotide sequences of a genome. This can involve the duplication or the deletion of a given DNA segment, or also its inversion. Much attention has been given in recent years to DNA inversion systems, the enzymes of which interact with specific DNA sequences. They thereby cut the DNA at these sites and rejoin the free ends in different ways so as to produce a DNA inversion. In rare cases, rather than to use a specific site on the DNA, the enzymes may also act on sequences widely deviating from the efficiently used sequences. This then can give rise to novel gene fusions. Such fusions can be within the sequences of two different genes and thereby produce a gene encoding a product composed of functional parts determined by the participating genes. This kind of mechanism can explain how the genetic information for a particular functional domain of a protein can become fused with information for other functional domains. As a matter of fact, many genes are known to be composed of several different functional domains. DNA inversion could thus be an evolutionary strategy to produce once in a while gene fusions. As long as such rearrangements affect only a few individuals in large populations and each of them in a different way, the population as such would not suffer from detrimental consequences but could still profit from very rare successful mutations providing improved fitness to the organism concerned. Although enzymes are involved in the generation of the mutations, none of the mutations is specifically programmed but rather appears to some degree at random.

Mobile genetic elements are defined as DNA segments able to change their location on DNA molecules. This phenomenon is called transposition and is catalysed by a specific enzyme produced by the element itself. The genomes of bacteria and higher organisms may each contain several different such elements. Generally speaking, mobile genetic elements have properties similar to those just described of DNA inversion systems. Indeed, each mobile genetic element has its characteristic strategy to select insertion target sites on DNA. Some elements prefer particular sites determined by a specific nucleotide sequence, but they can also use other targets although they do it with much lower probability. In contrast, other mobile genetic elements prefer certain regions on DNA molecules for insertion, but they may integrate practically at random into many different sites within the preferred region. This allows for a great number of structurally different transpositional DNA rearrangements. For the fitness of an individual suffering transposition, phenotypic alterations may be either negative or positive, depending on both the element involved and the target site used in transposition.

Mobile genetic elements which can also contain genes of chromosomal origin can transpose to so-called natural gene vectors which happen to be in a cell. A good example for a gene vector is the genome of a virus. Having suffered transposition, the virus can spread the mobilized genes of its host to other host strains. This is an example for another important mechanism that contributes to spontaneous mutagenesis of bacteria. The phenomenon is called horizontal gene transfer or acquisition of genetic information. Gene acquisition is also possible in direct contact between two genetically distinct cells in a process called conjugation. In addition, under particular conditions some bacteria are able to take up free DNA molecules from the environment upon transformation. All of these processes facilitate gene acquistion. Again, acquisition is to some degree random both in time and with regard to the informational content of the acquired DNA. But DNA acquisition does represent an important source of genetic information for bacterial strains which did not have a chance themselves to develop in the past an appropriate genetic function in need.

An impressive lesson on the ecological and evolutionary importance of DNA acquisition comes from studies of the reasons for the wide appearance of antibiotic resistances since antibiotic therapies have been introduced in the middle of our century. It has become clear that many resistance determinants which are today carried in enterobacteria had been acquired recently from other organisms. Horizontal gene transfer had of course always occurred at low rates and it still does today. What has changed, however, are the selection conditions for the benefit of rare cells having acquired resistance genes. This is a good example for our understanding the natural interplay between mutagenesis and selection. Upon drastic changes in the selection conditions, rare mutant forms in large populations can rapidly overgrow the large majority of nonselected cells.

As there is practically no action without counteraction in nature, gene acquisition also encounters natural barriers. These act at various levels, such as by inhibiting the penetration of foreign DNA or its establishment after uptake. In addition, functional compatibility between the products of the acquired genes and the residential biological activities of the recipient cell will be decisive for the success of an acquisition. Interestingly, practically all bacterial strains have one or several restriction systems able to specifically recognize if DNA penetrating into the cell is of foreign origin. In this case, the filamentous DNA is cleaved into fragments which are then rapidly degraded. But fragments may have some chance in their short lifetime to become incorporated into the cellular genome. This can represent an important evolutionary step for the involved bacterium. This type of acquisition in small steps (since relatively short DNA fragments are involved) is again to some degree aleatoric, and selection phenomena will eventually decide on the success of the process.

A tree has become the classical representation of biological evolution. Branching and the divergence of branches represents the increase in diversity and the accumulated differences between organisms of two branches. In this view, evolution progresses steadily as a result of the interplay between mutation and selection.

In view of the possibilities of gene acquisition which results from the horizontal transfer of bits of genetic information between two different branches, horizontal shunts should be drawn here and there between different branches of the evolutionary tree. In this way, it would symbolically become clear that future evolution, as in the past, will also depend not only on further internal development of existing branches, but also on genetic information which had become available in any other branch of the tree. The possibility of its horizontal transfer increases the chance for any type of living organism to make an important evolutionary step by sharing in the success of others. Indeed, as testified by the spreading of antibiotic resistances, gene acquisition seems to be a quite efficient evolutionary strategy, particularly if its efficiency is evaluated per event of transfer. This is related to the universality of the genetic language and to the fact that a biological function developed in one kind of organism may have a good chance also to serve similarly to another organism. As compared with gene acquisition, internal DNA rearrangement may be of lesser efficiency if again considered per event. This is due to both the relatively high chance of lethality of a DNA rearrangement and to the small probability to improve the fitness of the organism by the internal recombination. Finally, nucleotide substitution may be the least efficient of the three mechanisms of genetic variation which we have here specifically discussed. Single steps of nucleotide substitution are, however, important contributions to a longterm process of developing new biological functions. But it takes many consecutive steps to reach from a given DNA sequence a novel biological function.

We are aware that there exist many more molecular mutagenesis mechanisms than those described here and that each contributes in its specific way to the production of mutants. In most cases, the generation of mutants appears to be of fractal nature. However, the potential richness of fractal development is drastically limited by the effect of natural selection. Interestingly, selection is not constant. Rather, it is a function both of the physico-chemical environment and the activities of all living organisms present in the concerned ecological niche. Selection therefore varies with time and it of course also varies in space. Thus selection may favor a particular genetic constitution under some circumstances but disfavors it under other circumstances. This renders ecological studies so difficult. It is to be hoped, however, that progress in the fractal theorey will also open new approaches to better understand ecological equilibria and their evolutionary development.

Further readings:

- W. Arber: Mechanisms in microbial evolution. J. Struct. Biol. 104, 107–111 (1990)
- W. Arber: Elements in microbial evolution. J. Mol. Evol. 33, 4–12 (1991)
- W. Arber: Evolution of prokaryotic genomes. Gene (in press)(1993)

Error Propagation Theory of Chemically Solid Phase Synthesized Oligonucleotides and DNA Sequences for Biomedical Application*)

Zeno Földes-Papp, Armin Herold[1],
Hartmut Seliger and Albrecht K. Kleinschmidt[2]
Sektion Polymere
Universität Ulm
D-89069 Ulm (Donau)

1)   Schering AG, Zentrale Biologische Forschung, Berlin,

2)   Universität Ulm, Germany

Abstract. Our interest was focused on error propagation during the synthesis of single stranded oligonucleotides and DNA sequences. We have modeled the assumed degenerative effects of such $3' \rightarrow 5'$-driven syntheses, or vice versa $5' \rightarrow 3'$. Quantification in scaling the theoretically synthesized yields provided the rationale for our approaches to error propagation of synthesized oligonucleotides and DNA$_{ss}$ sequences. Our homeodynamical model is governed by nonlinear equations; they possess the relevant features of fractal dimensions. This is significant for implication of a new scale-invariant property of oligonucleotide synthesis. An inverse power law of driven multi-cycle synthesis on fixed starting sites is described with a model of growing oligonucleotides in fractal measures. It equates the constant coupling efficiency d_0 and the constant capping efficiency p_0 to the length distribution of sequences produced in the preparation set of oligonucleotides. Attractors in the dynamical model are included. Each sequence is produced randomly with the probability coupling function d and/or randomly with the probability capping function p; d and p are independent from one another. For this general problem we construct effective and computable recursive functions of the relation described by the inverse power law of driven multi-cycle synthesis. Nonlinear fractal dimensions $D(N)$ are computed for theoretical deviations from the constant coupling efficiency d_0. They offer a new access to the growth dependency on the nucleotides A, G, C, T.

1 Introduction

The chemical synthesis of oligonucleotides and single stranded DNA sequences is a powerful technique applied in genetic engineering, molecular biology and molecular medicine. Oligonucleotides as primers in polymerase chain reaction, for example, are essential for analytic, diagnostic and therapeutic purposes. The chemical synthesis of oligonucleotides and single stranded DNA using nucleoside-3'-phosphoramidites is a cyclic repetition of detrylation, coupling, capping, and oxidation in organic solvents [4]. We used mathematical formulations of monomers to build up a polymer structure as a model for directional growth. Multiplicative coupled, nonlinear equations for which the parameter search is made by experimental observations   were a way to formalize sequence errors. We studied the equivalent

---

*)   In memory of Severo Ochoa 1905–1993

of an algebraic instead of a geometric notation [13]. The mathematical model of a multi-cycle synthesis on fixed starting sites was tested by computer algorithms for experimental verification.

2   Chemical synthesis

Any mathematical formulation of chemical oligonucleotide synthesis on solid support is based on the following prerequisites (briefly described here, see also [4]): It is a driven growth process starting from 3' to 5' direction, or vice versa. The growing oligonucleotide or single stranded DNA is covalently bound to the support. We assume average constant conditions of fully tritylated nucleoside loading of the support. The cyclic repetition of coupling and capping steps during the growth process is considered as a linear undisturbed flow in first approximation. Error sequences are shorter than the target sequence, ending with a nucleoside; they can be separated according to their length. The error sequences are somehow truncated versions of a target sequence.

3   Mathematical formulation of a multi-cycle synthesis

The system described above acts as a variation generator. «1» represents an arbitrary (A, G, T, C) nucleotide/nucleoside in a sequence of oligonucleotides. As a coupling step either succeeds or fails, we use a second symbol for such missing nucleotides: symbol «0». Furthermore, we use a symbol for the capping group: symbol «2». Capping can only be done after a failed coupling step. Hence, symbol «2» is located only after symbol «0» as the last element of a sequence.

It is convenient to formulate the problem as follows: We consider the variation V with repetition w of two elements $(m = 2)$ «1» and «0» to $(N-1)$th power. N is the integer number of nucleotides/nucleoside of the target sequence. $N-1$ gives us the integer number of the last reaction cycle.

$$^{w}V_{m=2}^{N-1} = 2^{N-1} \tag{1}$$

Let us ask how many of these sequences for a given $N-1$ obtain either «1» or «0» as the last element. The answer is

$$\frac{1}{2}\left(^{w}V_{m=2}^{N-1}\right) = \frac{1}{2}\left(2^{N-1}\right) \tag{2}$$

From eqns. (1) and (2) follows that the number of sequences a_{N-1} for which the three elements «1, 0, 2» are mixed, is

$$a_{N-1} = \left(^{w}V_{m=2}^{N-1}\right) + \frac{1}{2}\left(^{w}V_{m=2}^{N-1}\right) \tag{3}$$

$$= 3\left(2^{N-2}\right) \tag{4}$$

$$= \sum_{l=0}^{N-2}\binom{N-2}{l} + \sum_{l=0}^{N-1}\binom{N-1}{l} \tag{5}$$

Here (Eqn. (5)), $l + 1$ is the integer number of nucleotides/nucleoside of target or error sequences. Since the first element of arbitrary sequences is «1», in fact, $N - 1$ nucleotides/nucleoside are varied. In other words, a_{N-1} is the number of nucleotides/nucleoside sequences of the last reaction cycle. They have been included in coupling and capping reactions during the last reaction cycle together with those that are left with reactive ends after the last reaction cycle.

Now we consider the number of truncated nucleotides/nucleoside sequences $a_{N-2,E}$ (errors, E) that have been left capped during all previous reaction cycles i

$$a_{N-2,E} = \sum_{i=1}^{N-1} \left(\frac{1}{2} \left({}^w V^i_{m=2} \right) \right) - \frac{1}{2} \left({}^w V^{N-1}_{m=2} \right) \tag{6}$$

$$= 2^{N-2} - 1 \tag{7}$$

$$= \sum_{l=1}^{N-2} \binom{N-2}{l} \tag{8}$$

Here (Eqn. (8)), l is the integer number of nucleotides/nucleoside of error sequences.

Hence (by Eqn. (6)), the number of all truncated, capped nucleotides/nucleoside sequences $a_{N-1,E}$ is

$$a_{N-1,E} = \sum_{i=1}^{N} \left(\frac{1}{2} \left({}^w V^i_{m=2} \right) \right) - \frac{1}{2} \left({}^w V^N_{m=2} \right) \tag{9}$$

$$= 2^{N-1} - 1 \tag{10}$$

$$= \sum_{l=1}^{N-1} \binom{N-1}{l} \tag{11}$$

including those of the last reaction cycle.

In view of eqns. (3) and (6), the total number of sequences $a_{N,S}$ (sum of error and target sequences, S) is presented

$$a_{N,S} = \left({}^w V^{N-1}_{m=2} \right) + \sum_{i=1}^{N-1} \left(\frac{1}{2} \left({}^w V^i_{m=2} \right) \right) \tag{12}$$

$$= 2^N - 1 \tag{13}$$

$$= \sum_{l=1}^{N} \binom{N}{l} \tag{14}$$

Because of identical serial arrangements of nucleotides/nucleoside, the total number of sequences $a_{N,S}$ is corrected to a lower value $a_{N,S(corr)}$ after cleaving off the capping group

$$a_{N,S(corr)} = \left({}^wV_{m=2}^{N-1} \right) + \sum_{i=1}^{N-1} \left(\frac{1}{2} \left({}^wV_{m=2}^i \right) \right) \tag{15}$$

$$- \left(\sum_{i=1}^{N} \left(\frac{1}{2} \left({}^wV_{m=2}^i \right) \right) - \frac{1}{2} \left({}^wV_{m=2}^N \right) \right)$$

$$= 2^{N-1} \tag{16}$$

$$= \sum_{l=0}^{N-1} \binom{N-1}{l} \tag{17}$$

These formulations make it possible to generate all integer numbers of sequences that are available in the synthesis. The nonlinear equations are typical for scale invariance [12] formally controlled by means of the scale exponent.

A system of equations is used for yields of sequences in dependence on the constant coupling efficiency d_0 as well as the constant capping efficiency p_0 in order to show the self-similarity of serial nucleotides/nucleoside arrangements. Here is the rule: Write all yields of sequences (A_{lN}) in dependence on the constant coupling (d_0) as well as capping (p_0) efficiencies until you get the yield of the target sequence (A_{NN}). Hence, we have eqns. (18) and (19)

$$A_{lN} = \mathbf{K}_{lN} \cdot \mathbf{a}_{lN} \tag{18}$$

where $l = \{1, 2, \cdots, N\}$, $N = \{2, 3, \cdots\}$ and

$$\mathbf{K}_{lN} = \left[\binom{N-1}{l-1} \quad \binom{N-2}{l-1} \quad \cdots \quad \binom{l-1}{l-1} \right] \tag{19}$$

The equations for yields of sequences are converted into a vectorial matrix format (eqns. (20) to (24))

$$\vec{a}_{11} = \begin{bmatrix} 0 \\ 0 \\ \vdots \\ 0 \end{bmatrix} \tag{20}$$

$$\vec{a}_{1N} = \begin{bmatrix} (1-d_0)^{N-1} \cdot (1-p_0)^{N-1} \\ (1-d_0)^{N-1} \cdot (1-p_0)^{N-2} \cdot p_0 \\ (1-d_0)^{N-2} \cdot (1-p_0)^{N-3} \cdot p_0 \\ \vdots \\ (1-d_0)^2 \cdot (1-p_0) \cdot p_0 \\ (1-d_0) \cdot p_0 \end{bmatrix} \tag{21}$$

$$\vec{a}_{NN} = \begin{bmatrix} (d_0)^{N-1} \\ (d_0)^{N-1} \\ \vdots \\ (d_0)^{N-1} \end{bmatrix} \tag{22}$$

$$\vec{a}_{lN} = \vec{a}_{l-1\,N-1} \begin{bmatrix} d_0 \\ d_0 \\ \vdots \\ d_0 \end{bmatrix} = \vec{a}_{l-1\,N-1} \cdot \vec{d}_0 \tag{23}$$

$$\begin{bmatrix} \vec{a}_{1N} \\ \vec{a}_{2N} \\ \vdots \\ \vec{a}_{NN} \end{bmatrix} = \begin{bmatrix} \vec{a}_{1N} \\ 0 \\ \vdots \\ 0 \end{bmatrix} + \begin{bmatrix} \vec{a}_{1N} \\ \vec{a}_{2N} \\ \vdots \\ 0 \end{bmatrix} + \cdots + \begin{bmatrix} 0 \\ 0 \\ \vdots \\ \vec{a}_{NN} \end{bmatrix} \tag{24}$$

Each vector element $\vec{a}_{l-1\,N-1}$ corresponds to a unique element $\vec{\sigma}\,(\vec{a}_{l-1\,N-1})$, where $\vec{\sigma}$ is a linear operator [5]. A model is obtained which is scale-invariant by vectorial definition. The constructed vector spaces have N dimensions. Each vector can be written according to Eqn. (25)

$$\vec{\sigma}\,(\vec{a}_{l-1\,N-1}) = \vec{a}_{l-1\,N-1} \cdot \vec{d}_0 = \vec{a}_{lN} \tag{25}$$

By serial nucleotides/nucleoside arrangements of target and error sequences this system embodies a higher-order fractal structure characterized by nonlinear fractal dimensions.

The moment of probability [7][11] of serial nucleotides/nucleoside arrangements $(M_{l,\,q=1})$ is derived from the total number of sequences $a_{N,S}$ presented after a given number of coupling cycles, and from the proportion of target and error sequences. The proportion of sequences depends on the constant coupling efficiency (d_0) and constant capping efficiency (p_0). The moment of probability of serial nucleotides/nucleoside arrangements during chemical synthesis of a target sequence is

$$M_{l,\,q=1} = \left[\binom{N-1}{l-1} \ \binom{N-2}{l-1} \ \cdots \ \binom{l-1}{l-1} \right]^{q=1} \cdot \mathbf{P}\,(l,\,N) \tag{26}$$

where $l = \{1, 2, \cdots, N\}$, $N = \{2, 3, \cdots\}$. $\mathbf{P}\,(l,\,N)$ is a probability matrix (not shown). Hence, the frequency distribution of nucleotides/nucleoside arrangements is described in the synthesis of a target sequence

$$M_{l,q=1} = f(l,N) \tag{27-a}$$

$$= \left(\frac{1}{d_0}\right)^{-(l-1)} \left\{ \left(\frac{1}{(1-d_0)p_0}\right)^{-1} \right.$$

$$+ \left(\frac{1}{p_0}\right)^{-1} \cdot \sum_{k=1}^{N-(l+1)} \left(\binom{N-(k+1)}{l-1} \left(\frac{1}{1-d_0}\right)^{-(N-(l-1+k))} \left(\frac{1}{1-p_0}\right)^{-(N-(l+k))} \right)$$

$$\left. + \binom{N-1}{l-1} \left(\frac{1}{(1-d_0)(1-p_0)}\right)^{-(N-l)} \right\}$$

$$M_{N-1,q=1} = f(N-1,N) = \left(\frac{1}{d_0}\right)^{-(N-2)} \left\{ \left(\frac{1}{(1-d_0)p_0}\right)^{-1} \right. \tag{27-b}$$

$$\left. + \binom{N-1}{N-2} \left(\frac{1}{(1-d_0)(1-p_0)}\right)^{-1} \right\}$$

$$M_{N,q=1} = f(N,N) = \left(\frac{1}{d_0}\right)^{-(N-1)} \tag{27-c}$$

where

$N =$ number of nucleotides/nucleoside of a target sequence

$d_0 =$ constant coupling efficiency

$p_0 =$ constant capping efficiency

An inverse power law of driven multi-cycle synthesis on fixed starting sites *(IPLCS)* is found (Eqn. 27-a,b,c). The linear operator $\vec{\sigma}$ of vector space S_{N-1} corresponds to the function F, where F is equal to

$$F = f(l, N) - f(l-1, N-1) \tag{28}$$

It is improbable that d and p are constants; d and p depend on internal conditions (e.g., concentrations of reactants, clustering, conformer variability, depurination, artificial branching, side reactions) and on external conditions (e.g., support fixed starting sites, temperature, pressure, time). We previously defined i as the integer number of reaction cycles ($i - \{1, 2, \cdots, N-1\}$), u and v are special internal or external conditions. Let $d = f(i, u)$ and $p = f(i, v)$ in the Euclidean plane $\mathbb{R}^2$, the coordinate plane of calculus, then $\{\mathbb{R}^2, f\}$ is a dynamical system of the iterated function system theory [1]. A simple example is given: Multivariate nonlinear analysis of factors d_0 and p_0 is equal to the «sums of exponentials» [6][3]

$$d = d_0 \left(\sum_{j=1}^{r} \alpha_j e^{-\beta_j(i_j-1)} \right) \tag{29}$$

$$p = p_0 \left(\sum_{j=1}^{s} \gamma_j e^{-\delta_j(i_j-1)} \right) \tag{30}$$

where $i_1 = \{1, 2, \cdots, g\}$, $i_2 = \{g + 1, \cdots, h\}$, $i_3 = \{h + 1, \cdots, N - 1\}$. Sum coefficients are r and s; α_j, β_j, γ_j, δ_j represent scaling factors. Parameters d_0 and p_0 are any constants of coupling and capping efficiency, respectively, in first approximation. Into the original relation described by the *IPLCS*, d and p are introduced as new functions. The analysis of this general case shows that the numbers of any nucleotides/nucleoside arrangements and the exponents of their mathematical expressions do not change. The inverse power law of driven multi-cycle synthesis on fixed starting sites has only to be modified in terms of integer numbers of reaction cycles and scaling factors. Assuming that «blocks in error» or «error clusters» influence d and p, phenomena of a multi-cycle synthesis on fixed starting sites may be related to the self-similar error cluster model of Mandelbrot [8], though for other purposes.

A priori, the regime of a multi-cycle synthesis on fixed starting sites is limited to values of d, or d and p less than 1.0 by acting of attractors. A fractal set is a distribution function (see [13]) which gives an inverse power law (see [13][1][9]). The *IPLCS* shows a logarithmic progression, which can be expressed in terms of fractal dimension D. The fractal dimension D depends on the errors of the growing chains.

Setting $\ln\left(\sum_{l=1}^{N-1} M_{l,\,q=1}\right) = Z + B \ln(N)$ and according to [2] $B = 2 - D$ then we have

$$\sum_{l=1}^{N-1} M_{l,\,q=1} = C\left(\frac{1}{N}\right)^{D-2} \tag{31}$$

Z is an intercept and B is the slope of a log-log plot $\sum_{l=1}^{N-1} M_{l,\,q=1}$ versus N [10]; C is given by e^Z. The values of fractal dimension D can be computed from Eqn. (31). The fractal dimension obtained from measuring data should allow to recognize all deviations from constant coupling efficiency d_0. This position can be made into a common expression. Substituting from Eqn. (29) into Eqn. (31) gives us

$$D = 2 - \frac{\ln\left\{1 - (d_0)^{N-1}\left(\sum_{j=1}^{r} \alpha_j^{N-1} \cdot \exp\left(-\frac{\beta_j}{2}(N-2)(N-1)\right)\right)\right\} + \ln(C)}{\ln(N)} \tag{32}$$

α_j $(0 < \alpha_j d_0 < 1)$, β_j are given by scaling procedures and $i_j = \{1, 2, \cdots, N-1\}$. When the distribution shown by Eqn. (31) is worked out for increasing values of N, the steps in the curve will become smaller and smaller until finally, with infinite N, a continuous curve is obtained. In continuous distribution we differentiate Eqn. (31) to eliminate C. This relation yields a general expression of D

$$D = 2 - N\left(\frac{\partial\left(\ln\left\{\sum_{l=1}^{N-1} M_{l,\,q=1}\right\}\right)}{\partial N}\right) \tag{33}$$

The fractal dimension $D(N)$ we discuss measures properties of sets of points determined by Eqn. (33) in a log-log plot $1 - M_{N, q=1}$ versus N. Hence, we discuss the functions which are defined by the independent variable N. The fractal dimension $D(N)$ is thus generalized. $\sum_{l=1}^{N-1} M_{l, q=1}$ can be directly measured by high performance liquid chromatography or capillary electrophoresis. Eqn. (33) gives us a new theoretical access to the growth dependency on the nucleotides A, G, C, T. Without substantial change we may consider sets of points that are defined by functions of l at $N = const.$

4 Conclusions

Error propagation of a multi-cycle synthesis of oligonucleotides on fixed starting sites can be theoretically regarded as a directed growth model with fractal measures. We use a model in which target and error sequences are formed by repetition of building blocks. Each sequence is produced randomly with the probability coupling function d and/or randomly with the probability capping function p; d and p are independent from one another. The inverse power law of driven multi-cycle synthesis on fixed starting sites has only to be modified in terms of integer numbers of reaction cycles and scaling factors. For this general problem we construct effective and computable recursive functions of the *IPLCS*. The theoretical yields of serial arrangements of sequences are indispensible for planning of experiments; the real growth process of synthesized short and long oligonucleotides is still arcane. The paper shows that the mathematical formulation of fractalistic error propagation in fractal dimensions D is a basis for new theoretical and experimental qualification (in preparation: I Theoretical considerations; II Application to experiments).

Acknowledgment:
Supported by DFG, Graduiertenkolleg «Biomolekulare Medizin» and Stiftung zur Förderung der molekularbiologischen Forschung, Univ. Ulm. For advice in mathematical symbols we thank Dr. Gerd Baumann, Dept. of Mathematical Physics, Univ. Ulm.

References

[1] Barnsley M: Fractals Everywhere. Academic Press, New York (1988)

[2] Batty M: Cities as fractals: Simulating growth and form. In: Fractals and Chaos. AJ Crilly, RA Earnshaw, H Jones (eds.), Springer, New York, pp. 43–69 (1991)

[3] Beechem JM: Global analysis of biochemical and biophysical data. In: Methods in Enzymol. **210**, 37–54 (1992)

[4] Caruthers MH, Beaton G, Wu JV, Wiesler W: Chemical synthesis of deoxy-oligonucleotides and deoxyoligonucleotide analogs. In: Methods in Enzymol. **211**, 3–20 (1992)

[5] Curtis ML: Abstract Linear Algebra. Springer, New York, pp. 9–46 (1990)

[6] Johnson ML, Faunt LM: Parameter estimation by least-squares methods. In: Methods in Enzymol. **210**, 1–37 (1992)

[7] Losa GA, Baumann G, Nonnenmacher TF: The fractal dimension of pericellular membrane from lymphocytes and lymphoblastic leukemic cells. Acta Stereol. **11**, Suppl. 1, 335–341 (1992)

[8] Mandelbrot B: Self-similar error clusters in communication systems and the concept of conditional stationarity. IEEE Trans. on Communication Technology, vol. **COM-13**, 71–90 (1965)

[9] Marek M, Schreiber J: Chaotic Behaviour of Deterministic Dissipative Systems. Cambridge Univ. Press, Cambridge (1991)

[10] Parker TS, Chua LO: Practical Numerical Algorithms for Chaotic Systems. Springer, New York, pp. 167–199 (1989)

[11] Small EW: Method of moments and treatment of nonrandom error. In: Methods in Enzymol. **210**, 237–279 (1992)

[12] Vicsek T: Fractal Growth Phenomena. World Scientific Publ., Singapore, pp. 9–46 (1989)

[13] West BJ: Fractal Physiology and Chaos in Medicine. World Scientific Publ., Singapore (1990)

Fractional Relaxation Equations for Protein Dynamics

Walter G. Glöckle and Theo F. Nonnenmacher
Department of Mathematical Physics
University of Ulm
Albert-Einstein-Allee 11
D-89069 Ulm, Germany

Abstract. Due to a large amount of conformational substates, relaxation processes in proteins are governed by many time constants and therefore, they decay more slowly than a Debye relaxation. For processes occurring on different time scales in a self-similar manner, we derive and solve a fractional order differential equation for the relaxation function. Solutions of this well-posed initial value problem are given in terms of a Mittag-Leffler function. Applications to ligand rebinding data of myoglobin are presented leading to a 3-parameter fractional model.

1 Introduction

In a large variety of biological systems, non-exponential relaxations are observed. Their relaxation function deviates from the Debye relaxation

$$\phi(t) = \phi_0 \exp\left(-\frac{t}{\tau}\right) \tag{1}$$

and it shows a decay which is slower than the exponential, e.g. in terms of a stretched exponential

$$\phi(t) = \phi_0 \exp\left(-(t/\tau)^\alpha\right) \tag{2}$$

or an inverse power law

$$\phi(t) \sim t^{-\beta} \tag{3}$$

for large t. In the last decades, among others, proteins were intensively studied from a physical point of view. According to their biological function, two states of a protein are distinguished, e.g. an ion channel can be open or closed, a hemoglobin or myoglobin protein can have bound oxygen or not. The transition between the two states is considered to be a relaxation process where the relaxation function is the probability of a protein to be in the «excited state».

The slow relaxations are often modeled by a superposition of Debye relaxations. If the two observable states consist of many substates, so-called conformational substates, with different relaxation times, the superposition of these single exponential processes leads to a slow relaxation. To model a power law decay, however, a large amount of such Markov processes is necessary. Therefore, in the last years concepts known from the description of polymers or glasses [1] have been applied to proteins. Doster, Cusack a. Petry [2] e.g. used the mode coupling

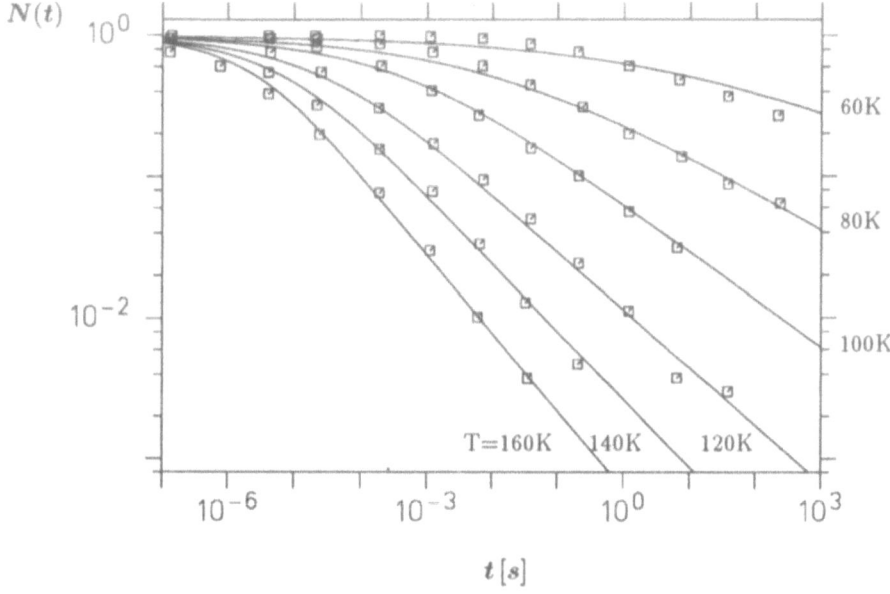

Fig. 1   Rebinding of CO to myoglobin after photodissociation [5]. The solid lines correspond to a fit with the solution (25) of the fractional equation (26) with $\beta(T) = 0.41T/120K$ and $\tau_0(T)$ given in fig.3.

theory of the glass transition to interpret inelastic neutron scattering data of proteins. Millhauser [3] applied the reptation model to estimate time constants of the dynamics of ion channels. Nonnenmacher [4] applied renormalization group ideas leading to the algebraic decay of the distribution function for open times of ion channels and to the observed oscillations around the power law trend.

Here we will consider the ligand rebinding in myoglobin after flash photodissociation [5]. Because of the well-known structure of this relatively simple protein, myoglobin serves as a model system to study protein motions and reactions. After the dissociation, the ligands are rebound to the heme iron and the fraction of the free ligands $N(t)$ decays. In fig.1 measurements at different temperatures [6] are shown. The decay is much more slowly than the exponential and the function

$$N(t) = (1 + t/\tau)^{-n} \qquad (4)$$

is often used to fit the data with temperature dependent parameters τ and n [5].

Alternatively the deviation from the exponential law is regarded to be a consequence of the large number of conformational substates leading to many different reaction rates or time constants [6]. Therefore, the reaction kinetics is described by using a distribution $g(E)$ of energy barrier heights. Then $N(t)$ is given by

$$N(t) = \int_0^\infty g(E) \exp(-k(E)t)dE \tag{5}$$

where for $k(E)$ an Arrhenius law

$$k(E) = A \exp(-E/RT) \tag{6}$$

is assumed. Instead of $g(E)$ the distribution $b(k)$ of reaction rates k

$$N(t) = \int_0^A b(k) \frac{RT}{k} \exp(-kt)dk \tag{7}$$

can be used. The two distributions are connected by $g(E) = b(A \exp(-E/RT))$. If we consider observation times t much greater than the time scale of molecular motions $1/A$, A can be substituted by ∞ in (7). Thus, $N(t)$ is the Laplace transform of $b(k)/k$:

$$N(t) = \mathcal{L}\left(\frac{RTb(k)}{k}, t\right). \tag{8}$$

Here we will use fractional order differential equations to describe the dynamics of proteins. Utilizing ideas of self-similar dynamical processes we will derive a fractional order initial value problem for the relaxation process.

2   Self-similar dynamics

Since many conformational substates are accessible to a protein, the protein system is not relaxing with a single time constant τ but with many time constants τ_n leading to

$$N(t) = \sum_n w_n \exp(-t/\tau_n) . \tag{9}$$

Self-similar dynamics means that the weights w_n and the time constants τ_n are not independent but correlated. If the time scale is changed by $t \rightarrow \lambda t$ ($\lambda > 1$), the behavior of the system should be the same apart from a renormalization of the statistical weight ($w \rightarrow \bar{w}$).

In case of the Bernoulli scaling

$$\tau_n = \lambda^n \tau_0 \qquad\qquad (\lambda > 1) \tag{10}$$

$$w_n = p^n p_0 = \left(\lambda^{-\beta}\right)^n p_0 \quad (\beta > 0, p < 1) \tag{11}$$

with infinitely many processes, one obtains in the continuum limit the relaxation function [4]

$$N(t) = \int_0^\infty p_0 \lambda^{-\beta x} \exp\left(-\frac{t}{\lambda^x \tau_0}\right) dx = \frac{p_0}{\ln \lambda} \left(\frac{t}{\tau_0}\right) \gamma(\beta, t/\tau_0) \tag{12}$$

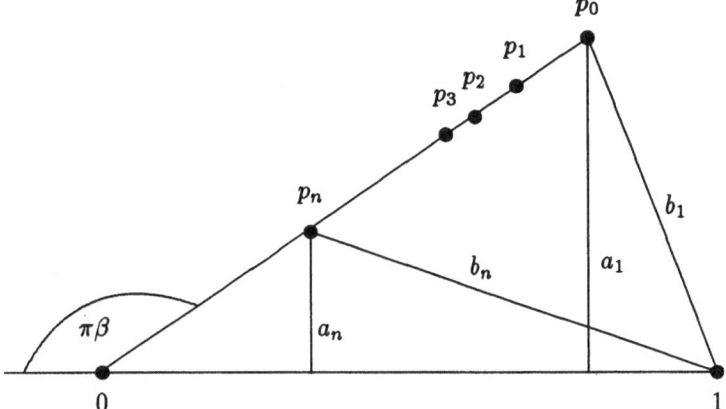

Fig. 2 Statistical weights $w_n = a_n/b_n^2$ of the fractional scaling model in comparison to p_n of the Bernoulli scaling model for $1/2 < \beta < 1$.

expressible by the incomplete γ-function. For large t an algebraic decay

$$N(t) \sim \frac{p_0 \Gamma(\beta)}{\ln \lambda} \left(\frac{t}{\tau_0} \right)^{-\beta} \tag{13}$$

is found. This continuous Bernoulli scaling can be translated to the description by energy barrier heights leading to

$$g(E) = p_0 \exp \left(-\frac{\beta E}{RT} \right) \tag{14}$$

which is independent of the temperature T if the parameter β is proportional to T.

In order to give a connection to the fractional relaxation equation, we consider a slightly different self-similar process, here called fractional scaling. Contrary to the Bernoulli scaling, the weight w_n is not proportional to $p^n = \lambda^{-\beta n}$ but modified in the form

$$\tau_n = \lambda^n \tau_0 \tag{15}$$

$$w_n = p_0 \frac{p^n \sin(\pi\beta)}{p^{2n} + 2p^n \cos(\pi\beta) + 1} . \tag{16}$$

In fig.2 the weights $w_n = a_n/b_n^2$ of the fractional scaling model are geometrically compared with those of the Bernoulli scaling model ($w_n = p_n$). For large values of n, b_n tends to 1 so that $w_n \sim p_n \sin(\pi\beta)$, i.e. for large n the two models are equivalent. In terms of the renormalization group theory, the two models belong to the same «class of universality» where the universality in these time processes is the asymptotic power law decay with the power $-\beta$.

For the fractional scaling, the energy barrier distribution is given by

$$g(E) = p_0 \frac{e^{-\beta E/RT} \sin(\pi\beta)}{e^{-2\beta E/RT} + 2e^{-\beta E/RT} \cos(\pi\beta) + 1} \tag{17}$$

and the corresponding distribution of reaction rates reads as

$$b(k) = p_0 \frac{(k\tau_0)^\beta \sin(\pi\beta)}{(k\tau_0)^{2\beta} + 2(k\tau_0)^\beta \cos(\pi\beta) + 1}. \tag{18}$$

3 Fractional equation

In order to derive the fractional order differential equation connected with the self-similar process, we calculate the Laplace transform

$$\frac{Q(p)}{p} = \mathcal{L}(N(t), p) \tag{19}$$

of the relaxation function $N(t)$. Because of (8), we can use the Stieltjes transform $\mathcal{S} = \mathcal{L}\mathcal{L}$ given by

$$\mathcal{S}(f(k), p) = \int_0^\infty \frac{f(k)}{p+k} dk \tag{20}$$

to calculate $Q(p)$ from the rate distribution $b(k)$. With (18),

$$Q(p) = N_0 \frac{1}{(\tau_0 p)^{-\beta} + 1} \tag{21}$$

is found with $N_0 = RT\pi p_0$, where the fractional power β indicates a fractional order differentiation or integration.

A proper initial value problem is obtained by considering the type of the relaxation process. With respect to the photodissociation experiment, there is a constant amount of free ligands because of an external force $E(t)$ for $t < 0$. At $t = 0$ the force is switched off and the number $N(t)$ decays. For this type of relaxation, (21) leads to the fractional integral equation

$$\tau_0^{-\beta} {}_0D_t^{-\beta} N(t) + N(t) - N_0 = \chi_0 \tau_0^{-\beta} {}_0D_t^{-\beta} E(t) \tag{22}$$

by inverse Laplace transform. In (22), ${}_0D_t^{-\beta}$ is the Liouville-Riemann fractional integral operator [7] with the order β being equal to the parameter β of the self-similar process (15,16). By applying the inverse operator, i.e. the fractional differential operator ${}_0D_t^\beta$, on (22), the fractional differential equation

$${}_0D_t^\beta N(t) - N_0 \frac{t^{-\beta}}{\Gamma(1-\beta)} = -\tau_0^{-\beta} (N(t) - \chi_0 E(t)) \tag{23}$$

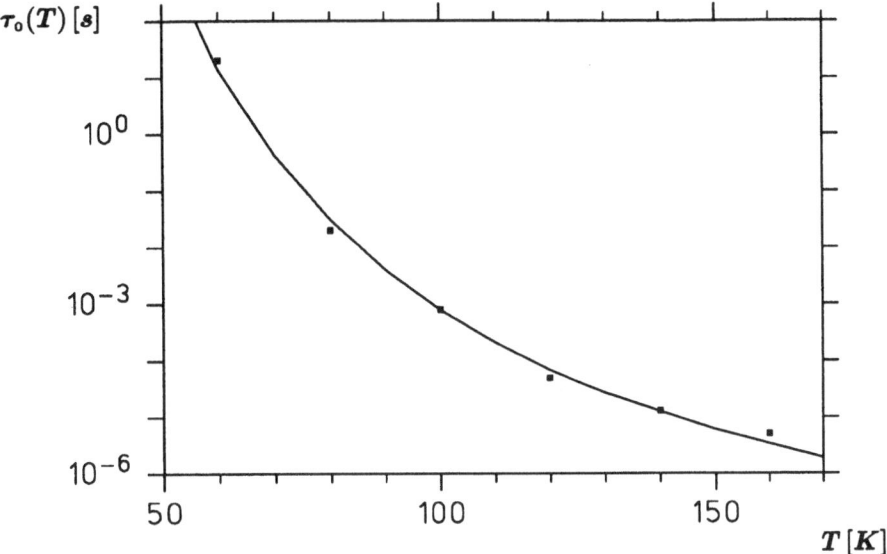

Fig. 3 Dependency of $\tau_0(T)$ used in fig.1. The solid line is an Arrhenius curve (28) with $E^* = 1470K$ and $\tau_m = 3.4 \cdot 10^{-10}s$.

is attained from which for $\beta \to 1$ the ordinary relaxation

$$\frac{d}{dt}N(t) = -\frac{1}{\tau_0}(N(t) - \chi_0 E(t)) \qquad (24)$$

results.

The solution [8]

$$N(t) = N_0 E_\beta\left(-(t/\tau_0)^\beta\right)$$
$$= N_0 \sum_{k=0}^{\infty} \frac{(-1)^k}{\Gamma(1+\beta k)}\left(\frac{t}{\tau_0}\right)^{\beta k} \qquad (25)$$

for times $t > 0$ follows from the initial value problem

$$\tau_0^{-\beta} \, _0D_t^{-\beta}N(t) + N(t) - N_0 = 0 . \qquad (26)$$

In (25), $E_\beta(x)$ is a Mittag-Leffler function [9] from which the asymptotic decay $N(t) \sim (t/\tau)^{-\beta}$ for $0 < \beta < 1$ can be derived. For $\beta = 1$ we recover the exponential function.

In fig.1 the fractional relaxation function (25) is applied to the ligand rebinding to myoglobin. The fractional function fits the data better than the power law (4). The temperature dependency of β is taken proportional to T

$$\beta(T) = \frac{0.41}{120K}T \qquad (27)$$

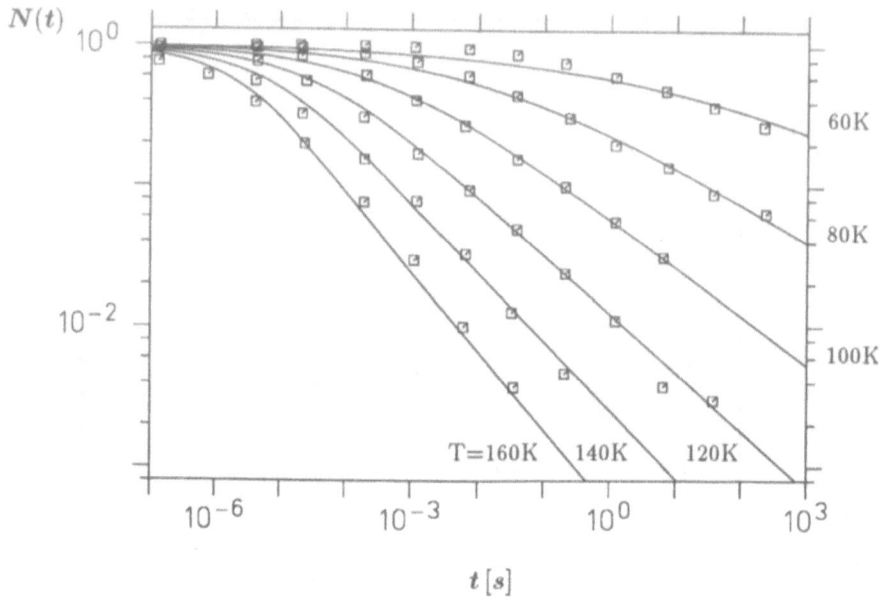

Fig. 4   3-parameter model (29) for the rebinding of CO to myoglobin with $\tau_m = 8.4 \cdot 10^{-10}s$, $\alpha = 3.5 \cdot 10^{-3}K^{-1}$, and $k = 130$.

and $\tau_0(T)$ is determined by a least square fit. As demonstrated in fig.3, $\tau_0(T)$ agrees well with an Arrhenius law

$$\tau_0(T) = \tau_m \exp\left(\frac{E^*}{T}\right) . \tag{28}$$

The activation energy $E^* = 1470K \hat{=} 12.3\,kJ/mol$ is of the order of a characteristic energy barrier height $E_{peak} \approx 10..11\,kJ/mol$ [10] and $\tau_m = 3.4 \cdot 10^{-10}s$ is a time constant of molecular motion.

Because of these temperature dependencies, the fractional equation (26) can be transformed to

$$k \, _0D_z^{-\beta}N(z) + N(z) - N_0 = 0 \tag{29}$$

with the dimensionless time $z = t/\tau_m$, the fractional order $\beta = \alpha T$, and a constant k which is independent of the temperature. In fig.4 this 3-parameter model is applied to the Mb-CO rebinding with $\tau_m = 8.4 \cdot 10^{-10}s$, $\alpha = 3.5 \cdot 10^{-3}K^{-1}$, and $k = 130 (\hat{=} E^* = 11.5\,kJ/mol)$. For temperatures T lower than the glass temperature $T_g \approx 180K$ [11] the dynamics and the temperature dependency are well-described by the 3-parameter model.

In this paper we considered self-similar processes in which each time scale occurs in a self-similar manner and therefore, no internal time scale exists. These processes form an analogue in the time domain to objects with fractal geometry in space. We demonstrated that fractional relaxation can be deduced from a self-similar process. Especially a connection between the statistical weight of the processes and the fractional order of the differential equation has been established. The fractional order initial value problem was applied to the kinetics of ligand rebinding to myoglobin after photodissociation. It turned out that for low temperatures the time and the temperature dependency is reproduced by a 3-parameter fractional model.

Acknowledgement:
This work has been supported by Deutsche Forschungsgemeinschaft (SFB 239, Project C8).

References

[1] A. Blumen, J. Klafter, a. G. Zumofen, in *Optical Spectroscopy of Glasses*, ed. by I. Zschokke, Reidel, Dordrecht (1986) 199.

[2] W. Doster, S. Cusack, a. W. Petry, J. Non-Cryst. Sol. **131–133** (1991) 357.

[3] G.L. Millhauser, Biophys. J. **57** (1990) 857.

[4] T.F. Nonnenmacher a. D.J.F. Nonnenmacher, Phys. Lett. A **140**, 323 (1989) and in *Stochastic Processes, Physics and Geometry* ed. by S. Albeverio, G. Casati, U. Cattaneo, D. Merlini, a. R. Moresi, World Scientific, Singapore (1990) 627.

[5] R.H. Austin, K.W. Beeson, L. Eisenstein, H. Frauenfelder, a. I.C. Gunsalus, Biochemistry **14** (1975) 5355.

[6] H. Frauenfelder, F. Parak, a. R.D. Young, Ann. Rev. Biophys. Biophys. Chem. **17** (1988) 451.

[7] K.B. Oldham a. J. Spanier, *The Fractional Calculus*, Academic Press, New York (1974).

[8] W.G. Glöckle a. T.F. Nonnenmacher, Macromolecules, **24** (1991) 6426.

[9] A. Erdélyi (edt.), *Bateman Manuscript Project, Higher Transcendental Functions* I–III, McGraw-Hill, New York (1953).

[10] R.D. Young a. S.F. Bowne, J. Chem. Phys. **81** (1984) 3730.

[11] I.E.T. Iben, D. Braunstein, W. Doster, H. Frauenfelder, M.K. Hong, J.B. Johnson, S. Luck, P. Ormos, A. Schulte, P.J. Steinbach, A.H. Xie, a. R.D. Young, Phys. Rev. Lett. **62** (1989) 1916.

Measuring Fractal Dimensions
of Cell Contours:
Practical Approaches and their Limitations

Gerd Baumann, Andreas Barth and Theo F. Nonnenmacher
Department of Mathematical Physics
University of Ulm
D-89069 Ulm

Abstract. We discuss practical methods to examine fractal properties of electron micrographs and outline a procedure to extract information from a gray scale image (EM-image) by using different filtering methods. The data obtained from this reduction of information is used as a basis for numerical calculations of fractal dimensions D. To determine D of a given object, we use the yardstick, the box counting and the probabilistic methods. These three approaches will be critically discussed and the finite scaling range for natural objects such as cells will be examined. The method of digital image analysis as discussed here incorporates an algorithm that detects self-similar domains in the structure of cells. We apply this method to determine the fractal dimension of cell contours and we discuss advantages and limitations of these methods. As prototype examples, we investigate cellprofiles of lymphocytes and lymphocyte leukemic cells.

1 Introduction

In close analogy to Richardson's examination of borders for different European countries [1], we discuss practical methods to examine fractal properties of electron micrographs and outline a procedure to extract information from a gray scale image (EM-image) by using different filtering methods. The data obtained from this reduction of information is used as a basis for numerical calculations of fractal dimensions D. To determine D of a given object, we use the yardstick, the box counting and the probabilistic methods. These three approaches will be critically discussed and the finite scaling range for natural objects such as cells will be examined. The method of digital image analysis as discussed here incorporates an algorithm that detects self-similar domains in the structure of cells. We apply this method to determine the fractal dimension of cell contours and we discuss advantages and limitations of these methods. As prototye examples, we investigate cell-profiles of lymphocytes and lymphocyte leukemic cells, as a first step into fractal analysis of cell profiles [2][3]. The characteristic property of boundaries Richardson and later on Mandelbrot found is the self-similarity of the border. For *true fractals* this means scaling with a common exponent on different length scales ranging from 0 to infinity. The corresponding scaling exponent, the fractal dimension D describes the deviation of the object from the euclidean dimension in which it is embedded. For some irregular and wrinkled curves in the plane one will determine the fractal dimension greater than 1 and less than 2. In our studies [5], we will restrict our considerations to sets in the plane because our experimental data are points of a two dimensional picture.

2 Image Extraction Procedure

Our examination is based on EM-pictures with a magnification of 8000 ×. The negative is turned into a positive at a magnification of 18400 ×. This print is used to extract the contour of a cell. Two different methods are taken to get the boundary of the cell: first, a computer based procedure and second, a manually based one.

Both procedures use a scanner and a contour tracing algorithm. The computer procedure is based on EM-prints. The print is first discretized at a resolution of 300 dpi in x and y directions. The amplitude of each picture element (pixel) is represented by a 8 bit data word. This representation discretizes the gray value of the original picture in 256 gray levels. The amplitude resolution produces about 4 MB of data depending on the size of the picture. The next step in deriving the cell's contour is the application of a filter. Two types of filter were used in our examinations: i) gradient filters and ii) band limited filters.

In a first test, we applied a standard gradient and a Laplace filter on the discretized picture and detected edges of steep, gray value changes. The results from these two filters were rather poor and contained many errors. By using a different filter operator, the so-called Sobel operator, we managed to get good results for our work. Depending on the print quality, the contours derived from this filter type may be broken in which case a further processing of the picture becomes impossible. Yet, a Sobel operator acting in x and y directions gives a good representation of the cell contour. A main problem with this filter which works above a certain gray value threshold is that the cell contour is not closed. This defect is the main obstacle in isolating the cell contour from a gray scale picture.

The second filter we used was a Marr-Hildreth or Mexican hat filter [4]. This type of filter uses a specific gray value distribution located around a pixel to detect edges in the picture. The results produced by this filter are as good as the results obtained by a Sobel filter. Again, the main problem is a broken contour line and a broadening of the edges. The broadening of the edges results in a small loss of information on the contour structure. This error originates from the assumed Gauß-distribution of gray values around a pixel.

Once a closed contour line is successfully extracted with one of the above filters, we can go on to the next step in processing the image. In the computer based procedure the object now needs to be isolated from its background. There is no standard method with which to isolate the cell under consideration. Normally, we autofill the objects by hand. The largest filled object is then selected by the computer program and used by the contour tracing algorithm. At times, it may be necessary to remove parts of the picture by using the delete functions of our program. If the contour is closed and if we succeeded in deleting the neighbouring cells from the picture, we can apply the last step of the contour extraction. This step is fully automatic and uses the contour tracing algorithm to isolate the (x, y)-coordinates from the binary picture. This step is necessary since both filter procedures widen the contours and consequently may cause errors in the ensuing analysis.

The second method which is in part a manually executed method is in comparison with the computer-based method a very simple, reliable yet time consuming method. In this procedure, we extract the contour from the EM-prints by means of a hand drawing. Parchment paper and a sharp Faber 2H pencil suffice to extract the contour from the print. Many different «artists» tested this method with good results. If the cell contours are drawn accurately, the results are as good as or better than from the computer-based method. EM-pictures always show some faded regions which a computer cannot interprete correctly. One of the benefits of the «artistic» method is a closed contour at the end of the drawing. Once the contour of a cell is drawn, we scan the drawing and apply the contour tracing algorithm on the picture to get the binary representation of the cell border. At this stage of the analysis, we extract the contour of the cell in binary form and the coordinates of the contour are known. Different methods can now be used to detect scaling behavior in the geometric structure of the cell contour.

3   Calculation of Fractal Dimensions

There are two main methods to find scaling behaviour in natural objects: first, the two metric methods and second, the probabilistic method.

The metric methods are the yardstick method and the box counting method. Both methods measure the perimeter of a cell by using the geometric properties of the contour. Each element in this measuring process has the same weight. Contrary to the property of equal weight, the probabilistic method allocates a certain weight to each measure. However, all three methods give approximations of the perimeter on the length scale of the yardsticks used.

3.1 Yardstick Method

One of the simplest and fastest methods is the yardstick method. The scaling or fractal dimension D for an object with a closed contour is derivable from the numbers $N(\epsilon)$ of yardsticks with a length ϵ needed to cover a cell border. Already Richardson applied this method for estimating border lengths between various countries [1]. The length is given by

$$L(\epsilon) = \epsilon N(\epsilon). \tag{1}$$

This perimeter tends to increase as ϵ decreases. For natural objects there exist lower and upper limits of the yardstick length ϵ. The lower limit in our computer studies is given by one pixel. One possible characteristic for the upper limit is given by the radius of the cell. If scaling behaviour is located between these two limits, such behavior is characterized by the fractal exponent D. For *true fractals* such upper and lower limits do not exist as ϵ ranges ideally from 0 to infinity. The

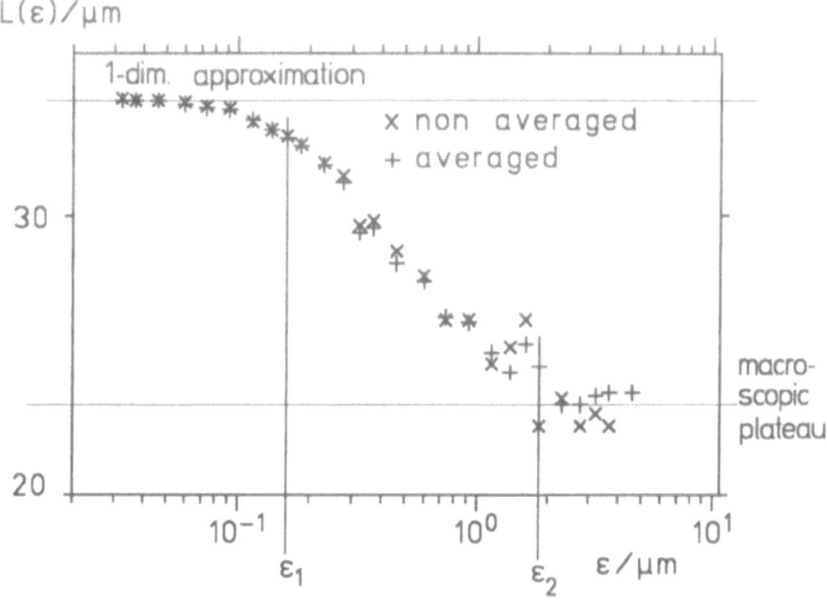

Fig. 1 Typical measurments with the yardstick method.

term *true fractal* also means that $N(\epsilon)$ diverges as $L \to \infty$ if $\epsilon \to 0$. For fractals, we thus get

$$N(\epsilon) \sim \epsilon^{-D} \tag{2}$$

and consequently

$$L(\epsilon) \sim \epsilon^{1-D} . \tag{3}$$

In order to estimate the scaling exponent for a *true fractal*, it suffices to count the number of yardsticks at different yardstick lengths ϵ and to find the limit

$$D = \lim_{\epsilon \to 0} \frac{\log N(\epsilon)}{\log(1/\epsilon)}. \tag{4}$$

In practical applications, the number of yardsticks $N(\epsilon)$ at a fixed length ϵ ranging between lower and upper limits has to be counted. Then the logarithm of both quantities $N(\epsilon)$ and ϵ is taken and their values are plotted in an (x, y)-plot. If the natural object possesses the scaling property, a straight line in the log-log-plot can be detected. The slope of the scaling region is directly related to the scaling exponent or fractal dimension D. It is clear from figure 1 that this procedure is only reliable if a scaling region is detected in the plot. To find the scaling region, we used an automatic fit procedure in our studies to check the confidence level within which this region is being located. Checking the confidence level guarantees the reproducability of the results. Using a standard linear regression program, all points of the calculation have the same weight in the fit procedure and as a result, the total ϵ range is assumed to possess scaling behavior. A second source of errors in this method is the closing gap at the end of the tracing. The closing error is a typical error of the yardstick method. The error occurs if the yardstick length does not match the contour length exactly. This error is very large for large yardstick lengths and shows up in large fluctuations in the number of yardsticks for larger ϵ. To eliminate this effect, we use all discrete points of the contour as starting points when measuring the length. The calculation of an averaged number $\bar{N}(\epsilon)$ by

$$\bar{N}(\epsilon) = \frac{1}{N_{max}} \sum_{i=1}^{N_{max}} N_i(\epsilon) \tag{5}$$

will suppress these fluctuations. N_{max} is the total number of contour points. $N_i(\epsilon)$ corresponds to the number of yardsticks for each starting point i. The scaling exponent D is then calculated by

$$D = \lim_{\epsilon \to \epsilon_1} \frac{\log \bar{N}(\epsilon)}{\log(1/\epsilon)}, \tag{6}$$

with ϵ_1 the lower limit of the scaling region. While the yardstick procedure is a very fast method, its problems are the detection of the scaling region and the closing error at the end of the contour. However, the closing error can be avoided and the scaling range extended by using the box counting method.

3.2 Box Counting Method

The box counting method is based on the division of a plane into squares of edge length ϵ. The box counting method also delivers an estimate of the length of a contour by counting the number of boxes $N(\epsilon)$ of a given size ϵ. Each box containing at least one pixel is counted in $N(\epsilon)$. This number is roughly proportional to the number of N needed by the yardstick method. Starting with the smallest ϵ scale (one pixel), the grid length ϵ is increased successively to about 1300 pixels depending on the size of the EM-print. Again in a log-log-plot of $N(\epsilon)$ versus ϵ, a scaling range for self-similar structures is obtained (see Fig. 2). The scaling dimension D is again given by (4). Since the covering of the plane with a grid of edge length ϵ is somewhat arbitrary, we have to use several locations of the grid's origin to eliminate this arbitrariness. For each edge length ϵ, which is an integer value, we use ϵ^2 points as starting points to calculate a mean number $\bar{N}(\epsilon)$ by

$$\bar{N}(\epsilon) = \frac{1}{\epsilon^2} \sum_{i=1}^{\epsilon^2} N_i(\epsilon), \tag{7}$$

whereby $N_i(\epsilon)$ counts the number of occupied boxes of the grid for each starting point i. Thus the dimension D can be estimated by using the described fit procedure in the sense of equation (6). It is clear from this procedure that after selecting the

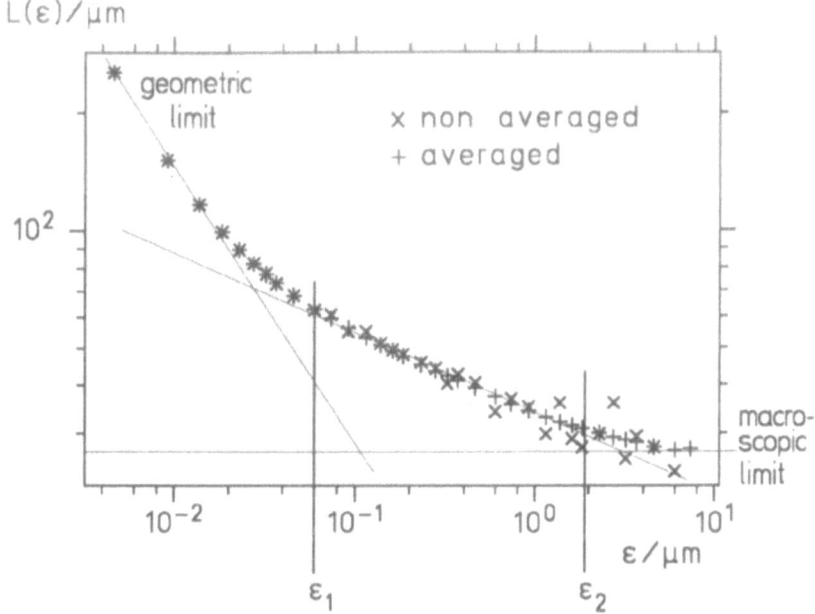

Fig. 2 Measurments taken by the box counting method for a contour line with a finite width. If the width of the boundary is reduced to a one pixel line, we observe the same behavior as shown in figure 1.

scaling range, only reliable estimates for L are obtained. The box counting method does not show any closing errors and thus the scaling range is larger than in the yardstick method. The upper scaling limit ϵ_2 is again given by the macroscopic extension of the cell. The lower scaling limit ϵ_1 depends on the method of analysis and the details of preparation.

As seen in figures 1 and 2 both methods possess for large and small yardsticks ϵ limiting values. As mentioned above the plateau value for large ϵ's is determined by the macroscopic extension of the object (cell). The limit for small ϵ depends on the method used. For the yardstick method one observes a plateau which is typical for the one dimensional approximation of the contour line. In case of the box counting method the limiting behaviour for small ϵ's is also a plateau if the width of the contour line is one pixel (not shown in fig. 2). This plateau changes to a straigth line with a slope near one if the width of the contour line is some pixels. Such a situation is shown in figure 2. The line with slope around one is infact a consequence of the two dimensional approximation of the contour line. The limits for small ϵ's occur since the structural variance of the boundary disappears for these yardstick lengths. The scaling range between ϵ_1 and ϵ_2 is thus a function of the roughness of the contour line too.

3.3 Probabilistic Method

The third method used in our experiments is the probabilistic method. In determining the scaling dimension D by the two methods discussed so far, all squares or yardsticks needed to cover the cell border are equally weighted even if the number of pixels with which a border visits a box may be different. Lacking other information, any point of the contour is assumed to be equivalent and is an equally probable origin for our analysis. The spatial arrangement of the points in the discretized image along the cell border is used to determine the probability $P(m, \epsilon)$. The distribution $P(m, \epsilon)$ gives the probability to locate m points within a square of size ϵ which is centered on an arbitrary point along the contour line. Again, the origin of the grid which covers the picture is arbitrary and thus we determine the number of boxes of edge length ϵ containing m pixels as an average over different locations of the origin. The scaling behaviour of the moments

$$M_q = \sum_{m=1}^{N} m^q P(m, \epsilon) \tag{8}$$

of the probability distribution $P(m, \epsilon)$ can be used to determine scaling exponents D_q which are given by

$$D_q = \frac{1}{q} \frac{\log M_q(\epsilon)}{\log \epsilon}. \tag{9}$$

Different D_q values for the moments indicate multi-fractal behaviour of the cell profile.

4   Results

First, we tested the three methods as to their accuracy and reliability in determining the fractal dimension D of an exact self-similar construct. In these checks, we changed the resolution of representation on a grid of the fractal. The yardstick method was tested with a Koch curve. The result is that with increasing resolution R (number of pixels) this method approaches the exact scaling dimension $D_{exact} = 1.261\ldots$ from below. For instance, taking $R = 1200 \times 1200$ pixels, we get $D = 1.22$ and for $R = 2400 \times 2400$, we get 1.238 as scaling exponents. The box counting method and the probabilistic method were tested with a Sierpinski triangle ($D_{exact} = 1.5849\ldots$). The results are collected in table 1. Two points become obvious: i) the higher the resolution R, the more accurate the value D, ii) the probabilistic method approaches D_{exact} from above while the box counting method approches D_{exact} from below. The results listed in table 1 are clearly in favour of the probabilistic method since the errors are about a factor of 10 smaller than for the box counting method. However, the computer time for one run is larger by a factor of three.

To test the scaling hypothesis for cell boundaries, we examined lymphocytes in four different states: stimulated, unstimulated, healthy, and disease states. The

R	D_{PM}	error(PM)	D_{BCM}	error (BCM)
1024×1024	1.63	2.84%	1.49	-5.7%
1124×1124	1.63	2.84%	1.51	-4.6%
1526×1526	1.62	2.21%	1.53	-3.3%
1824×1824	1.60	0.95%	1.54	-3.0%
2048×2048	1.59	0.321%	1.54	-3.1%
2524×2524	1.58	0.309%	1.54	-2.6%

Table 1 Determination of the fractal dimension by the probabilistic method (PM) and the box counting method (BCM) for a Sierpinski triangle for different resolutions R. R counts the number of pixels on a square lattice in x and y directions. The Sierpinski triangle was constructed with an iterated function system with $n = 10^6$ iterations. The exact theoretical value for this fractal is $D_{exact} = 1.5849\ldots$.

fractal dimensions collected in the work [3] are results from the probabilistic method. As a system of reference we took 52 normal T-lymphocytes. The mean value of all individual fractal dimensions for this cell type is $D = 1.20 \pm 0.05$. The same value within the standard deviation is obtained for integer moments of q up to $q = 5$. In [3] there are also some other types of cells with which various influences like preparation, cell line etc. on the scaling exponents are discussed.

Acknowledgement:
The work has been supported by the Swiss National Science Foundation under the grant No. 31-25702.88.

References

[1] B. Mandelbrot, The fractal geometry of nature, Freeman, San Francisco, (1983).

[2] G.A. Losa, G. Baumann, and T.F. Nonnenmacher, Path. Res. Pract. **188**, 680–686 (1992).

[3] G.A. Losa, G. Baumann, and T.F. Nonnenmacher, Acta Ster. **11**/Suppl I, 335–341 (1992).

[4] T.G. Smith Jr., W.B. Marks, G.D. Lange, W.H. Sheriff Jr., and E.A. Neale, J. Neuro. Sci. Meth. **26**, 75–82 (1988).

[5] T.F. Nonnenmacher, G. Baumann, and G. Losa, Trends in Biological Cybernetics **1**, 65–73 (1990).

Fractal Properties of Pericellular Membrane from Lymphocytes and Leukemic Cells

Gabriele A. Losa

Laboratorio di Patologia Cellulare

Istituto Cantonale di Patologia, 6600 Locarno

and Faculté des Sciences, Université de Lausanne

1000 Lausanne, Switzerland

Abstract. Blood mononuclear cells examined on two-dimensional electron microscopy pictures display plasmalemmal membranes with irregular contour due to the alternance of microprotrusions and invaginations at the cell surface. Surface irregularity can be quantified in terms of the fractal dimension D because cell contours manifested the property of statistical self-similarity. A probabilistic method which defined the scaling behaviour through five scaling exponents or moments of order q was applied for evaluating the fractal dimension D of the plasma membrane of distinct cell populations. Resting CD8 suppressor T-lymphocytes showed a monofractal dimension D (the five moments of D had a similar value) of their plasma membrane higher but than the D of plasma membrane from CD4 helper T-lymphocytes. In contrast, blast cells of acute lymphoblastic leukemia of B and T cell lineage were characterized by a smaller fractal dimension D. This indicates that the examined leukemic lymphoblasts were proliferating cells with surface pattern smoother (i.e. less irregular) than those of mature immunocompetent lymphocytes. Similar smooth surface patterns showed T-blasts raised by culturing mononuclear cells with mitogenic phytohaemagglutinin. In contrast, concanavalin A-triggered T-cells displayed a plasma membrane of more irregular contour and consistently higher fractal dimension, close to that of control T lymphocytes. Finally, established lymphoblastoid cell lines of B lineage with progressive stage of differentation showed a corresponding increase of their surface fractional dimension.

1 Introduction

The immune system is constituted of several populations of circulating and resident cells, including lymphocytes, granulocytes, mononuclear phagocytes, which generate in the bone marrow from cellular precursors. Through successive stages of differentation and selection final cells reach the immunologic competence. The current working hypothesis postulated that cells derived from precursors may not proceed through maturation steps but remain in effect frozen in a particular point of the differentiation process. Those elements that fail to become differentiated into mature cells due to the neoplastic transformation, can eventually expand into a dominant population of undifferentiated or partially differentiated cells [with a phenotype close to the corresponding normal counterparts] and thus give rise to leukemias of various type. The quantitative evaluation of morphological patterns of surface and nuclear membranes is of great importance not only to establish correlations with biochemical functional mechanisms but also in histopathology in order to identify the types of leukemia and to assess grading systems wich may help in predict cancer progression, invasiveness, and malignancy [1]. A certain number of morphometric shape descriptors integrated by a series of mathematical transformations [2] are used to quantitate structural and morphological features of cells and tissues [3]. The application of conventional morphometric methods to measure size, shape and area reposes on the assumption that biological structures approximate geometrical objects and conventional dimensions adequately describe morphological features and peculiar changes of cellular and nuclear patterns. In practice, this implies to disregard the real form and shape of cells and cellular organelles. It however offers an explanation why ambiguous and divergent results may often occur from morphometric estimates recorded at a single scale [8]. Even computer-assisted quantitative analysis which utilizes the ratio of investigated area to the area of the surrounding rectangular figure recorded at a unique scale might not be adequate to quantify nuclear irregularity [4]. In fact, cellular structures are almost irregular and therefore reveal more details about their morphological configuration by closer inspection at increasing scale of magnification. Hence the degree of morphological irregularity and the entity of geometrical parameters cannot be evaluated at a single scale of length in reason of the property of the statistical self-similarity on multiple scales, a characteristic of fractal objects [5], which connotated also subcellular ultrastructures and organelles of many physiological competent and neoplastic cells as well [6,7,8]. In the present study, a probabilistic method was applied [9,10,11,12] to evaluate the degree of irregularity of cell surface contour in terms of fractal dimension D of several cell populations, namely immunocompetent human blood lymphocytes, T-lymphocytes stimulated by mitogenic lectins in vitro, lymphoid blasts isolated from patients with acute lymphoblastic leukemia of different phenotype and finally, established lymphoblastoid cell lines which constituted a neoplastic system at progressive stages of differentation along the B lineage.

2 Material and methods

2.1 Immunophenotyping of mononuclear cells
Resting mononuclear cells isolated by gradient centrifugation from blood samples of normal donors [21], mononuclear cells cultured in vitro in presence of mitogenic lectins Phytohaemagglutinin or Concanavalin A, established lymphoblastoid cell lines and blasts cells isolated from patients with acute lymphocytic leukemia were identified on the basis of the membrane surface immunophenotype.

Cells were stained with fluorescent conjugated monoclonal antibodies directed against surface membrane antigens CD2, CD3, CD4, CD7, CD8, CD10, CD19, CD20, CD33, CD34, and positivity assessed by flow cytometry (FACScan, Becton Dickinson).

2.2 Lectins stimulation and cell cultures
Aliquots of peripheral blood mononuclear cells (2×10^6/ml) were incubated at $37°C$ for five days in a humidified 5% CO_2 atmosphere with phytohaemagglutinin (PHA) and Concanavalin A (ConA) at the concentration of $10 \mu g$/ml.

Stimulated cells were then collected, washed with 0.05 M Tris-K-Mg buffer solution (TKM), counted and processed for both conventional and immunoelectron microscopy.

Cultures of human lymphoblastoid cell lines, REH-6, Nalm-1, Raji and 6410, were maintained at 37°C in a humidifed 5% CO_2 atmosphere in plastic flasks. The medium consisted of RPMI 1640 supplemented with 2 mM L-Glutamin, 40 mg/l folic acid, 2 g/l sodium bicarbonate, 50 mg/l gentamycin and 10% FCS. Cells were examined for surface markers phenotype by flow cytometry and processed for electron microscopy.

2.3 Conventional and immunoelectron microscopy

Cell Samples ($1 - 5 \times 10^6$) were fixed during 5–6 hours at 4° with a 30 g/l of glutaraldehyde, calcium chloride in 0.1M sodium cacodylate containing solution, pH 7.4. After washing, cell samples were postfixed for 1–2 hours at room temperature with a 2g/100ml of osmium tetroxide in 0.1M sodium cacodylate containing solution. Dehydratation with ethanol and propylene oxyde preceded embedding in Epon.

Thin sections having an interference color of gray to silver were cut with an LKB 2128 microtome, mounted on 200-mesh grids coated with a formwar film and finally stained with uranyl acetate (5g/100ml) for 1 hr and successively with lead citrate for 30 min. Micrographs were recorded on plastic plates with a Philiphs 400-ET microscope at the magnification of 8000 $\times$ and 12500 $\times$ in order to visualize completely a single cell. The cell phenotype was identified in EM preparations by the peroxidase-antiperoxidase technique after staining cell sample for 30 min at 4°C, with unconjugated monoclonal antibodies directed to surface antigens previously mentioned. After washing with phosphate buffered NaCl solution, cells were fixed for 30 min at 4°C in a 0.1g/100 ml glutaraldehyde solution, thereafter incubated for 30 min at room temperature with rabbit anti-mouse IgG-antibody (diluition 1/100), rinsed and finally reacted for 30 min with a mouse peroxidase-antiperoxidase complex. The enzymatic activity was activated by adding diaminobenzidine-H_2O_2 as substrate which transformed in a electron dense complex.

3   Fractal Dimension of the Plasmalemma

The fractal dimension D_q of cell surface perimeter was estimated by applying a probabilistic method exhaustively described in previous reports [10,11] and in the present volume [13,14]. The boundary profile of a cell projected into a plane was picked up by a scanner and the data stored into a memory of a computer in the form of a two dimensional array. Some twenty EM pictures of each cell population have been analyzed with a NEC SX-3 of the Swiss Scientific Computing Center (CSCS), Manno (Ti), using a frame of 2048 $\times$ 2048 pixels. Experimental image analysis is dealing with a geometrical object defined by a set S of points at position $x = (x_1, x_2, \ldots x_\epsilon)$ in a ϵ-dimensional Euclidean space. The spatial arrangement of pixels along the cell border was then used to define a probability $P(m, \epsilon)$ which counts the number of points m within a square of size ϵ centered about an arbitrary point of the image S. $P(m, \epsilon)$ is directly related to other probability measures as introduced by Mandelbrod [5]. The moment $M_q(\epsilon)$ of the probability $P(m, \epsilon)$ is defined by

$$M_q(\epsilon) = \sum_{m=1}^{N} m^q P(m, \epsilon) \qquad (1)$$

where N counts the number of boxes which at least contain one pixel. These $M_q(\epsilon)$ are related to the fractal dimension D_q by logarithmic derivatives

$$D_q = \frac{1}{q} \left(\frac{d \log M_q(\epsilon)}{d \log \epsilon} \right) \qquad (2)$$

This particular definition of the probabilty $P(m, \epsilon)$ has been introduced by [15] and it is very efficient to implement on a computer [14]. The range of ϵ used in our study was established in three orders of magnitude, namely from $\epsilon_{min} = 0.003 \mu$m to $\epsilon_{max} = 3 \mu$m. If all moments give the same value of D_q, then the fractal set is considered uniform (monofractal) while for a non-uniform fractal set (multifractal object) the fractal dimension D_q would take different values. Our findings indicated that distinct cell populations [20 elements for each population were examined] had D_q values for the five moments q almost similar within a maximum standard deviation of SD = 0.06, what excludes the presence of multi-fractal structures.

4 Results

Differentiated resting lymphocytes isolated from peripheral blood of normal donors share a plasmalemma with a definite degree of irregularity as documented by the dimension D_q of 1.21 ± 0.04 (table 1). Among T-lymphocytes distinct subsets could even be identified by the fractal dimension: namely, CD4 T-helper lymphocytes with a $D_q = 1.17 \pm 0.03$ which revealed a less irregular surface border than CD8 T-suppressor lymphocytes ($D_q = 1.23 \pm 0.03$). In another series of experiments, isolated mononuclear cells were cultured for five days with T-cell mitogenic lectins, namely PHA and ConA at the unique concentration of $10 \mu g/ml$. The fractal dimension D_q was found reduced in PHA rather than in ConA transformed T-cells, i.e $D_q = 1.11 \pm 0.02$ versus 1.27 ± 0.02 respectively, indicating that the former proliferating T-cells lose morphological irregularity of their plasma membrane (table 1). We have also examined peripheral blood cells isolated from patients affected by acute lymphoblastic leukemia (ALL) of distinct immunophenotype, identified by the expression of surface markers specific of B and T cell lineage. Seven out of nine cases of acute leukemia of B cell lineage, i.e four cases of common acute lymphoblastic leukemia (c-ALL) (no 1,7,8,9) and three cases of pre-B-ALL (no 2,4,5), both undifferentiated subtypes of leukemia, expressed D values ($D = 1.12-1.17$ and $1.17-1.20$ respectively) lower than the

Cells	D_1	D_2	D_3	D_4	D_5	$\pm SD$
Monomolecular cells	1.19	1.20	1.21	1.21	1.22	0.04
CD3 T-lymphocytes	1.20	1.20	1.20	1.20	1.21	0.03
CD4 T-lymphocytes	1.17	1.16	1.16	1.16	1.17	0.03
CD8 T-lymphocytes	1.22	1.23	1.23	1.23	1.23	0.03
CD19 B-lymphocytes	1.19	1.20	1.20	1.19	1.20	0.03
PHA-stimulated mononuclear cells	1.11	1.11	1.12	1.12	1.12	0.03
ConA-stimulated mononuclear cells	1.25	1.26	1.26	1.27	1.27	0.02

Table 1 Fractal dimension D_q for five moments of plasmalemma from human lymphocytes. *SD*: standard deviation.

fractal dimension value which characterized the plasma membrane of mature B lymphocytes. (fig. 1, table 1). Two other cases, namely 3 and 6, were characterized by a plasmalemmal contour more irregular with a value of $D_q = 1.23$, i.e. higher than the D_q of normal B cell counterpart. Both cases were recognized as acute lymphoblastic leukemias of the early B-immunophenotype (B-ALL), which corresponds to a more differentiated stage of acute leukemias according to the hypothetical scheme of B cell differentiation [16]. Accordingly, these cells were found devoid of CD10 and slightly positive for CD34 antigen, which are usually expressed at the surface of undifferentiated lymphoid cells and on leukemic blasts of c-ALL and pre-B-ALL (figure 1).

As a matter of comparison cells of Hairy cell leukemia (case 10), a chronic type of leukemia, were characterized by a plasma membrane with numerous microvilli and protrusions wich consistently yielded an high fractal dimension, namely $D_q = 1.33 \pm 0.03$.

Our investigation has also included four cases of acute leukemia of the T-cell lineage (T-ALL).

Blasts with an immature T phenotype (pre-T ALL), i.e. expressing CD7 and CD3 thymic antigens, were found to have a very low fractal dimension of 1.10 ± 0.03, with an order of magnitude of D_q close to that evaluated on proliferating cells triggered in vitro by mitogenic PHA (table 2).

Two other cases of T-ALL with cells of thymic medullar origin bearing surface markers of more mature T-phenotype (CD5) have an higher fractal dimension of 1.20 ± 0.02, whilst the highest D_q value of 1.23 ± 0.03 was recorded on T-ALL blasts with a phenotype similar to the one characterizing normal thymus cells of the cortical zone.

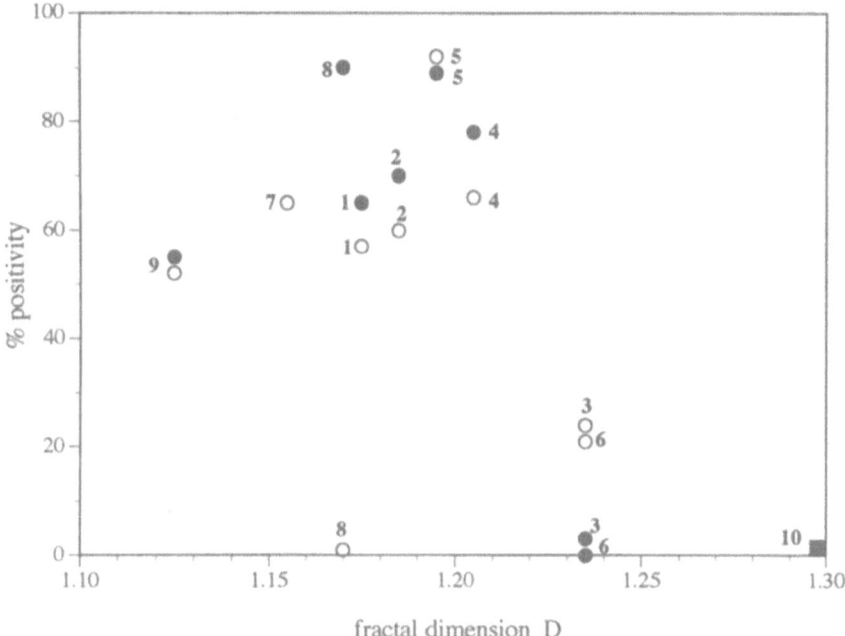

Fig. 1 Common Acute Lymphoblastic Leukemia (c-ALL): cases no 1,7,8,9. Surface immunophenotype: TdT, HLA-DR, CD10, CD19, (CD34). Pre-B-Acute Lymphoblastic Leukemia (pre-B-All): cases no 2,4,5. Surface immunophenotype: Tdt, HLA-DR, CD10, CD19, (CD34), Cytopasmic immunoglobulin μ (Cyμ). B-Acute lymphoblastic Leukemia (B-ALL): cases no. 3,6. Surface immunophenotype: HLA-DR, CD19, CD20, (CD10), surface membrane immunoglobulin (SmIg). Hairy cell leukemia: case no 10. TdT, deoxynucleotidyl terminal transferase; HLA-DR, human leukocyte antigen class II. CD = Antigen cluster differentation. Cells positive to fluorescent anti CD10(o) and CD34(●) monoclonal antibodies were evaluated by flow cytometry as indicated in the Material and Methods section.

Cells	D_1	D_2	D_3	D_4	D_5	$\pm SD$
pre T-ALL (CD7; CD3;TdT-)	1.10	1.11	1.11	1.11	1.11	0.02
T-ALL medullar (CD2; CD3; CD5; TdT-)	1.21	1.20	1.20	1.20	1.20	0.02
T-ALL medullar (CD3; CD5; TdT+)	1.20	1.20	1.20	1.20	1.20	0.02
T-ALL cortical (CD2; CD5; TdT+)	1.25	1.23	1.24	1.24	1.24	0.03

Table 2 Fractal dimension D_q for five moments of plasmalemma from Acute Lymphoblastic Leukemias (ALL) expressing thymic antigens. SD: standard deviation.

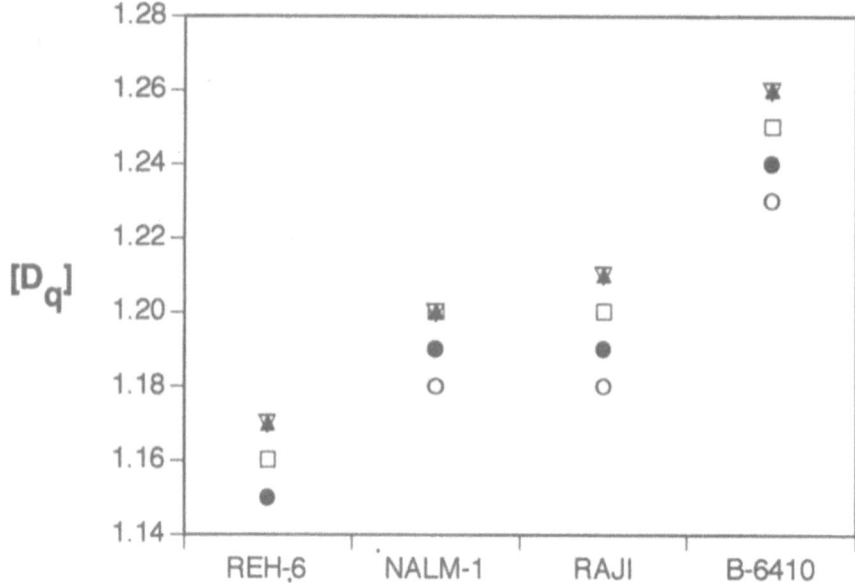

Fig. 2 Fractal dimension D_q for q: 1(o), 2(●), 3 (□), 4 (△) and 5 (▽) of plasmalemma from lymphoblastoid cell lines; REH-6 (HLA-DR, CD10); NALM-1 (HLA-DR, CD10, Cyμ); RAJI (HLA-DR, CD10(+/−), SmIg) and B-6410 (HLA-DR, SmIg, K).

Human lymphoblastoid leukemic cell lines with an immunophenotype corresponding to a progressive stage of differentation of the B lineage were examined at the end of the growth exponential phase (fig. 2).

The lowest D_q value (1.16) characterized the cell membrane of REH-6 cells, the less differentiated among the four cell lines examined, while the highest D (1.25) pertained to the plasma membrane of the most differentiated B-cell line 6410 (fig 3). Nalm-1 and Raji cells could not be distinguished from each other on the basis of the fractal dimension; interestingly, they presented a similar intermediate surface immunophenotype.

Indeed, the range of D_q was 1.18–1.20 and 1.18–1.21 for Nalm-1 and Raji cells respectively. This indicates that the plasma membrane of these cell lines has a fractional border with a similar degree of irregularity (fig.3) although inferior to that of the differentiated 6410 B cell line.

5 Discussion

In the present study we could demonstrate that mature blood mononuclear cells like lymphocytes, immature cells isolated from patients with acute lymphoblastic leukemia and cultured human lymphoblastoid cells as well, have plasma membrane boundaries of irregular pattern. Cell surface borders manifest the property

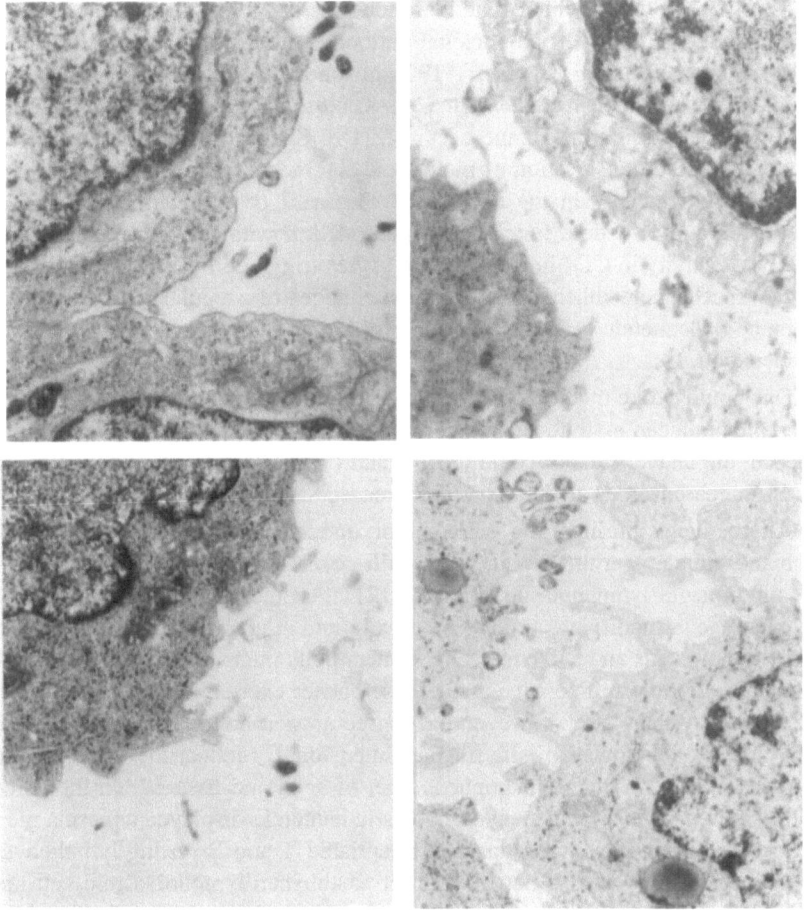

Fig. 3 Electron micrograph of plasmalemma portion from lymphoblastoid cell of a) REH-6; b) NALM-1; c) RAJI; and d) B-6410. Final magnification 18400 x; bar: 1cm= 540 nm.

of statistical self-similarity within a definite scaling range and could be portrayed as fractal objects properly described in terms of fractal rather than topological dimensions. According to fractal dimension values of the plasma membrane it turns out that neoplastic lymphoid cells from patients with acute leukemia of the undifferentiated type (c-ALL, pre-B-ALL and T-ALL) were in general smoother than the corresponding mature T or B lymphocytes [20]. Similar average fractional dimensions have been reported for human lymphocytes from whole blood [17]. Contour smoothness characterized also the plasma membrane of mononuclear cells bearing T-antigens which were triggered for proliferation with PHA mitogenic lectin. In contrast, the plasma membrane of cells proliferating with ConA displayed an higher degree of irregularity confirmed by the fractal dimension value.

Recently, we could demonstrate that mononuclear cells stimulated with either PHA or mitogenic anti-CD4 monoclonal antibodies underwent a partial translocation of signal transducing enzymes (PIP2 phospholipase C) bound to the plasma membrane toward nuclei, whereas in ConA cultured cells a reversed translocation toward the plasma membrane occurred [18]. Such a transmembrane process which enabled the translocation of membrane fragments from the surface into the cell interior might explain the decreased roughness of cell membrane in PHA treated cells and provide a functional-structural link with the apparent reduction of the PLC activity in CD4 helper T-lymphocytes triggered by PHA. The true biochemical mechanisms inherent the observed changes of irregularity on the plasma membrane of leukemic cells and of cells proliferating in vitro are however far to be understood.

One could evoke modifications in the configuration and composition of membrane lipids induced by mitogens and/or neoplastic factors and cytokines. As consequence, the bilayer fluidity could change and create zones within the membrane with an increased order of lipids but a decreased structural complexity [19].

Alternatively, proliferating cells might undergo an enhanced shedding of membrane fragments from cell surface as still observed with undifferentiated cells of common acute lymphoblastic leukemia [21]. All these membrane phenomena may intervene in modifying surface complexity and structural richness of proliferating lymphoid cells and leukemic cells. Cancer cells might eventually exploit one or the other mechanism to escape the defense barrier exerted in vivo by the reticuloendothelial system. Does, however, the degree of morphogeometrical irregularity of cell surface quantitated by the fractal dimension D_q reflect the differentiation level, in terms of surface immunophenotype, of cells and tissues? On the whole, undifferentiated cells from acute lymphoblastic leukemia displayed a plasma membrane less irregular in comparison to differentiated T and B resting lymphocytes. Such a relationship was also noticed in four established lymphoblastoid cell lines which showed a similar high proliferation rate or S-phase but a specific degree of differentiation. Consistently, the higher fractal dimension was measured on cell line with the most differentiated surface immunophenotype.

The fractal dimension D_q constitutes a quantitative descriptor of the real cell surface morphology and other cellular organelles as well. So far, the level of cellular differentiation expressed in numerical term might help, together with other quantitative cellular parameters, in view of a comprehensive description of the biological profile of neoplastic tissues. The closer the fractal dimension to the reference D_q of membranes from structural and functional mature cells the more differentiated appear to be the neoplastic cells. Finally, this study aimed to underline that by taking into account the property of self-similarity and the scaling domain beyond which morphological measures remain invariable one could estimate the true length, size, surface and volume of irregular living structures [22].

Acknowledgments:
The present study has been supported by grant no. 31-25702.88 of the Swiss National Science Foundation.

References

[1] Baak J.P.A., Chin D., Van Diest P.J., Ortiz R., Matze-Cok p., Bacus S.S. 1991. Comparative long-term prognostic value of quantitative HER /neu protein expression, DNA ploidy, and morphometric and clinical features in paraffin embedded invasive Breast cancer. Lab.Inv. 64, 215–223.

[2] Pienta K.J.Coffey D.S. 1991 Correlation of Nuclear morphometry progression of Breast Cancer. Cancer 68, 2012–2016

[3] Sorensen F.B. 1992. Quantitative analysis of nuclear size for objective malignancy grading: a review with emphasis on new, unbiased stereologic methods. Lab. Inv. 66, 4–23.

[4] Wydner K.S, Godyn J.J, Lee M.L, Sciorra L.J, 1991. A new approach to the computer-assisted quantitative analysis of nuclear shape. Modern Pathology 4, 154–160.

[5] Mandelbrot BB. 1982. The fractal geometry of the nature. Freeman, San Francisco.

[6] Nonnenmacher TF. 1987. A scaling model for dichotomous branching processes. Biol Cyber 56, 155–159.

[7] Nonnenmacher TF. 1988. Fractal shapes of cell membranes and pattern formation by dichotomous branching processes. In: Lamprecht I, Zotin AI,eds. Thermodynamics and Pattern Formation in Biology, Walter de Gruyter, Berlin-New York, 1988, 371–394

[8] Paumgartner D., Losa GA., Weibel ER. 1981. Resolution effect on stereological estimation of surface and volume and its interpretation in terms of fractal dimensions. J Micros 121, 51–63.

[9] Baumann G., Nonnenmacher,T.F. 1989. Determination of fractal dimensions. In Losa GA, Merlini D, Moresi R, eds. Gli oggetti frattali in astrofisica, biologia, fisica e matematica. Cerfim, Locarno, 93–104.

[10] Nonnenmacher T. 1989. Fractal structures in biomedical systems and morphogenetic scaling processes. In: Losa GA, Merlini D., Moresi R., eds. Gli Oggetti Frattali in Astrofisica, Biologia, Fisica e Matematica, Cerfim, Locano, 64–92.

[11] Nonnenmacher TF., Baumann G., Losa GA. 1990. Self-organization and fractal scaling patterns in biological systems. In Trends in Biological Cybernetics, Menon J.,ed., Publications Manager, Research Trends, Council of Scientific Research Integration, Trivandrum India, 1, 65–73.

[12] Nonnenmacher TF,Baumann G., Barth A., Losa G.A. 1993 A quantitative structural analysis of self-similar cell profile. Proceedings of the Royal Society (London) Series B (Biological Sciences) (in press).

[13] Nonnenmacher TF. 1993. Spatial and temporal fractal patterns in cell and molecular biology. In Fractals in Biology and Medicine. Eds. Nonnenmacher TF., Losa GA., Weibel ER., Birkhäuser Verlag, Basel

[14] Baumann G, Barth A., Nonnenmacher T.F. 1993. Measuring Fractal dimensions of cell contours: practical approaches and their limitations. 1993 Fractals in Biology and Medicine. Eds. Nonnenmacher TF., Losa G.A., Weibel ER., Birkhäuser Verlag Basel

[15] Voss RF. 1985. Random Fractals: Characterization and Measurements. In: Pynn R., Skjeltrop A., eds. Scaling Phenomena in Disordered Systems, Plenum Press, New York, 1–11.

[16] Van Dongen J.J.M., Adriaansen H.J., Hooijkaas H. 1988. Immunophenotyping of leukemias and non-Hodgkin Lymphomas. Neth.J.Med. 33, 298–314

[17] Keough K.M.W., Hyam P., Pink D.A, Quinn B. 1991. Cell Surfaces and fractal dimensions. J.Microscopy 163, 95–160.

[18] Graber R., Leoni L., Carrel S., Losa G.A. 1993 Lectins and anti-monoclonal antibodies induced changes of second messengers generating enzymes in human peripheral blood mononuclear cells. Cellular Molecular Biology. 39, 45–54.

[19] Graber R., Losa G.A. 1993 Subcellular localization of inositide enzymes in established T-cell lines and activated lymphocytes. Analytical Cell. Pathology, 5, 1–16.

[20] Losa G.A., Baumann G.,Nonnenmacher TF. 1992. Fractal dimension of pericellular membrane in human lymphocytes and lymphoblastic leukemic cells. Path.Res.Practice 188, 680–686.

[21] Losa G.A., Heumann D, Carrel S, Von Fliedner V, Mach JP.1986. Characterization of membrane vesicles circulating in the serum of patients with common acute lymphoblastic leukemia. Lab. Investigation 55, 573–579.

[22] Weibel ER. 1991. Fractal geometry: a design principle for living organisms. Am J Physiol, 261, L361–L369.

Cellular Sociology: Parametrization of Spatial Relationships Based on Voronoi Diagram and Ulam Trees

Raphaël Marcelpoil, Franck Davoine and
Michel Robert-Nicoud
Equipe de Reconnaissance des Formes et Microscopie Quantitative
Université Joseph Fourier
CERMO, BP 53X. 38041
Grenoble Cedex, France

1   Introduction

At all levels (molecules, cells, organisms, . . .), biological systems are made of structural and functional units. Those units are interacting with each other and, tend to use space in an optimal way with respect to their specific function and environmental constraints. Hence, cells can be defined as being the smallest structural and functional units capable of auto-reproduction. The rapid technological advances during the 20th century have made possible to describe the inner architecture of cells at different levels of organization. Our approach is an attempt to study cellular populations at the «sociological» level, i.e. at the level of their spatial organization and interrelationship in a given tissue. The topographies of cellular population have to be considered linked to morphogenesis, structural stabilities and functional state of a given tissue. This approach makes use of the relations that links form to disorder and is based on space partition constructed from a set of points defining the position of cells. This spatial partition consisting of a set of individual forms (Voronoi paving) associated to its dual (Delaunay triangulation) permits the calculation of Ulam trees which are descriptors of the local surrounding of a given cell and the search for characteristic features of the normal and pathological state.

The invariance by dilation, exact or statistic, is a characteristic of all fractals assumed figures. For an object statistically self-similar like an aggregate, we can demonstrate this invariance by tracing a circle of radius R around a point X of the aggregate. We note that the mass of the object $M(R)$ within the circle is proportional to $R^{D(X)}$, $D(X)$ being the local scale exponent at the point X. In the case of the Ulam tree (characteristic of the local topography around a given cell), the figure growths with the iteration number I. We observe that the number of segments (branches and leaves of the tree) is proportional to I^D, where D is the fractal dimension ($D \leq 2$). The Ulam tree can thus be used to give a consistent description of the area covered by the figure.

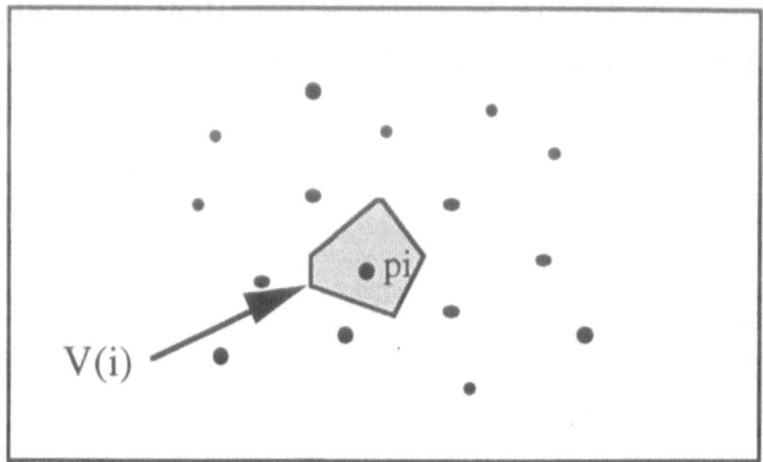

Fig. 1   Bidimensional distribution of points and the Voronoi polygon V(i) associated to p_i.

This method requires three main steps in its development, (i) a step of construction of the Voronoi diagram and Ulam trees which associates a polygonal form and a Ulam tree to each point of the population, (ii) a step of elimination of points whose associated polygonal form or Ulam tree has been altered due to border effects, and (iii) a final step of parametrisation and quantitation of topographical informations.

2   Construction of the Voronoi diagram

The Voronoi diagram is the space partition containing the most information (Toussaint, 1980). Let us briefly recall the basic definitions of this space partition. Let S, be a set of N points in the plane, i.e. the nucleus barycenters of cells. For each point pi in S what is the locus of points (x, y) in the plane that are closer to pi than to any other point of S?

The solution to the above problem is to partition the plane into regions (each region being the locus of points (x, y) closer to a point of S than to any other point of S). Given two points p_i and p_j, the locus of points closer to pi than to p_j is the half-plane containing p_i that is defined by the perpendicular bisector of p_ip_j. Let us denote this half-plane by $H(p_i, p_j)$. The locus of points closer to p_i than to any other point, which we denote by $V(i)$ (Fig. 1), is the intersection of $N - 1$ half-planes, and is a convex polygonal region having no more than $N - 1$ sides, that is,

$$V(i) = \cap_{i \neq j} H(p_i, p_j). \tag{1}$$

Where ∩ denotes the intersection and $V(i)$ is the Voronoi polygon associated with p_i (Preparata & Shamos, 1985). The construction of the Voronoi partition associated to S, denoted $Vor(S)$, follows an incremental method by local modification of the diagram after each insertion of a point of S. (Bowyer, 1981).

3   Construction of the Ulam trees

After the study of the tissular structure at the level of its global topography, we now present a method for the analysis of the local environment of each cell. This method is based on the analogy between a cell in a tissue, and a tree in a forest. Ulam trees are objects growing in time and space under given induction rules (Ulam, 1962). One of the aims of Ulam's work is to throw light on the question of how much «information» is necessary to describe the seemingly enormously elaborate structures of living objects.

Let us briefly recall the basic definitions used to construct those trees.

The basic idea is a fixed division of the plane (or space) into regular elementary figures. For example, the plane may be divided into squares or into equilateral triangles, or into regular hexagons, the space into cubes. An initial configuration will be a finite number of elements of such a subdivision and the induction rule will define successive accretions to the starting configuration.

Our purpose is to quantify the environment of every cell. The initial configuration we use is the Voronoi graph of the population, and a particular point of the population on which the tree is to be constructed.

A large variety of Ulam trees can be generated depending on the choice of the induction rule used to grow the tree. We chose an Ulam rule which respects the following conditions: (i) the rule does not favor any direction in the graph, (ii) the tree covers as much space as possible to increase its statistical significance, (iii) a strong physical significance can be given to the resulting process.

Given a number of Voronoi regions in the nth generation, the regions of the $(n + 1)$th generation will be all those which are adjacent to the existing ones but with the following proviso: the regions which are adjacent to more than one region of the preceding generations will not be taken into account. (Fig. 2)

Remark: The neighborhood notion is the Delaunay neighborhood notion (dual to the Voronoi diagram). This Ulam rule may be compared to a wave which originates on one point in the graph, propagates from node to node, and dies everywhere the wave intersects itself. The propagation rate is one graph distance per unit time.

4   Elimination of marginal points

Due to the properties of the Voronoi partition, some regions of the paving should be modified if other points outside the analysis window were acquired. The points

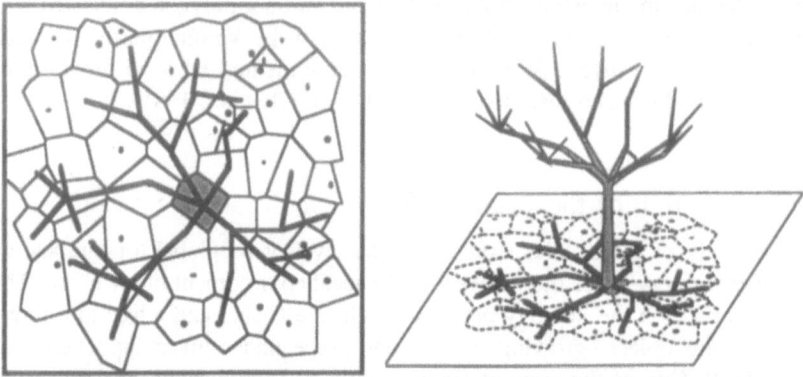

Fig. 2   Part of a Voronoi diagram and the Ulam tree (bold lines) of depth 3 associated to the gray
polygon. A true like tree can be generated from the Ulam tree. A given population of cells
thus can be transformed to a forest.

of S which belong to such regions are considered to be marginal. The subset of
all the marginal points is called the Marginal Subset (MS). It can be demonstrated
that these points are associated to regions having one or more summits which
are outside or closer to the analysis window than to the considered point. All the
marginal points of S are not taken into account in the subsequent calculations.
For example, let us consider a bidimensional distribution of points and the associated Voronoi diagram. Border-effect-free polygons (in white) are conserved for
further calculations and marginal polygons (in gray) are eliminated according to
the previous rule (Fig. 3).

Selection of the subset of points which are mathematically significant for the
Ulam tree construction:

This step consists in the selection of points for which the associated Ulam
trees are strictly included in the set of points which are not marginal. This step
is performed with the help of mathematical morphology. Let us consider an Ulam
tree depth of seven (i.e., a length of seven in terms of graph, separates the furthest
leaf of the tree from its root). To be mathematically significant, this tree must not
intersect the subset of marginal points (MS). If a marginal point is included in a
tree, this tree can be biased due to border effects of points that would be outside
the analysis window. Therefore, a morphological dilatation of seven (equal to the
Ulam tree depth) of the MS subset is performed before any Ulam tree is calculated.
The points of S which are eliminated during the dilatation of MS make up the
Intermediary Subset (IS). This dilatation ensures mathematical significance of all
the trees associated to the final subset of points (FSP), and the following relation
can be derived:

$$FSP = S - (MS + IS) \qquad (2)$$

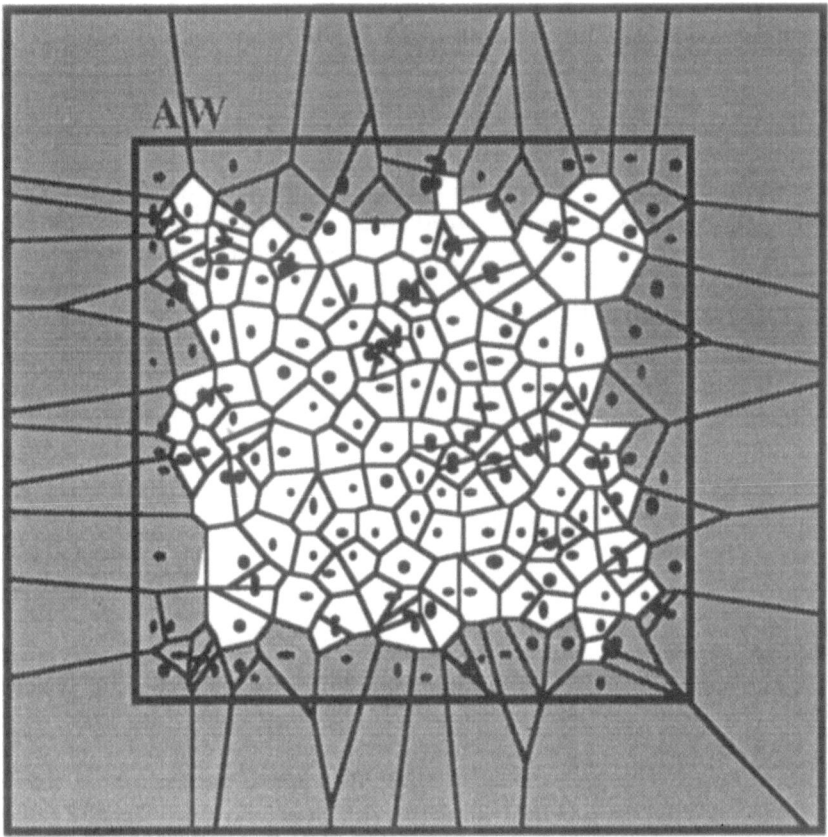

Fig. 3   Bidimensional distribution of points and the associated Voronoi diagram. Polygons in white
are conserved for further calculations and marginal polygons, in dark gray, are eliminated
according to the marginal rule and the analysis window AW.

5   Topographical parametrisation and quantitation

5.1   Quantition of the global topography

This attempt to quantitate cellular topography is based on the strong relationship
which obviously links form to disorder. Therefore, we defined parameters that
are descriptors of both form and disorder. These parameters are calculated on the
polygonal form which has been associated to each cell of the population during
the construction of the Voronoi partition.

Determination of the average type of the spatial occupation: for a convex set,
X, where $A(X)$ is the area and $L(X)$ is the perimeter, it is possible to demonstrate
the following isoperimetric inequality:

$$L(X)^2 - 4\pi A(X) \geq 0 \tag{3}$$

Since Voronoi polygons are convex, the average type of S population spatial occupation is well characterized by the average roundness factor $(RFav)$

$$RFav = \frac{1}{N}\sum_{i=1}^{N}\frac{4\pi A(X_i)}{L(X_i)^2} \qquad (0 < RFav \leq 1) \qquad (4)$$

Example: the roundness factor $(RF=4\pi A(X)/L(X)^2)$ of a n sided Reuleaux polygon (Regular polygon with all angles and sides equals) is $RF(n)=\pi/(n\tan(\pi/n))$. Thus, this formula makes it possible to predict the RF value of simulated polygons. The RF of a circle is 1, the RF of a line is 0.

The intrinsic disorder of the population has been expressed as two prime factors, the disorder concerning the area heterogeneity of the population and the disorder of the geometrical properties.

Determination of area heterogeneity and geometrical homogeneity of the spatial occupation inside S can be quantified by the following two parameters, area disorder AD, and roundness factor homogeneity RFH

$$AD = 1 - \left(1 + \frac{\sigma_a}{A_{av}}\right)^{-1} \qquad\qquad RFH = \left(1 + \frac{\sigma_{RF}}{RF_{av}}\right)^{-1} \qquad (5)$$

where σ_A denotes the area standard deviation, σ_{RF} the roundness factor standard deviation, A_{av} and RF_{av} the mean area and the mean roundness factor. Using two types of invariant, geometrical one and area one, it can be demonstrated that RF_{av}, AD and RFH are uncorrelated. Thus RF_{av}, RFH and AD define a three axis graph which represents topographical informations. We tested this model with different theoretical populations.

5.2 Quantitation of the Ulam trees
The first step in the quantitation of the cell neighborhood from the Ulam tree is to transform the tree into a mathematical object. It is possible to transpose the tree into a matrix M. This matrix is composed of two orthogonal properties which are respectively: the integration and the topological properties of the tree.

The integration property is the number of points added to the tree at a given expansion level. The topological property of the tree is the number of simple, double, triple, etc., junctions in the tree.

Thus the $M(i, j)$ cell of the matrix is set to the number of junctions of jth order at the ith level in the tree.

For example, in a population ordered at the nodes of a triangular mesh, the root of any tree is the level 0 of the tree and is a 6 order junction node. Thus the $M(0, 6)$ cell of the matrix is set to 1. The first level of the tree is made of 6 nodes which are all 2 order junctions. Thus the $M(1, 2)$ cell of the matrix is set to 6. The tree depth is 7 and the maximal junction order is 6. Thus this tree matrix is a (7+1) by 6 matrix.

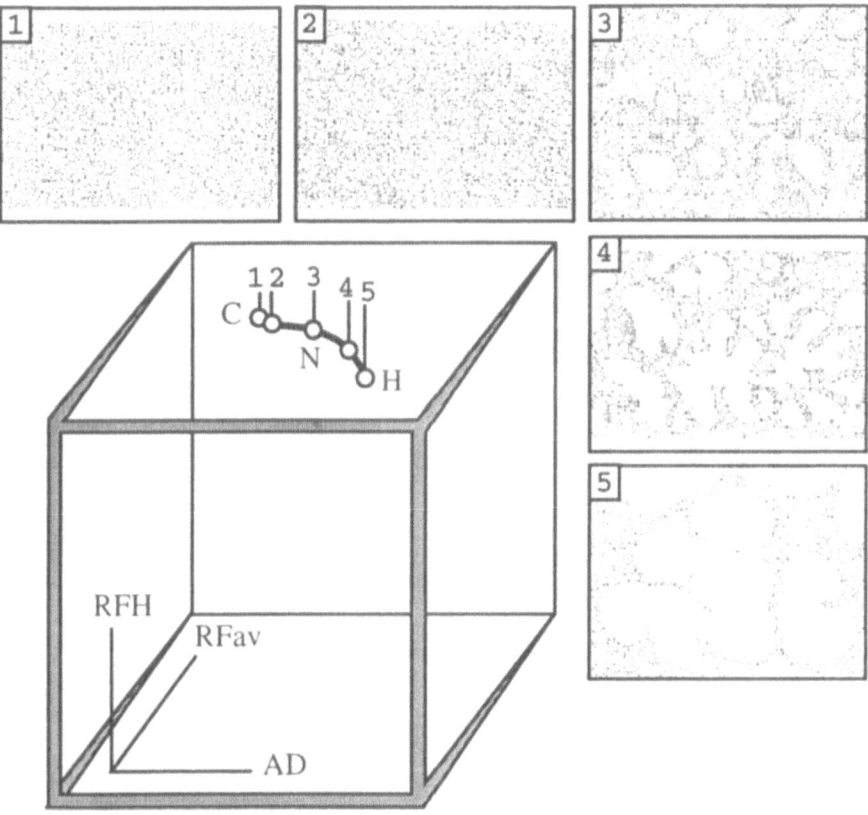

Fig. 4 Each prostatic tissue section can be located in the topographical space on a particular way which shows the topographic evolution of the prostatic tissue from the normal homeostasy (N) to the cancer (C) or to the hyperplasia (H). RF_{av} = average Roundness Factor, RFH = Roudness factor homogeneity and AD = area disorder.

From each Ulam tree, we determine 2 parameters which are characteristic of the cell environment.

The first factor, DORT is expressed as the integrated absolute difference between the matrix of an ordered reference tree constructed on a perfectly triangular ordered population, and the matrix of the given tree. The second one is expressed as the branch length heterogeneity, ELH of the tree.

6 Results

6.1 Global topography quantitation

The topography of different prostatic tissue sections of various pathological grades

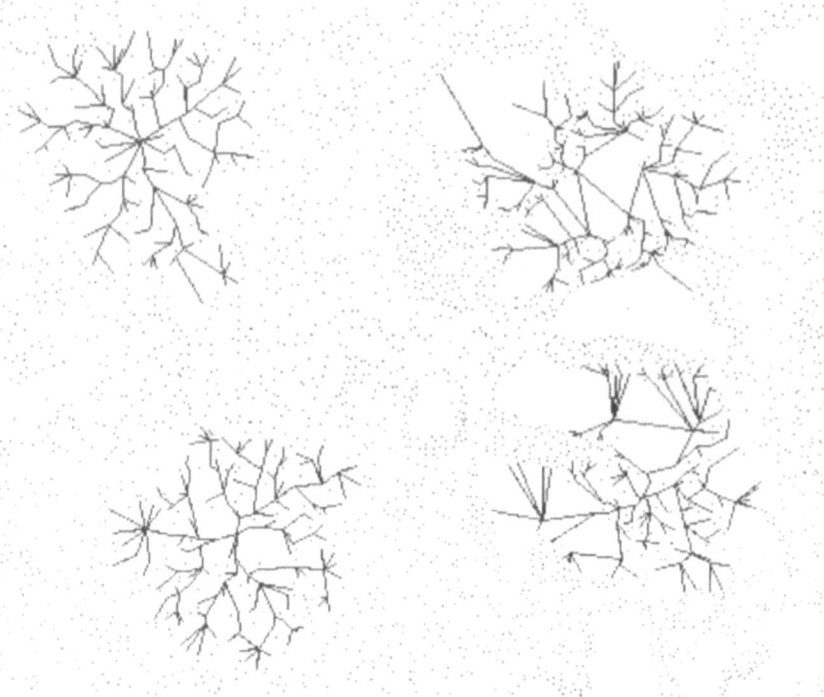

Fig. 5   Prostatic carcinoma on the left side and hyperplasia on the right side. The form of the Ulam
trees grown on the cancer and hyperplasic parts of the tissue are very different and thus can
help diagnosis.

have been mesured from their Voronoi diagram and located in the topographical
space as shown in figure 4.

6.2 Local topography quantitation
Using the Ulam trees, it is possible to caracterize the environment of each cell in
a prostatic tissue for example (Fig. 5).

7   Conclusion

This approach should make it possible to describe and quantify abnormalities of
the cellular topography characteristic of pathological states, and to objectively
grade tumours by measuring the amount of perturbations compared with normal
population, and in a more general way, to study cellular interactions and sociology.

Finally, this model appears to be useful for analyzing the effects of the cell
surrounding on a given cellular function and vice-versa.

References

[1] Bowyer A: Computing Dirichlet tesselation. Comput J, 24: 162–166, 1981.

[2] Preparata FP, Shamos MI: Computational geometry. Springer-Verlag, 1985.

[3] Toussaint GT: Pattern recognition and geometrical complexity. In: Proceedings 5th International Conference on Pattern Recognition, IEEE Catalog NO. 80CH1499-3, Miami Beach 1980, pp. 1324–1347.

[4] Ulam S: Patterns of growth of figures: mathematical aspects. Proceedings of symposia in applied mathematics, XIV: Mathematical problems in the biological sciences, American Mathematical Society. 64–75, 1962.

A Fractal Analysis of Morphological Differentiation of Spinal Cord Neurons in Cell Culture

Tom G. Smith, Jr.
Laboratory of Neurophysiology, NINDS and
Laboratory of Developmental Neurobiology, NICHD, NIH
Bethesda, MD., U.S.A.

Abstract. Cell cultures of murine spinal cord neurons, selected at discrete intervals after plating over a 5–7 day period, were fixed, stained and classified on the basis of the number of large, primary dendrites emanating from the cell body (2; 3/4; 5 or greater) and dendritic branching patterns (5A; 5B). Images of individual cells were captured from a light microscope via a video camera to an image processor. Gray scale images were converted to binary ones with the Marr-Hildreth convolution algorithm. The fractal dimension (D) of individual images was determined by Flook's dilation logic method. The two main features that contribute to D are the profuseness of branching and the ruggedness of the border. Plots of mean D vs. time for all of the four cell types (2; 3/4; 5A; 5B) could be well fitted to a model of the form: $D(t) = A + B \exp(t/t)$, where A and B are constants, t is time (hours) in culture and t is a time-constant (hours). Each cell type had a distinct final, plateau D (ascending rank order: 2 = 1.28; 3/4 = 1.32; 5B = 1.37; 5A = 1.41) and a different time course as measured by t (ascending rank order: 5B = 12.6; 5A = 14.5; 3/4 = 16.4; 2 = 20.4). The time-courses of the changes in D for all four groups were significantly different ($p < 0.05$) from one another (ANOVA). We conclude that the fractal dimension is a useful, quantitative and unbiased measure of the complexity of neuronal borders and branching patterns and that its time course of development can be described by a simple equation with a characteristic time constant and final plateau value. We suggest that D is also a useful measure of morphological cellular differentiation.

1 Introduction

As the Proceedings of First International Symposium on Fractals in Biology and Medicine [14] adequately and convincingly demonstrate, Mandelbrot's fractal dimension of fractal geometry [8] is a useful measure or statistic of images and other types of data. The definition and application of fractal concepts were demonstrated abundantly in the various papers to be found in these proceedings. In this paper we show how the fractal dimension can be employed as a useful, quantitative and unbiased measure of the complexity of neuronal borders and branching patterns and that its time course of development can be described by a simple equation with a characteristic time constant and final plateau value. We also suggest that it is a useful measure of morphological cellular differentiation.

2 Methods

The raw materials for our research are grey scale images of individual neurons, mainly grown in dissociated cell culture. Figure 1 is a photograph of a mouse

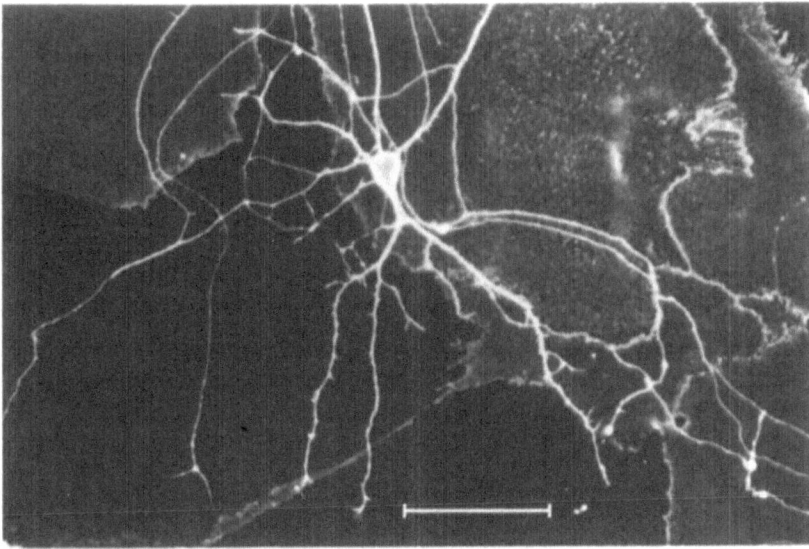

Fig. 1 A spinal cord neuron in dissociated cell culture for four days and stained immunohistochem-
ically using Fragment C of tetanus toxin and a monoclonal antibody against Fragment C.
Neuronal surface membranes are stained intensely with this procedure. The stained neuron is
a pyramidal Type 3/4 neuron. Bar = 100 μm. Reprinted with permission from Neale, Bowers,
and Smith, 1993.

spinal cord neuron stained after 4 days in culture. One particularly important technical point is that it is necessary to have an original image with as much contrast
as possible between the object of interest (cell) and the background. In addition,
since the initial goal is to obtain the border of the object, no intracellular structures
should be stained or visible, elsewise they produce unwanted borders. Finally, the
analog TV signal used to produce the digitized image should match the dynamic
range of the A/D converter of the image processor's frame grabber (e.g., 8 or
16 bits). These objectives are achieved by proper histological preparation (high
contrast) and/or adequate illumination. In addition, an analog video amplifier, with
gain (contrast) and DC (background) controls placed between the video camera
and the frame-grabber can help considerably in obtaining the desired high-contrast
digitized image [21]. To measure the fractal dimension (D), we need a binary image. To that end we employ the Marr-Hildreth algorithm, which is a convolution
operation employing a large (7×7 to 13×13 pixels), two-dimensional, circularly
symmetrical Laplacian of a Gaussian kernel («Mexican Hat» filter) [21]. The principal merit of this technique is that it almost always produces unbroken borders
which are ideal for filling to obtain a cell silhouette. Most cells illustrated are
silhouettes since they are easier to visualize than borders, particularly where dendritic branches are numerous and crossing. But the measurements for the fractal
dimension are made on borders-only images. The main drawback of the method

is that, with large kernels, there is some blurring of detail; however, this is not serious if, as here, the images and their measures are used for comparative purposes only. In our hands, the size of the kernel needed is inversely related to the degree of contrast in the original image.

3   Measuring the Fractal Dimension

We have employed three different methods of measuring the fractal dimension of the borders of individual cells [22]. Two of these, the yardstick and box-counting methods, have been discussed fully by the Ulm group in these proceedings [2]. The third method, called the dilation algorithm, was developed by Flook [4]. It is a convolution operation, where the kernels are single-valued (Boolean 1) circular discs of differing diameters. These kernels are applied, one at a time, to a border image and each resultant area is divided by the relevant disc diameter to give an equivalent perimeter. Then, the log of that result is plotted against the log of the diameter of the disc. A straight line, with slope $= -S$, is plotted and D calculated from: $D = 1 - S$. For images lying in a plane, $1 < D < 2$ [8].

All of these three measurements are related to lengths, and the filtering action of the larger measuring elements remove the higher spatial frequencies and produce smaller equivalent perimeter lengths. We calibrated our D-measuring techniques against borders of known fractal dimension (e.g., Koch islands, snowflakes, etc.) and found that our measurements consistently underestimated the true value of the known fractal dimension by about 5%. This may be due to the fact that our digitized images (512×512 pixels) do not contain the degree of resolution inherent in a true fractal. That was of little consequence to us, since we used the numbers for comparative purposes only.

As has been discussed elsewhere [2,22], the yardstick and box-counting methods can have technical problems in securing unambiguous data. For example, with short yardsticks and images with relatively smooth borders (straight lines), the log-log curves tend to deviate from a straight line and approach a Euclidian slope of zero. On the other hand, with the box-counting method and large boxes, the box count of a poorly centered image is underestimated and the log-log slope steepens. Both these cases make for difficulty in drawing an objective, unambiguous straight line. The dilation method, being a convolution operation rather than a counting one, is not as sensitive to these problems over the range of discs of 2-128 pixels in diameter — we consistently get unambiguous log-log straight lines over the entire range (6 octaves or 1.26 decades, the range required to obtain useful results). Thus, all of the data illustrated were obtained by the dilation method.

4   Measuring Neuronal D's

We obtained our original biological measurements from spinal cord neurons which had been grown in cell culture for more than 6 weeks [22]. Some selected results

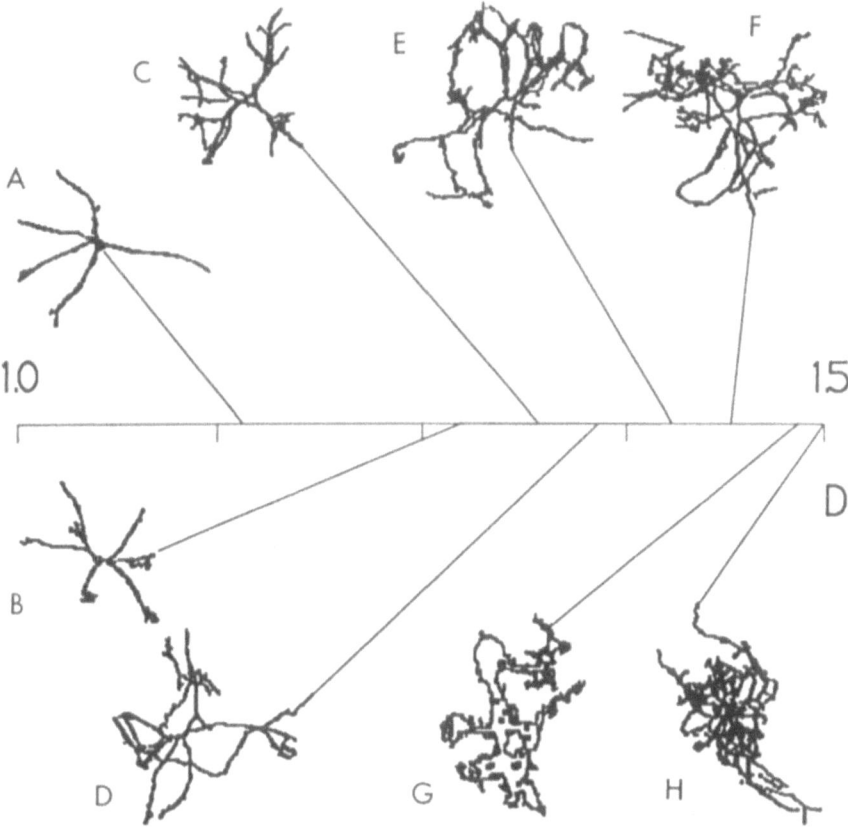

Fig. 2   A set of neuronal images ordered by increasing fractal dimension. The value of D for each image is indicated by its position on the D axis. Visually perceived morphologic complexity correlates well with estimated fractal dimension. Reprinted with permission from Smith et al., 1989.

are shown in Fig. 2 as a one-dimensional plot in D, where a line is drawn from each cell to its corresponding measured D. This illustrates that the fractal dimension of a cell's border is a measure of the complexity of that border. As cellular morphology becomes more complex, D increases. We find that two characteristics are important in determining D: 1) the ruggedness of the border and 2) the profuseness or degree of dendritic branching. In addition, a particular D does not uniquely define a border. This is illustrated in Fig. 3, where two cells have approximately the same D but very different morphological characteristics. Cell A has few branches but a very rugged border, while B has smooth borders but lots of branches. Having developed a «tool» for quantifying the complexity of an image's border, we next attempted to apply this tool to an experimental situation to determine its utility.

Fig. 3   Two neuronal images of differing morphology but with similar high values for D. The high
D for cell A is related to its spiny surface, whereas the extent of branching accounts for the
high D of cell B. Reprinted with permission from Smith et al., 1989.

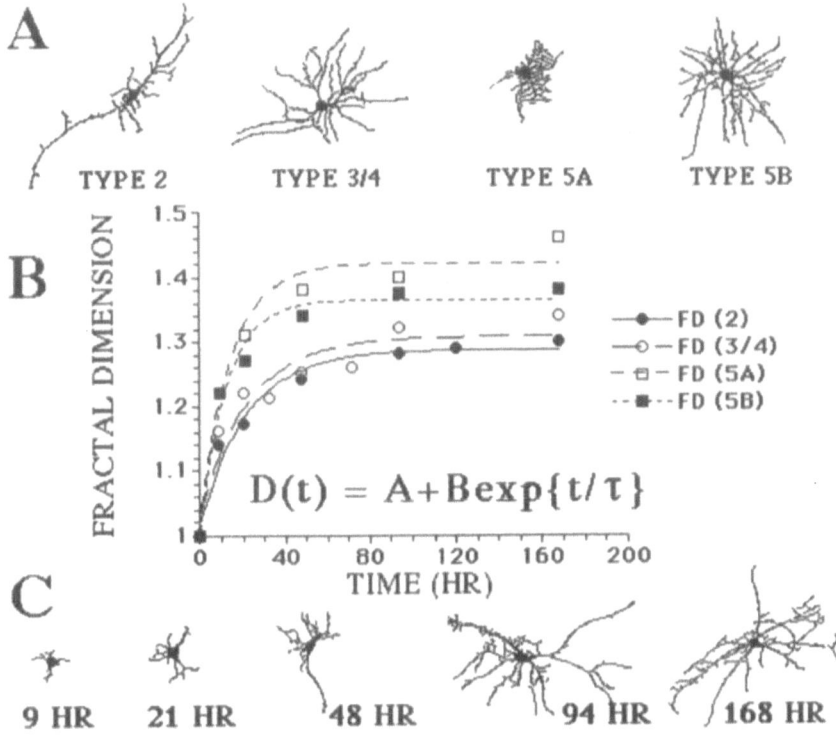

Fig. 4 Change in fractal dimension and morphology of spinal cord neurons with time in culture. Spinal cord neurons were grouped (see text) into four types, all illustrated in (A). Fractal dimensions were analyzed at various times during the first week in culture and plotted for each morphologic type in (B). The development of morphology over this same interval is shown for Type 3/4 neurons in (C). Reprinted with permission from Neale, Bowers, and Smith, 1993.

5   Results and Discussion

To provide a frame of reference, we grouped 258 spinal cord neurons on the basis of the number of primary dendrites growing directly from the cell body (Fig. 4A). The neurons can be recognized in this way even when the cells are quite immature. We grouped our population as Type 2 (two primary dendrites), Type 3/4 (three or four primary dendrites) and Type 5 (five or more primary dendrites). This latter group breaks down to Type 5A or 5B, depending upon the pattern of dendritic branching [12,13].

   We start a batch of sister cultures at time zero and halt growth of individual cultures at given times by fixing and staining the cells. These cultures are then examined under the microscope and discrete cells of each cell type are chosen for analysis. After digitization, the fractal dimension of each individual neuronal image is measured by the dilation method [4,22].

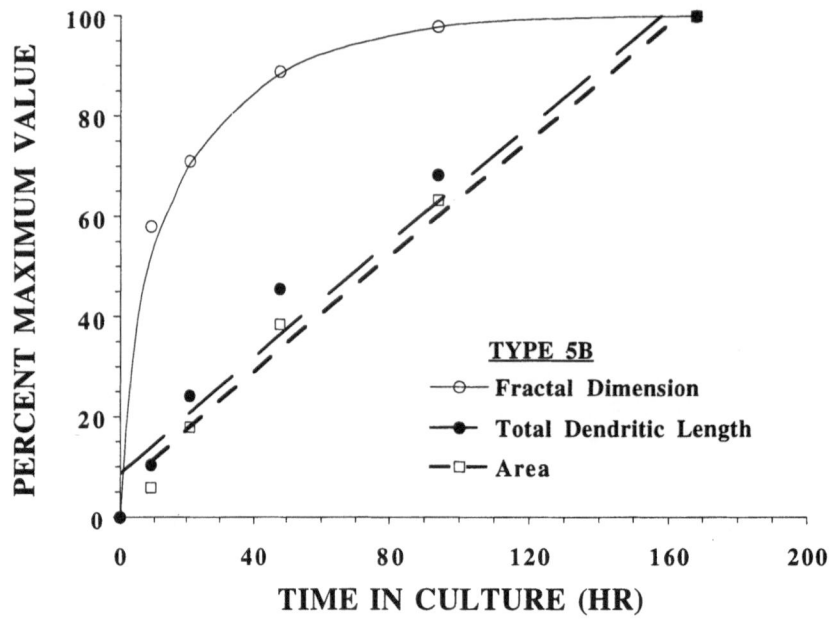

Fig. 5 Change in neuronal growth compared with change in D. Measures of total dendrite length and area covered by the dendrite arbor (territory) are shown along with changes in D for Type 5B neurons. Whereas fractal dimension increases rapidly during the first 2–3 days in culture and then plateaus, other measures of dendrite growth continue to show linear increases for the first week after plating. Reprinted with permission from Neale, Bowers, and Smith, 1993.

What is clear, as shown in Fig. 4C, for Type 3/4 cells, is that the cell borders become progressively more complex, over time, during growth and differentiation in culture. Figure 4B shows the time courses for all 4 cell types, starting with a D of 1 for spherical cells plated at time zero. As dendrites emerge and branch, D increases progressively and plateaus at a maximum value for each cell type. The curves that best fit the time course of the mean value for each of the cell types is the model shown in Fig. 4B, center. The correlation coefficients for the fit of all the curves shown in Fig. 4B to the data were greater than .95. This equation-model, which describes how D varies with time, t, has two constants, A and B, and is characterized by a time constant, t. We should emphasize that we find it quite remarkable that so complex a process, with its many-fold changes in complexity, can be described so well by such a simple expression. It suggests to us that complexity is an important aspect of cellular morphological growth and differentiation. Table I shows the actual values for the maximum D and the t's for the 4 cell types. For our population of cells, the maximum D rises from a low value of 1.28 for Type 2 cells to a high value of 1.41 for Type 5A, while the time constants run from a low value of 12.6 hours for Type 5B to a maximum of 20.4 hours for Type 2. All these curves are significantly different from one another ($p < 0.05$; ANOVA).

We would note in passing that similar developmental data have been obtained from glial cells grown in culture, and the same D(t) model provides the best fit for those data also [20].

Figure 5 illustrates how the change in D with time (t), for Type 5B cells, relates to other more conventional or Euclidian forms of cellular measurements (growth area, total dendritic length), all normalized to the maximum value for each variable. It is immediately apparent that, while the Euclidian measures all increase nearly linearly, D changes non-linearly with time. We would note that if one takes the conventional definition of «fractal growth» as growth with constant D, then the change in area, etc. from time zero to about 80 hours is not fractal growth, but that the period after about 80 hours, when D has reached its stable value, does represent fractal growth. We would note the changes in D with time are clearly not fractal growth.

Some authors have proposed that neuronal growth and differentiation represent a special case of diffusion limited aggregation-like growth (DLA) [3]. The evidence cited for such a conclusion is that some neurons have a fractal dimension of 1.70, the hallmark of DLA. The results from our group [13,20,21] and from others [7,9,10,11] suggest that such an assignment of DLA properties may be premature since most neurons, either in vivo or in vitro, do not in fact have a $D = 1.70$.

We have some observations that we would like to mention that may relate to the meaning of «fractal growth». In order to produce images of known fractal dimension to calibrate our methods (like the Koch islands), we used commercial software that employs the so-called L-systems method, named after Landenmayer, a Dutch botanist, who devised it [15]. It employs a system much like that discussed in Mandelbrot's book, «The Fractal Geometry of Nature» [8], except Mandelbrot's «initiator» and «generator» are called «axiom» and «production», respectively, in L-systems. We undertook an exercise [8; p.51], beginning with a Euclidian square (initiator, axiom) and applied the generator or production through successive interations to generate progressively more complex images. Then, we took the result of each iteration of a number of deterministic fractals and calculated the fractal dimension for that image. The results are shown in Fig. 6 for six different deterministic fractals, listed in the figure, where the fractal dimension is plotted as a function of iteration. Then we fitted these points with the same equation we had used for neurons demonstrated in this paper and, as mentioned, for glia.

It is clear that this fractal generation begins with what are Euclidian objects (initiator, axiom) like lines, squares or triangles and have a D of one. As they progress, they become more complex and D increases to reach some plateau value like our neurons and glia. But is this fractal growth? And, how do these lattermost results relate to the similar changes found with cultured neurons and glia?

The notion that biological tree-like, fractal structures represent an optimal design for a particular function is generally accepted. For example, such ideas have been proposed for the lung for the flow of air [1] and the vascular bed for

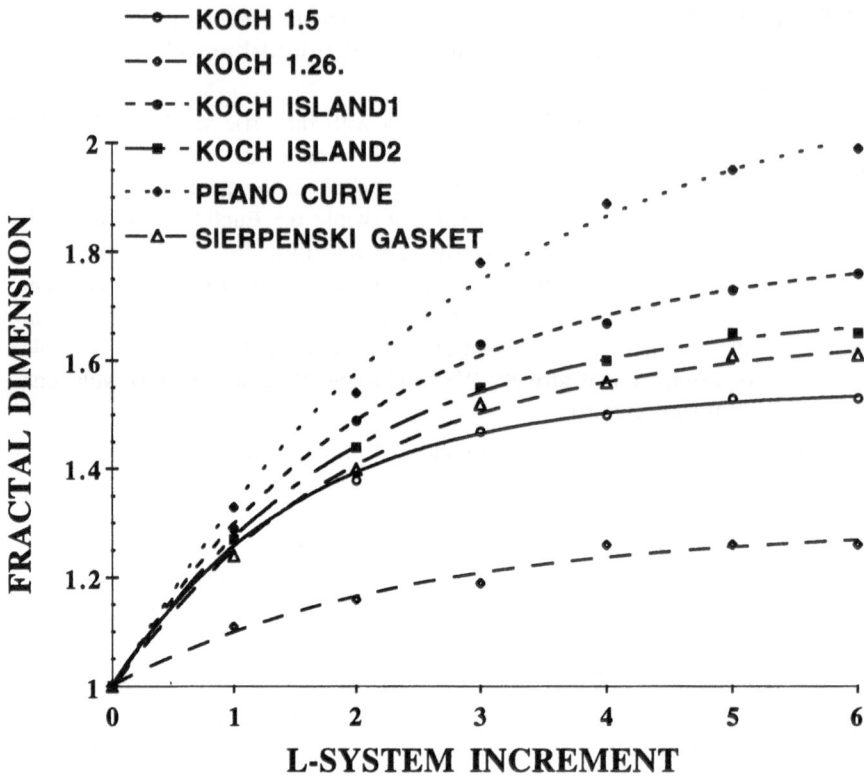

Fig. 6   A plot of the fractal dimension as a function of iteration number for a number of known fractal
objects listed in the upper, left part of the figure. The curves of the points fit an expression
of the form: $D(i) = A + B\exp(i/n)$ (cf. equation given in Fig. 4B).

the flow of blood [23]. It may well be that the fractal, dendritic trees of neurons
are also optimally designed, but in this case for the flow of their most important
commodity: information.

Indeed, there is a considerable corpus of theory and data which indicates
that some neurons are optimally structured for the integration of their graded
input signals (synaptic potentials) which results in a pulsatile output signal (action
potential) [5,6,16,17,18,19]. In cases where the organization is known, excitatory
synaptic inputs are located mainly on distal dendritic branches whereas inhibitory
synaptic inputs are located on the cell body and proximal dendrites. Since the
action potential generating membrane is located at the axon hillock or the axon
itself, the excitatory synaptic potentials (EPSPs) must depolarize that membrane to
a threshold value in order to generate an action potential. If however, prevention
of such action potentials is desired, then an increase in inhibitory conductances
can shunt the excitatory currents flowing from the distal dendrites toward the axon
hillock and block the generation of an action potential.

6   Conclusion

Finally, we conclude that the fractal dimension is a useful, quantitative and unbiased measure of the complexity of neuronal borders and branching patterns and that it is often more useful and insightful than conventional Euclidian measures. Moreover, its time course of development can be described by a simple equation with a characteristic time constant and final plateau value. We suggest that D is also a useful measure of morphological cellular differentiation, in the sense that Webster's Collegiate American Dictionary defines differentiation as «a change from the simple to the complex» [24].

References

[1] Bassingthwaighte JB: News Physiol. Sci. 3, 5–10 (1988).

[2] Bauman G, Barth A and Nonnenmacher TF: In Proceedings of the First International Symposium on Fractals in Biology and Medicine, (ed. Nonnenmacher TF, Losa GA and Weibel ER), Birkhäuser-Verlag, Basel, pp. (1993).

[3] Caserta F, Stanley HE, Eldred WD, Daccord G, Hausman RE and Nittmann J: Phys. Rev. Lett. 64, 95–98 (1990).

[4] Flook AG: Powder Technol. 21, 295–298 (1978).

[5] Jack JJB, Nobel D and Tsein RW: Electric Current Flow in Excitable Cells, Clarendon, Oxford, U.K., (1975).

[6] Jack JJB and Redman SJ: J. Physiol. (Lond.) 215, 321–352 (1971).

[7] Kniffki KD, Pawlak M and Vahla-Hinz C: In Proceedings of the First International Symposium on Fractals in Biology and Medicine, (ed. Nonnenmacher TF), Birkhäuser-Verlag, Basel, pp. (1993).

[8] Mandelbrot BB: The Fractal Geometry of Nature, W.H. Freeman, New York, (1982).

[9] Montague PR and Friedlander MJ: Proc. Natl. Acad. Sci. (U.S.A.) 86, 7223–7227 (1989).

[10] Montague PR and Friedlander MJ: J. Neurosci. 11, 1440–1457 (1991).

[11] Morigiwa K, Tauchi M and Fukuda Y: Neurosci. Res. Suppl. 10, S131–S140 (1989).

[12] Neale EA, Bowers LM and Smith TG, Jr.: Soc. Neurosci. Abst. 17, 36 (1991).

[13] Neale EA, Bowers LM and Smith TG, Jr.: J. Neurosci. Res. 34, 54–66 (1993).

[14] Nonnenmacher TF, ed.: Proceedings of the First International Symposium on Fractals in Biology and Medicine, (Basel: Birkhäuser-Verlag) (1993).

[15] Prusinkiewicz P and Landenmayer A: The Algorithmic Beauty of Plants, Springer Verlag, New York, (1990).

[16] Rall W: In Handbook of Physiology, Vol. 1, Pt. 1, The Nervous System, Cellular Biology of Neurons, (ed. Brookhart JM, Mountcastle VB, Kandel ER and Geiger SR), American Physiological Society, Bethesda, Maryland, pp. 39–97, (1977).

[17] Rall W: Biophys. J. 9, 1483–1508 (1985).

[18] Rall W, Burke RE, Smith TG, Nelson PG and Frank K: J. Neurophysiol. 30, 1169–1190 (1967).

[19] Redman S and Walmsley B: J. Physiol. (Lond) 343, 117–133 (1983).

[20] Smith TG, Jr., Behar TN, Lange GD, Marks WB and Sheriff WH, Jr.: Neuroscience 41, 159–169 (1991).

[21] Smith TG, Jr., Marks WB, Lange GD, Sheriff WH, Jr. and Neale EA: J. Neurosci. Methods 26, 75–82 (1988).

[22] Smith TG, Jr., Marks WB, Lange GD, Sheriff WH, Jr. and Neale EA: J. Neurosci. Methods 27, 173–180 (1989).

[23] West BJ and Goldberger AL: American Scientist 75, 354–365 (1987).

[24] Woolf HB, ed.: Webster's New Collegiate Dictionary, (Springfield, Massachusetts, U. S. A.: G & C Merriam Company) (1976).

Fractal Dimensions and Dendritic Branching of Neurons in the Somatosensory Thalamus

Klaus-D. Kniffki, Matthias Pawlak and
Christiane Vahle-Hinz[1]
Physiologisches Institut
Universität Würzburg
D-97070 Würzburg, Germany

1)   present address: Physiologisches Institut, Universitäts-Krankenhaus Eppendorf, Universität Hamburg, D-20246 Hamburg

1   Introduction

The investigation and modelling of irreversible growth phenomena has become a topic of considerable interest in the last decade, stimulated by the introduction of the concept of fractality by B.B. Mandelbrot. This concept provides a quantitative framework to study in particular biological growth phenomena of complex shapes, such as the branching structures of trees, of bronchial trees and of blood vessels. Recently, it was shown that the shapes of 2-dimensional retinal neurons are fractal objects, and hence may be characterized by their fractal dimension $D = 1.68 \pm 0.15$ (Caserta et al., 1990). The authors proposed an explanation of certain stages of neuronal development by a diffusion-limited particle-cluster aggregation (DLA) model, which predicts in 2-dimensional space $D = 1.70 \pm 0.1$.

2   Fractal dimension of thalamic neurons

The fractal dimension of neurons stained with a Golgi impregnation method located in the cat's thalamic ventrobasal complex (VB) and its ventral periphery (VB_{vp}) was examined using 2-dimensional projections of the fractal structures embedded in Euclidean space E^3 (Kniffki et al., 1991; Kniffki et al., 1993).

Fractal structures embedded in E^d usually do not have a finite d-dimensional volume and a D-dimensional Hausdorff measure has to be used in order to assign well-defined contents to such objects. $D \in [0, d]$ is the Hausdorff dimension of a fractal point set $\mathbf{X} \subset E^d$. A very good estimate can be achieved by computing the so-called box counting dimension D_B (Mandelbrot, 1983). It is defined as

$$D_B = \lim_{\epsilon \to 0} \frac{\ln N(\epsilon)}{\ln(1/\epsilon)}, \qquad (1)$$

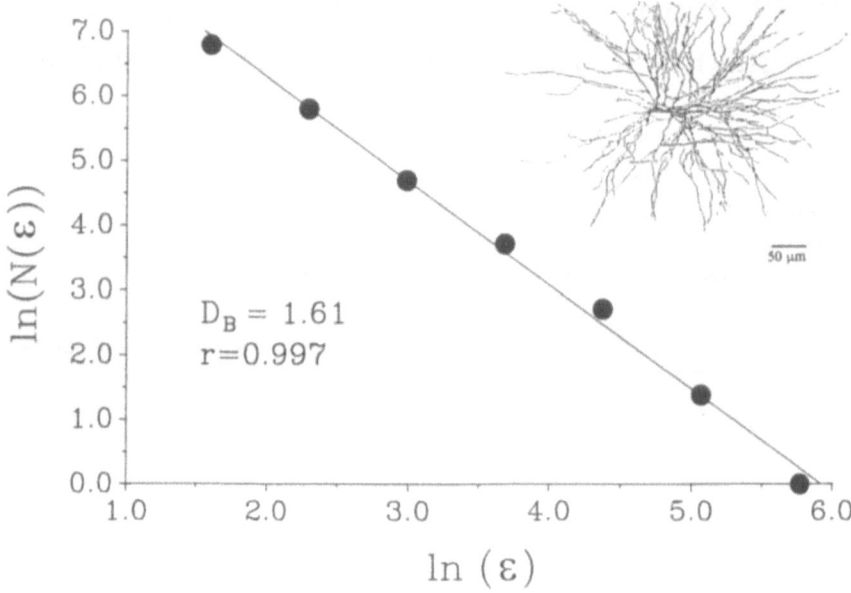

Fig. 1   Fractal analysis (box-counting method) of the dendritic branching of a neuron (inset) located in the cat's ventrobasal complex. A box covering the dendritic tree is successively dissected into smaller and smaller boxes, where ϵ is the edge length of the boxes. N is the number of boxes of a particular edge length, which are necessary to cover the whole dendritic field: N vs ϵ in ln-ln plot. The logarithmic straight line implies the indicated fractal dimension D_B (Eq. 1).

where ϵ is the lattice constant of d-cubic covers of **X** and $N(\epsilon)$ is the number of cubes contained in the minimal cover. Numerically, D_B is calculated by counting the number $N(\epsilon)$ of boxes of edge length ϵ, which are necessary to cover all points of the data set and calculating the slope of $\ln N(\epsilon)$ versus $\ln \epsilon$ for decreasing values of ϵ. Two methods of analysis were employed in the present study. Firstly, the drawings of the thalamic cells were placed under grids of different sizes ϵ. These grids of different scales were constructed by successively dissecting the largest grid which covered the whole cell. A typical example is shown in Fig. 1. As indicated by the straight line (least square fit), this particular cell exhibited $D_B = 1.61$ with a correlation coefficient $r = 0.997$. Secondly, the images of the thalamic neurons were digitized by a scanner with a resolution of 200 dpi in 2 grayscales and the obtained data were automatically processed using a box counting algorithm implemented on an AT personal computer (Block et al., 1990).

A main result of new study (Kniffki et al., 1993) is the fact that dendritic branching patterns of thalamic cells have fractal structures. The mean fractal dimension ($\langle D_B \rangle$) of neurons within *VB* in general is larger than that of neurons within VB_{vp}. In particular, neuron of the ventral posteromedial nucleus (*VPM*) of *VB* have $\langle D_B \rangle = 1.51 \pm 0.05$, those in the ventral posterolateral nucleus

(VPL) of VB have $\langle D_B \rangle = 1.44 \pm 0.05$. The corresponding values of neurons located within the ventral peripheries of these nuclei are: ventral layer of VPM_{vp}, $\langle D_B \rangle = 1.39 \pm 0.04$; VPL_{vp}, $\langle D_B \rangle = 1.39 \pm 0.08$.

It is important to emphasize that we have determined the fractal dimension of the camera lucida drawings of thalamic cells (see inset of Fig. 1) which are projections of the original 3-dimensional objects. Nevertheless, this fact does not affect the estimation of the fractal dimension, if the fractal dimension of the 3-dimensional object is smaller than 2. In general, for the dimension of a fractal structure S of dimension D projected upon an Euclidean subspace of dimension d the projection S^* satisfies:

$$\text{dimension } S^* = \min \ [d, D] \qquad (2)$$

(Mandelbrot, 1983; Takayasu, 1990). In our case this means $D^* = D$, i.e. the fractal dimensions which we have estimated from the 2-dimensional camera lucida drawings are the true fractal dimensions of the neurons embedded in E^3. As a consequence, simple DLA models are not suitable for describing the fractal pattern formation of dendritic outgrowth of thalamic neurons (Kniffki et al., 1993). Since the estimated fractal dimensions are significantly smaller than predicted for neurons described by a DLA model in 3-dimensional space, which would give $D = 2.5$, it is suggested that the neuron's morphology is not a result of purely stochastic processes but of strong interactions between the cell and its environment. This is reminiscent of, e.g., aggregation with interaction, which leads to growth structures that are similar to DLA products at first glance, yet possess fractal dimensions significantly smaller than 2.5 (Block et al., 1991).

3 Thalamic neuronal branching

When a biological tree structure is a fractal object and exhibits self-similarity over a certain range of length scales, in addition to its fractal dimension D another parameter is involved, termed the diameter exponent n. This exponent n is part of the scaling law for a bifurcation:

$$(d_0)^n = (d_1)^n + (d_2)^n, \qquad (3)$$

where d_0 is the diameter of the parent branch and d_1 and d_2 are the diameters of the two daughter branches, the thicker and the thinner, respectively. For the bronchial tree, e.g., Mandelbrot (1983) assumed a simple self-similar growth process, which does not require the branching ratio $d_0/d_i (i = 1, 2)$ to be encoded genetically. The branching pattern of the bronchial tree, which at first sight appears to be very complex, is fully determined by two parameters: the width/length ratio of the branches and the diameter exponent $n = 3$.

Assuming a similar simple rule for the growth of neurons, it was the aim to quantify the dendritic branches of neurons being located within VB and VB_{vp}

A   B

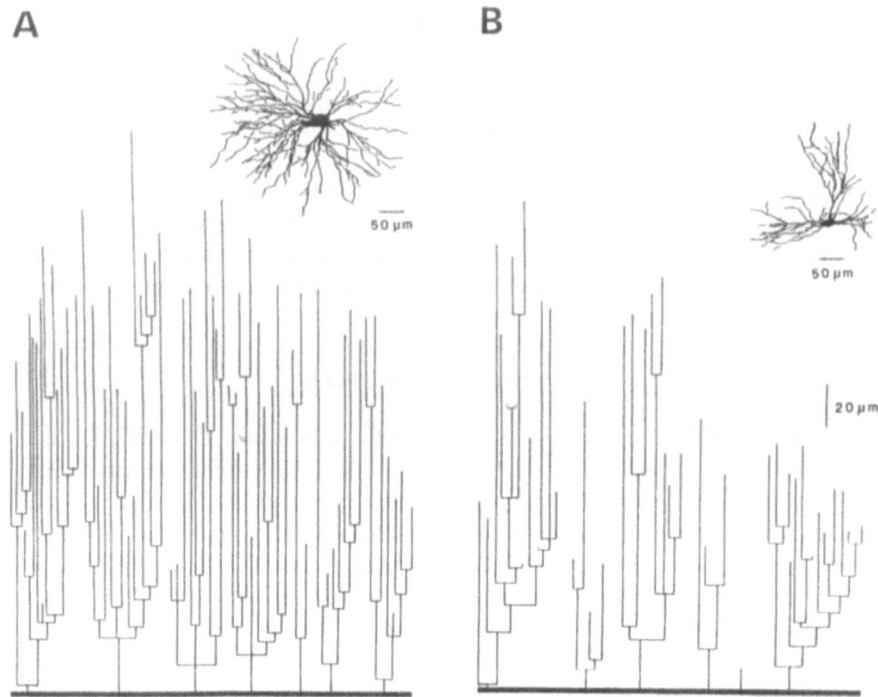

Fig. 2   Dendritic morphology of a thalamic neuron located in the ventrobasal complex (A), and of a
neuron of its ventral periphery (B), as well as their schematic reconstructions. The diagrams
represent the topological branching pattern. The lengths of the branches have been corrected
for the travel through the planes of section. The diameters of the individual dendritic branches
were measured directly from the histological sections and were not drawn to scale in this
figure.

(Kniffki et al., 1993). The morphology of neurons in these two thalamic regions
has been described previously (Vahle-Hinz et al., 1993). Here we wanted to test
in particular, whether the dendritic arborizations of the neurons in both regions of
the lateral thalamus may be describable by the scaling law [Eq. (3)] with a single
value of the diameter exponent n.

    The quantitative analysis was carried out on representative neurons from VB
and VB_{vp} (Kniffki et al., 1993). The diameters and the lengths of the primary
dendrites and the dendritic branches of the neurons were measured directly from
the histological sections using a 1000 × magnification and an eyepiece graticule.
Two-dimensional schematic drawings were made for each neuron representing
the topological pattern of dendritic branching and the lengths of the individual
branches. In Fig. 2 the dendritic morphology of a thalamic neuron located within
VB (A) and of a VB_{vp} (B) neuron as well as their schematic reconstructions are
shown.

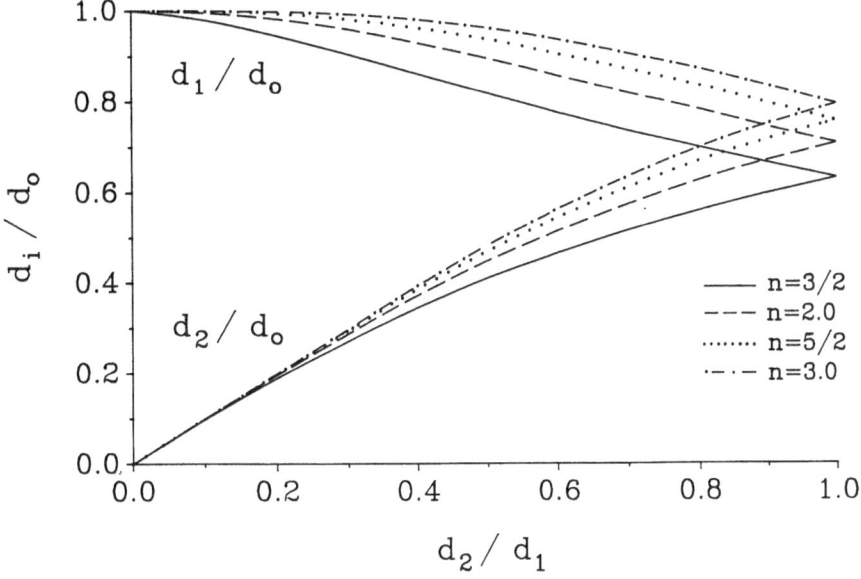

Fig. 3 Representation of Eqs. (4) and (5) with the indicated diameter exponents n.

To test the scaling law [Eq. (3)], some transformations were carried out. After division of Eq. (3) by d_0 and with the introduction of the asymmetry ratio $a = d_2/d_1 \leq 1$, Eq. (3) is equivalent to:

$$d_1/d_0 = (1 + a^n)^{-1/n} \tag{4}$$

and

$$d_2/d_0 = a(1 + a^n)^{-1/n}. \tag{5}$$

When both interbranching segments have the same diameter, i.e. $a = d_2/d_1 = 1$, the d_i/d_0 results in

$$d_1/d_0 = d_2/d_0 = 2^{-1/n}. \tag{6}$$

For $n \to \infty$ this leads to an increasing value of $2^{-1/n}$ up to the value of 1.

This method of representing the scaling law [Eq. (3)] has the advantage that d_1/d_0 for $n > 1$ shows only a weak dependency as a function of $a = d_2/d_1$, while d_2/d_0 shows a stronger relationship. The functional behaviour of d_i/d_0 for four different values of n is shown in Fig. 3. For the experimental data, a nonlinear least square fit routine was used for Eqs. (4) and (5) to determine the exponent n for d_1/d_0 and d_2/d_0 separately.

The diameters of the dendritic segments of neurons within VB and VB_{vp} were pooled according to the order of branching with respect to the soma and the mean d_i/d_0 were plotted against the asymmetry ratio d_2/d_1 after the d_2/d_1 values

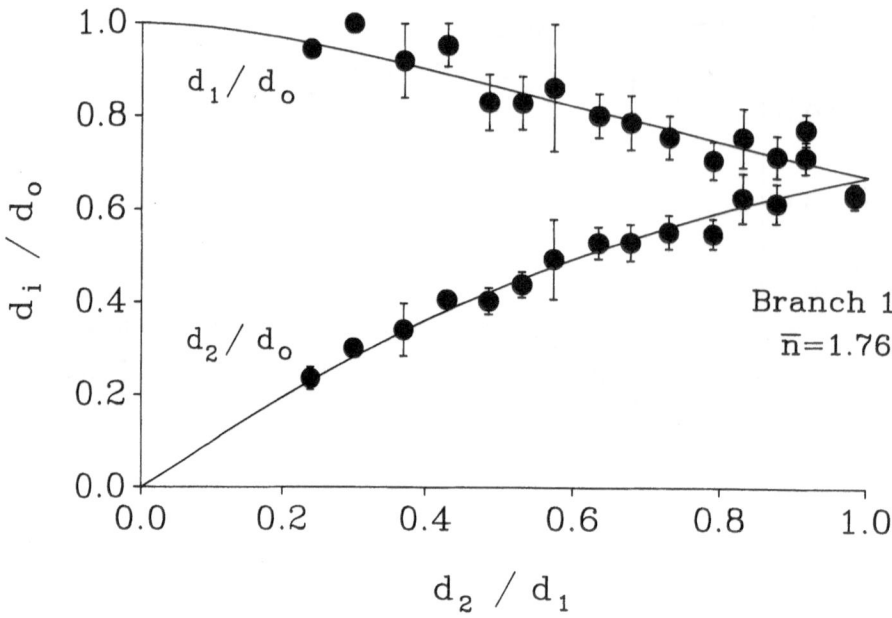

Fig. 4 Scaling behaviour for thalamic dendritic bifurcations for the 1st order of branching. The filled circles show the averaged measured values, whereas the continuous lines represent Eqs. (4) and (5), respectively, with the indicated values of the diameter exponent resulting from a nonlinear least square fit procedure. The error bars larger than the size of the symbols indicate the standard error of the mean.

had been grouped into classes with a width of 0.05. No statistically significant difference was found for the diameter exponent n for d_1/d_0 and d_2/d_0 of VB and VB_{vp} neurons (Kniffki, et al., 1993a). Therefore, the data of both groups of neurons were pooled and in Figs. 4 and 5 the filled circles represent the pooled data for branch order 1 and 3. For branch order 1, $\bar{n} = 1.76$ and for branch order 3 the corresponding value $\bar{n} = 2.93$ was found. Near the soma, i.e. for branch order 1 the lowest value of $\bar{n}$ was obtained. For increasing order of branching the diameter exponent increased up to its highest value of $\bar{n} = 3.92$ for the 7th (and last) order of branching. The corresponding values of $\bar{n}$ are plotted against the order of branching in Fig. 6.

In the thalamic neurons studied, the scaling relation [Eq. (3)] was fulfilled for the ramifications of the dendritic trees; the diameter exponent n, however, was not constant throughout the dendritic trees, but varied systematically with the order of branching.

4   Conclusion

During their outgrowth, neuronal dendrites exhibit a morphology typical for the kind of neuron within a particular nucleus in the central nervous system. To char-

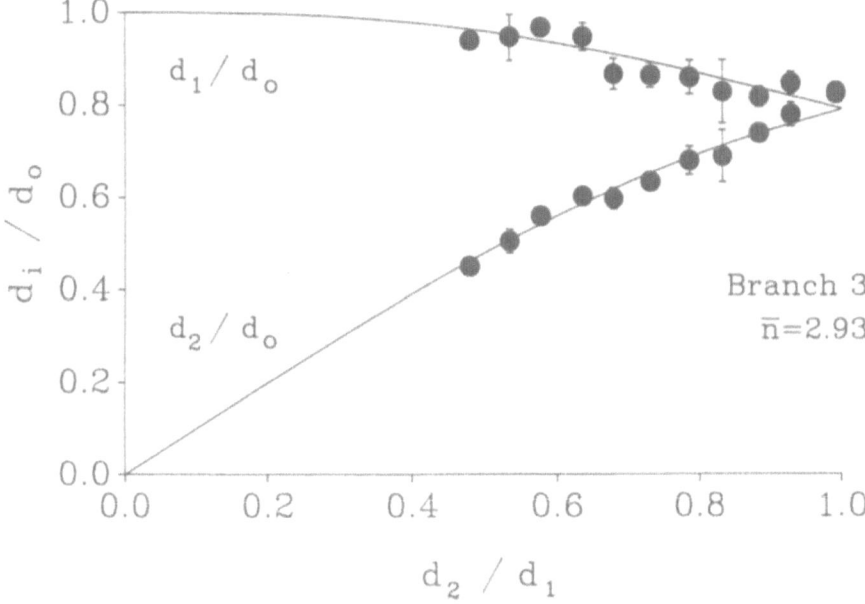

Fig. 5 Scaling behaviour of thalamic dendritic bifurcations for the 3rd order of branching (filled circles) with the indicated diameter exponent. The continuous lines represent Eqs. (4) and (5); the error bars larger than the size of the symbols indicate the standard error of the mean.

acterize the space-filling properties of the neurons' dendrites, the fractal dimension D might be a useful parameter (Smith et al., 1989). In addition, the fractal dimension might be a useful parameter to describe differences in the dendritic trees of neurons in different nuclei. Differences between VB and VB_{vp} neurons with respect to their fractal dimension were shown, D of VB_{vp} neurons being smaller or equal to D of VB neurons (Kniffki et al., 1993). Since D of both types of neurons is significantly smaller than predicted for DLA model neurons, which would give $D = 2.5$, the morphology of thalamic neurons is not a result of purely stochastic processes but of strong interactions between the outgrowing cell and its local environment, e.g., by specific chemical gradient fields (Hamilton, 1993). Models of aggregation with interaction, which leads to growth structures similar to DLA structures, exhibit fractal dimensions significantly smaller than 2.5 (Block et al., 1991). However, the functional significance of D for neurons is currently unknown. It is speculated to describe some aspects of the neuronal integrative properties, including aspects of the synaptic connectivity.

In order to study the dendritic integrative properties of synaptic signal conduction, i.e. the passive spread of depolarization towards the axon hillock, Rall (1959) introduced an ideal neuron with an equivalent-cylinder dendrite. At each dendritic bifurcation this ideal equivalent-cylinder dendrite requires a diameter exponent $n = 3/2$ in the scaling law [Eq. (3)]. Agreement with and deviations from

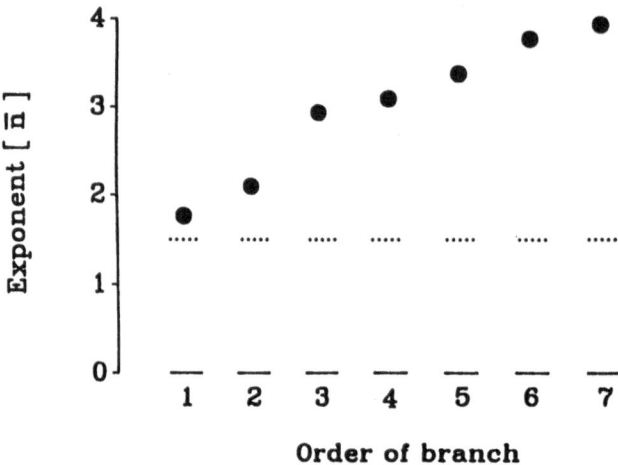

Fig. 6 Values of the diameter exponent as a function of the order of dendritic branching of thalamic neurons. The small dots represent $n = 3/2$, the value for the ideal equivalent-cylinder model dendrite.

the ideal equivalent-cylinder model have been noted in neurons of several species (for reviews, see Schierwagen and Grantyn, 1986; Kernell and Zwaagstra, 1989) resulting in values for n ranging from 1.0 to 3.0. The present results also show clear deviations from $n = 3/2$. In dendritic trees of thalamic neurons, up to the 7th order of branching n increased systematically from 1.76 for the 1st order of branching near the soma to 3.92 for the 7th order of branching (Kniffki et al., 1993a). That means, that the description of the detailed integrative properties of these neurons requires a segmental non-uniform equivalent cable model instead of Rall's equivalent-cylinder approximation (Schierwagen, 1989).

The outgrowth of the branching dendrites should be studied by modeling the dynamic behaviour of the growth cones, which are the active structures of the outgrowing dendritic tips. The basic actions of these growth cones comprise elongation, branching and guidance as a result of the cell's local environment (van Veen and van Pelt, 1992). The dependence of the diameter exponent n on the order of the dendritic branching is judged as an indication for the involvement of more than one intrinsic rule in the neuronal growth process, assuming that the dendritic branching ratio for bifurcations is not required to be encoded genetically within the DNA-sequences.

References

[1] Block, A., von Bloh, W., Schellnhuber, H.J. Efficient box-counting determination of generalized fractal dimensions. *Phys. Rev. A.* **42**: 1869–1874 (1990).

[2] Block, A., von Bloh, W., Schellnhuber, H.J. Aggregation by attractive particle-cluster interaction. *J. Phys. A. Math. Gen.* **24**: L1037–L1044 (1991).

[3] Caserta, F., Stanley, H.E., Eldred, W.D., Daccord, G., Hausman, R.E., Nittmann, J. Physical mechanisms underlying neurite outgrowth: A quantitative analysis of neuronal shape. *Phys. Rev. Lett.* **64**: 95–98 (1990).

[4] Hamilton, P. A language to describe the growth of neurites. *Biol. Cybern.* **68**: 559–565 (1993).

[5] Kernell, D., Zwaagstra, B. Dendrites of cat's spinal motoneurons: Relationship between stem diameter and predicted input conductance. *J.Physiol.* **413**: 255–269 (1989).

[6] Kniffki, K.-D., Chialvo, D., Vahle-Hinz, C., Apkarian, A.V. Fractal dimensions of neurons located in the cat's thalamic ventrobasal complex (VB) and its ventral periphery (VB_{vp}). *Soc. Neurosci. Abstr.* **16**: 622 (1991).

[7] Kniffki, K.-D., Pawlak, M., Vahle-Hinz, C. Scaling behavior of the dendritic branches of thalamic neurons. *Fractals* 1/2: 171–178 (1993a).

[8] Kniffki, K.-D., Vahle-Hinz, C., Block, A., von Bloh, W., Schellnhuber, H.J. Fractal scaling characteristics of 3-dimensional thalamic neurons. *In preparation* (1993).

[9] Mandelbrot, B.B. *The fractal geometry of nature.* Freeman, New York (1983).

[10] Rall, W. Branching dendritic trees and motoneuron membrane resistivity. *Exp. Neurol.* **1**: 491–527 (1959).

[11] Schierwagen, A., Grantyn, R. Quantitative morphological analysis of deep superior colliculus neurons stained intracellularly with HRP in the cat. *J. Hirnforsch.* **27**: 611–623 (1986).

[12] Schierwagen, A. A non-uniform cable model of membrane voltage changes in a passive dendritic tree. *J. theor. Biol.* **141**: 159–179 (1989).

[13] Smith, T.G., Marks, W.B., Lange, G.D., Sheriff Jr., W.H., Neale, E.A. A fractal analysis of cell images. *J. Neurosci. Methods* **7**: 173–180 (1989).

[14] Takayasu, H. *Fractals in the physical sciences.* Manchester University Press, Manchester and New York (1990).

[15] Vahle-Hinz, C., Pawlak, M., Kniffki, K.-D. Morphology of neurons in the ventral periphery of the thalamic ventrobasal complex (VB_{vp}) of the cat. *J. Comp. Neurol.* (in press, 1993).

[16] van Veen, M. van Pelt, J. A model of outgrowth of branching neurites. *J. theor. Biol.* **159**: 1–23 (1992).

Fractal Structure
and Metabolic Functions

Organisms as Open Systems

Manfred Sernetz
Institut für Biochemie und Endokrinologie
Justus-Liebig-Universität Giessen
D-35392 Giessen, Germany

1 Introduction

Organisms are open, energy-dissipative driven systems which build up and sustain their complex structure and functional organization in a multiplicity of steady states far from thermodynamic equilibrium by continuous exchange of mass and energy. Both from theoretical reasoning and from experimental evidence they are as living systems comparable with continuous bioreactors as technological equivalents (Fig. 1). This applies to scale dependence of reaction kinetics and of structure and can be demonstrated on a broad variety of common features with far reaching correspondences (Table I). Application of the concepts of heterogeneous catalysis and of fractal structure helps to understand scaling phenomena and the organization of metabolism on the level of single cells as well as of metazoa.

2 The Reduction Law of Metabolism

It is a well known phenomenon in biology that absolute metabolic rates $\dot{S}$ of similar organisms of different size, say mammals, are not directly proportional to their volume V (or body mass, at constant average density ≈ 1), but scale according to an empirical power law

$$\dot{S} = a \cdot V^b \tag{1}$$

with an non-integer exponent $b \approx 0.75$. This relation is experimentally well documented over a range of $1 : 10^6$, that is from about 3 ml to $3\,m^3$ (mouse to elephant). Thus the specific rates $\dot{S}/V$ (in the conventional definition) decrease with body volume

$$\dot{S}/V = a \cdot V^{b-1} \approx a \cdot V^{-0.25} \tag{2}$$

leading to the allometric reduction law of metabolism. This means that small animals metabolize at appreciably higher volume-specific rates than bigger ones.

We can achieve, however, constant specific rates a, if we relate the absolute metabolic rates $\dot{S}$ to the intuitively strange, but proper fractal entity V^b of organisms, namely

$$\dot{S}/V^b = a = \dot{S}/L^{3b} \approx \dot{S}/L^{2.25} = \text{constant} \tag{3}$$

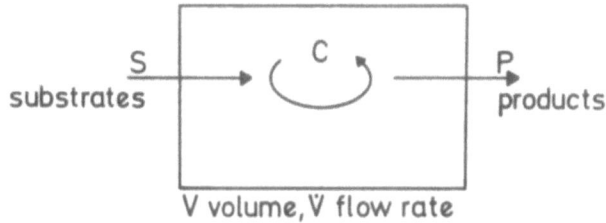

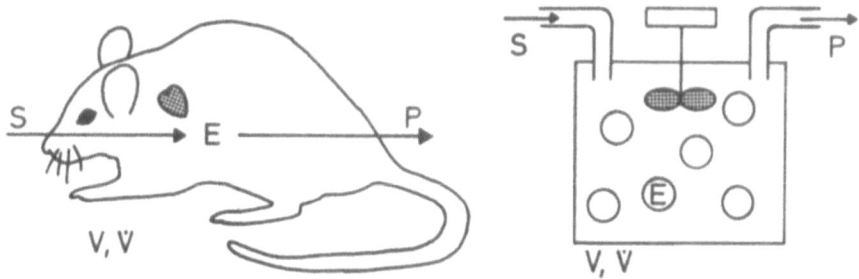

Fig. 1 Correspondence between organisms and continuous bioreactors as open systems. S substrates, P products, E constraint enzymes, C concentrations, V volume, $\dot{V}$ dilution rate.

This indicates that the body is not adequately represented by its mere mass or its unstructured volume $V = L^3$, but behaves as a whole or with its rate limiting parts rather like a 2.25-dimensional object (Sernetz et al. 1985, 1986, 1989a). It is noteworthy that isolated cells under identical experimental conditions exhibit equal turnover rates irrespective of the species of origin. Thus allometry of metabolism is not due to genetic differences, as previously suspected, but is a «technical» consequence of transport limitation of turnover rates in scaling volumes.

Allometric scaling can be observed for clearances Cl (e.g. of kidneys) and many other first order elimination processes with rate constants ^1k or relaxation times $\tau = ^1k^{-1}$.

$$\frac{Cl}{V^b} = \frac{V_v \cdot ^1k}{V \cdot V^{b-1}} = a \tag{4}$$

Since the relative volume of distribution V_v/V of a compound in the body can be regarded as a constant for similar organisms of different size, rearrangement gives

$$^1k = a' \cdot V^{b-1}$$
$$\tau = 1/a' \cdot V^{1-b} \tag{5}$$

and shows that these cases reduce to the scaling merely of rate constants ^1k, frequencies or time constants τ in bodies of different size. Only by reference to the proper fractal entity V^{1-b} the so defined specific time constants τ_{spec} become independent of size

$$\tau_{spec} = \tau/V^{1-b} = \text{constant} \tag{6}$$

Organisms	Bioreactors
open, energy-dissipative, driven, continuous, catalytic and structured systems	
living animal, experimental analysis of physiological functions	technical device, analysis of reaction kinetics and energy balance
catalytic, compartimented, multiphase systems (solid, liquid, gaseous). Stationary phase as the carrier of catalytic activity (enzymes)	
Cells as basic, catalytic units. Tissues, parenchyma with intracellular constraint enzymes	Suspensions of porous enzyme carrier particles or enzyme membranes with immobilized enzymes
Liquid phase, driven for transport of reactants	
Blood, intercellular liquid driven by the heart	Bulk solution of substrates and products driven by a stirrer
Kinetics: heterogenous catalysis, turnover determined by enzymatic reaction and convective and conductive transport, transport limited turnover	
Metabolic rates	Turnover rates in steady state
Size dependent volume specific rates and frequencies	
Allometry of metabolism	Scaling of turnover
Kinetics of mixing between stationary and liquid phase	
«Almost turbulent» mixing via a fractal vessel system. Clearances and rate constants of first order processes. Allometry of characteristic times and frequencies	Turbulent mixing by stirring Dilution rates, exponential residence time distributions. Scaling of characteristic times
Optimization of efficiencies E by reduction of transport resistances	
Fractal structure	
Fractal branching of transport vessels, fractal folding of tissues, fractal borders between stationary and liquid phases Ergodic trajectories of blood cells in the vascular system. Percolation of stationary and liquid phase from macroscopic to molecular, intracellular scale	Turbulence as the dynamic equivalent of fractal structure of stirred bioreactors. Ergodic trajectories of carrier gel particles. Percolation of stationary and liquid phase in the open porosity of polymer enzyme carrier gels

Table 1 Correspondence between organisms and continuous bioreactors (CSTR)

3 Scaling of Turnover in Continuous Bioreactors

We can apply the same arguments to describe the turnover of bioreactors, especially the type of continuous stirred tank reactors (CSTR) (Fig. 1c). Under turbulent stirring its characteristics are again the dilution rates D

$$D = \dot{V}/V =^1 k \tag{7}$$

and clearances Cl

$$Cl =^1 k \cdot V \tag{8}$$

Here too the clearances or relative turnover rates follow scaling power functions with non-integer exponents of the reactor volume V_r.

Its proportional part is the catalytically active stationary phase, the gel volume V_g with immobilized enzymes

$$Cl = \dot{S}_{het}/S = k \cdot V_g^b = f \cdot V_v^b \tag{9}$$

Turnover in bioreactors is governed by the laws of heterogeneous catalysis, where the catalytic activity is constrained to the Volume V_g of the stationary phase. The efficiency E is defined as the ratio of the actual turnover rate of the substrate in the heterogeneous system S_{het} to the maximum possible rate S_{hom} in the homogeneous case of solute enzymes.

$$E = \dot{S}_{het}/\dot{S}_{hom} = f(\mu, \sigma) \tag{10}$$

The efficiency depends on the influence of two modules of rate limitation, namely the module μ of external transport resistance between stationary and liquid phase, and the module σ of internal diffusion of the substrates within the porous stationary phase. The module μ describes the influence of a quasi unstirred Nernst-layer δ at the surface of the stationary phase. Its influence can be reduced by rigorous, turbulent stirring. The module σ represents the influence of diffusional transport limitation within the stationary phase. It can be minimized by reducing the radius R of the carrier gel particles or the thickness of catalytic membranes.

Thus the efficiency E of a bioreactor can be optimized by reducing these two transport resistances, that is of the distance δ between liquid and stationary phase by stirring and of the distances R within the stationary phase by using small particles or thin layers.

By transferring this terminology from bioreactor kinetics, we can define organisms of increasing size as bioreactors with constant resistance of internal diffusion, since the size R and the density of cells remain constant, independent of the size of an animal, but with increasing external transport resistance. The resulting decrease of efficiency with size must be counterbalanced by optimizing the distribution characteristics of a growing vessel system and by maximizing the exchange surfaces between liquid phase and the tissue or parenchyma as the stationary catalytic phase.

The scaling of turnover rates has been demonstrated with CST-bioreactors on the basis of suspensions of mammalian cells and bacteria in continuous fermentations (Sernetz et al. 1990). The turnover rates of bioreactors as purely technical, analytical systems scale with reactor volume and with the particle density as experimental parameter according to power functions with non-integer exponents in complete analogy to the allometric reduction law of the metabolism of organisms. The value of the exponent however depends very much on the cell or particle number densities in the reactor. At low cell densities the total turnover of the reactor is merely the sum of the individual turnover of independent cells of the suspension. In this case the specific turnover $\dot{S}/V$ is independent of size and the system behaves isometrically. At high cell densities however, the Nernst-layers of cells overlap, the total turnover now is the sum of concurrent, interacting cells and the specific turnover decreases allometrically with the size of the system (Sernetz et al. 1990).

A comparison between the value of the scaling exponents of bioreactors and the allometric exponent of organisms leads to an important qualitative difference. The specific turnover rates of bioreactors reach scaling exponents in the range of $b - 1 = -0.3$ at cell densities of about $10^9/l$. In contrast the allometric exponent of specific metabolic rates of organisms is only $b - 1 \approx -0.25$, yet for average cell densities of about $10^{12}/l$. Therefore we postulate that our body exhibits excellent mixing characteristics in spite of its extremely high packing density of the stationary phase by virtue of the peculiar properties of its fractal vessel systems.

4   Fractal Structure

Over a broad range of scales organisms are characterized by fractal structure of tissues and vessels or transport systems. It is the intricate interlacing of a stationary, catalytic phase with a liquid, driven phase, which minimizes distances and resistances of transport down to subcellular resolution and enables optimum turnover through all scales and sizes of organisms. In fact we can define the structural organization of the body as that of an effective percolation of two phases, solid and liquid, with their common fractal border.

The comparison with a percolation can already be applied to the gel-state of polymers in the subcellular and molecular range. The fractal properties have been established experimentally e.g. for the gel matrix of hyaline cartilage and for the intercellular permeabilities of corneal tissue by gelfiltration (Sernetz et al. 1989b, Sernetz 1992). Biological gels and tissues are characterized by the open porosity of the interlacing phases. Open porosity means that both in the solid phase and the liquid phase a wanderer can reach any point of the entire space.

The broad pore size distributions of gels can be determined experimentally by size exclusion or permeation equilibrium techniques. Classical size exclusion chromatography or gelfiltration, however, is a typical fractal analytical approach, since it measures the scaling of the gel pore volume accessible to testmolecules of different size as yardsticks, thus corresponding to the mass-radius- or box-counting-method in image and structure analysis (Wlczek et al. 1992, Sernetz et al. 1992). The fractal dimension of the border between both phases of the gel can be determined from the scaling of the complementary, namely non-accessible volume of the polymer network, which includes the fractal border (Sernetz et al. 1989b).

Within the body the cells are only constant, autarc units of almost common size, above which the fractal architecture of tissues and transport systems continues in an even more conspicuous manner up to macroscopic scales. Thus e.g. for the lung the scaling of bronchial branching (Kitaoka, this volume) and that of the alveolar surface as a phase border (Weibel, this volume) has been measured by serial sectioning and image analysis.

In my group we devote special attention to the fractal properties of the kidney arterial tree, which we derived from arterial cast preparations according to the corrosion cast technique and by 3-dimensional, scale-dependent image analysis of 3D-data sets. These data sets were both gained from NMR-computertomography and from fine serial sectioning of the entire organ with an isotropic resolution of 0.1 mm (Sernetz et al. 1992). The mass-radius-analysis yielded a scaling of the mass with about $D = 2.2$ to 2.5, which corresponds well with the overall scaling exponent of organisms for metabolic rates as a function of body size (see above).

The scaling of the surface of the kidney arterial tree varies from about $D_S \approx$ 2.5 at coarse resolution (> 3 mm) to about $D_S \approx 2.0$ at fine resolution (0.3 mm). Due to natural limits exponents in the body represent scale dependent, local dimensions and an average fractal behaviour in a certain range and change towards approaching these limits. It is possible that at the scale of diffusing molecules the exchange surfaces (say in alveoli or capillaries) may become two-dimensional again. Thus non-limiting structures may therefore exhibit isometric, i.e. non-scaling behaviour in a proper range of resolution.

With the maximum resolution achieved experimentally in the kidney arterial tree (0.1 mm) we are still within the transport and thus turnover limiting part of arteriolae and far from the capillary bed. A technical consequence of mass-radius analysis of arterial surfaces is, that at high resolution it is not necessarily the fine structure of the smallest vessels, but rather the «flat» surface of the bigger ones that is tested (Sernetz et al. 1992). Non-integer exponents can be interpreted as direct evidence of the fractal organization and in comparison with the allometric exponents of specific turnover rates as a measure of the effective transport-limiting border in a catalytic system (Sernetz et al. 1985).

Now what does fractal organization mean and what are the fractal borders in the body? It can easily be demonstrated with any two arterial and venous casts of the vessel tree of an organ (e.g. of kidneys, Sernetz et al. 1985) or even of the entire body that each of them has a definite volume (blood) and that in its finest branchings the vessels reach almost any point and depict the form of the entire organ or body. Moreover, the two systems of defined volume fit exactly between each other, yet are still merely negatives of the tissue. Only the remaining inter«space» as a highly folded fractal «area» and border between the two volumina of the vessels (liquid phase) actually represents the tissue, literally the parenchyma or stationary phase of the organ or body.

Thus from both our present knowledge on the fractal exponents of vessels and from the allometric reduction law of metabolism, organisms should be represented in fractal terms as 2.25-dimensional rather than voluminous objects.

5 Turbulence in the Body

The determination of the fractal dimension of tissues or vessels by means of 3D-image analysis techniques and by 3D-computer simulation (Bittner and Sernetz 1991, Bittner 1991) serves not only for the morphological description as a fractal structure, but also yields the key to the dynamics of turnover and turbulent mixing by means of the vessel system. In bioreactors high efficiency E (eq. 9) is achieved by turbulent mixing, which keeps Nernst-layers δ small and resistance of external diffusion μ low. Turbulence itself is a dynamic self-similar phenomenon, namely a recursive regress of eddies in eddies through a wide range of scales. Turbulent systems continuously dissipate energy and must be driven (periodically) to maintain their structure. In turbulent flow the paths of two initially neighboured particles or volume elements diverge exponentially and unpredictably into entirely different regions, after sufficiently long, periodical or stroboscopic observation their paths however fill the entire region and depict its internal structure.

There is ample kinetic evidence, although it is apparently paradoxical, of turbulent distribution and transport in the organism as the basis of the high efficiencies of metabolic reactions. Although all the flow in the blood vessels is locally and everywhere laminar, as known from the low Reynolds-numbers, the mixing characteristics of the organism are practically turbulent, as expressed by the exponential residence time distributions which are the basis of the definition of clearances in physiology as first order distribution processes. The fractal organization of tissues and transport systems is the structural equivalent and morphological support of turbulence.

Nevertheless the turnover remains to some extent transport limited as can be derived from actual deviations of ideal clearance or pure first order distribution. Traditionally this is taken into account e.g. in pharmacokinetics by defining so-called superficial and deep compartments of tissues. In our context by way of fractal folding and branching of vessels all the cells within the body share an average reactive surface of dimension 2.25.

Although blood flow is locally laminar within the entire blood vessel system, due to its fractal structure it imposes turbulent mixing characteristics. The path of single cells, say erythrocytes, within the vessels is unpredictable and ergodic. After sufficiently long, stroboscopic observation the pathway of a single cell would however fill the entire volume of the vessels and with the resolution of its diameter depict its structure and the anatomy of the entire body as its complement. The vessel system behaves as a «strange attractor», and the tissue as a «strange repellor» of a cells trajectory. Any real anatomical cross-section x, y of the body is also a two-dimensional x, y-section of the complete Poincaré-map of the ergodic pathway of an erythrocyte. The coordinates of its repetitive transitions through this plane of section after sufficiently long iterations (or simultaneously for all erythrocytes the instantaneous «bleeding» of the section) fill and depict with increasing resolution the anatomical hierarchy, the resolution dependent interlaced architecture both of the liquid (vessels) and stationary (tissue) phase of the body and also the areas of nested common residence time distributions (Sernetz et al. 1989a).

6   Limited selfsimilarity

Up to now allometric relations in biology have usually been formulated as power law equations without setting limits of validity. Instead, as in any natural fractal system, it is however realistic to assume limited ranges of allometric or fractal behaviour. This can be achieved by two different approaches: One can either define a fractal or selfsimilar region by truncation of the power law within upper and lower bounds (cut off) ϵ_{max} and ϵ_{min} of the range of resolution ϵ.

As opposed, selfsimilarity can also be limited by upper and lower limits y_{max} and y_{min} of the measured quantity y, with asymptotically fading fractal properties of the measured system towards approaching these limits as a function of the resolution ϵ. By introducing such limits, the selfsimilar power function, which is a straight line in a graph with logarithmic coordinates, modifies into a log-logistic distribution function with sigmoidal course in logarithmic coordinates (Sernetz et al. 1985, Sernetz and Bittner 1991) and with the fractal dimension related to the parameter of dispersion of the distribution function.

Viewing the body as a fractally structured system allows a new, functional interpretation of the allometric power law of metabolism. The relation of turnover rates to the proper fractal entity, namely the 2.25 dimensionality instead of a volume, yields equal efficiency for small and large animals and thus explains the broad size range of realization of similar metazoa.

Acknowledgement:
This work was supported by grant of Deutsche Forschungsgemeinschaft (Se 315/13-1 to 4).

References

[1] M. Sernetz, B. Gelleri and J. Hofmann: The organism as bioreactor. Interpretation of the reduction law of metabolism in terms of heterogeneous catalysis and fractal structure. J. Theor. Biol. 117 (1985) 209–239.

[2] M. Sernetz, H.R. Bittner and H. Willems: Organismen als Bioreaktoren: Fraktale Struktur und heterogene Katalyse, Umschau 86 (1986) 582–587.

[3] M. Sernetz, H Willems and H.R. Bittner: Fractal organization of metabolism, in: Energy Transformations in Cells and Organisms, W. Wieser, E. Gnaiger (eds.), Thieme, Stuttgart 1989a, pp. 82–90.

[4] M. Sernetz, H. Willems and K. Keiner: Dispersive analysis of turnover rates of a CST-reactor by flow-through microfluorometry under conditions of growth. Ann. N.Y. Acad. Sci. 613 (1990) 333–337.

[5] M. Sernetz, H.R. Bittner, H. Willems and C. Baumhoer: Chromatography, in: The Fractal Approach to Heterogeneous Chemistry, D. Avnir (ed.), Wiley, Chichester 1989b, pp. 361–379.

[6] M. Sernetz: Fractal structure of gels determined by size exclusion chromatography or equilibrium techniques, in: Proc. IUPAC-Symp. Macromolecules 1992, J. Kahovec (ed.), VSP, Zeist 1993 pp. 423–429.

[7] P. Wlczek, A. Odgaard and M. Sernetz: Fractal 3D analysis of blood vessels and bones, in: Fractal Geometry and Computer Graphics, J.L. Encarnação et al. (eds.), Springer, Berlin 1992, pp. 240.-248.

[8] M. Sernetz, J. Wübbeke and P. Wlczek: Three-dimensional image analysis and fractal characterization of kidney arterial vessels. Physica A 191 (1992) 13–16.

[9] H.R. Bittner and M. Sernetz: Selfsimilaritiy within limits: Description with the log-logistic function, in: Fractals in the Fundamental and Applied Sciences, H.-O. Peitgen et al. (eds.), Elsevier, Amsterdam 1991, pp 47–58.

[10] H.R. Bittner: Modelling of fractal vessel systems, in: Fractals in the Fundamental and Applied Sciences, H.-O. Peitgen et al. (eds.), Elsevier, Amsterdam 1991, pp. 59–71.

Transfer to and across Irregular Membranes Modelled by Fractal Geometry

Bernard Sapoval
Laboratoire de Physique de la Matière Condensée
C.N.R.S. Ecole Polytechnique
91128 Palaiseau, France

Abstract. We apply results from the study of fractal electrodes to diffusion across an irregular passive membrane with finite permeability. Diffusion efficiency is more specially discussed and seems to explain semi-quantitatively the final step of respiratory physiology i.e. the diffusion and capture of oxygen in the acinus region of the lung. We define a «best possible acinus» for which the acinus cut by a plane has a length equal to the ratio of the diffusivity to the membrane permeability. The observed anatomy of the acinus of several animals corresponds roughly to this optimized geometry. This close relation between morphometry and transport parameters like diffusivities and permeabilities could help to understand the design of biological organisms.

1 Introduction

Transport across rough or porous surfaces is a basic problem in the study of several natural or industrial processes. For instance, in the design of high current batteries it is natural to consider porous electrodes as a means of increasing the net output current. Any process that is limited by transport across a surface or interface can be enhanced in this way. Many systems like plant roots, villi in the human intestine or lung alveoli are found that have a very irregular or ramified geometry that can be considered as approximate examples of large area «fractal» structures [1–3]. In the case of a membrane, neutral reacting species are brought to the membrane surface by diffusion currents whereas in a battery the ions are transported by the electric field. The transport across the membrane plays the same role as the redox reaction on the battery electrode [4]. The problem is best stated in the electrochemical case and was studied extensively in this context because frequency dependent transport can easily be studied experimentally by impedance spectroscopy. A variety of studies have indicated that very often the impedance of electrochemical cells behave as

$$Z = R_0 + k(j\omega)^{-\eta} \tag{1}$$

where R_0 represents the electrolyte resistance, ω is the frequency of the a.c. applied voltage, j is $(-1)^{1/2}$. This behavior is called «constant phase angle» (CPA) response and η is the CPA exponent which depends on surface electrode roughness (k is a constant). A number of groups have worked on the fractal impedance problem and a number of papers have been published on this subject by A. Le Mehauté, L. Nyikos and T. Pajkossy, T. Kaplan, L.J. Gray and S.H. Liu, M. Blunt,

R. de Levie, W. Geertsma, J.E. Gols and L. Pietronero, A.M. Marvin, F. Toigo and A. Maritan, W.H. Mulder and J.H. Sluyters [5–8]. A complete list of references up to 1990 is given in the review given in reference [5].

Up to very recently, there existed apparent contradictions in the literature on that subject but we believe that these contradictions can now be lifted and a single quantitative model, confirmed by numerical simulations and by experiments, can now be considered as giving a proper understanding of the transfer across self-similar interfaces [9]. The scaling form of this theory is presented in this volume by R. Gutfraind and B. Sapoval [10]. Our purpose here is to apply what we know from electrodes to the problem of diffusion to membrane in a biological or physiological context. Here diffusion could be diffusion of oxygen or CO_2 in air at the level of the pulmonary acinus or diffusion of Na^+ or other ions or drugs in the kidney or of some nutrient in the intestine.

It was shown that an exact equivalence exists between the response of an electrode under d.c. conditions and the net diffusion across a membrane of the same geometry. There exists also a si mple and direct connection between the a.c. and the d.c. response of an electrode [4,5,8]. The electrode results that we use here have been obtained by several methods: numerical simulation, scaling arguments and a continuous fraction expansion study of the problem. They allow to explain quantitatively the impedance measurements on model electrode in $d = 2$ [9].

We present briefly the 2d results and their extension to more real 3d systems and uses these results to understand why and to what extent fractal exchangers can be more efficient. The reader is refered to [10] to find the scaling arguments that justify the following results.

2   The Diffusion Results in $D = 2$

We consider here the self-similar fractal membrane shown in Fig.1. The simplified diffusion system that we consider has a diffusion source at the bottom of the figure where the concentration of diffusing particles is constant. We then consider systems in which particles have to diffuse before crossing the membrane. The problem that we consider is then a problem of diffusion to and across the membrane. Diffusing particles (think of oxygen in air) emitted at the diffusion source execute random walks. If the random walker collides with the membrane it is absorbed with a finite probability σ (also known as the sticking probability) which corresponds to the existence of a finite permeability of the membrane. In a diffusion situation most of the particles which leave the diffusion source return to it. This is due to the fact that the net flux, from Fick's law, is proportional to the gradient of the concentration and not to the concentration itself. The net diffusion transfer between the source and the membrane is due to the very few particles that are absorbed before returning to the source. There are then two limitations to the transfer efficiency. First the particles have to diffuse to the membrane and second they have to be absorbed by the membrane. In much the same situation, in an

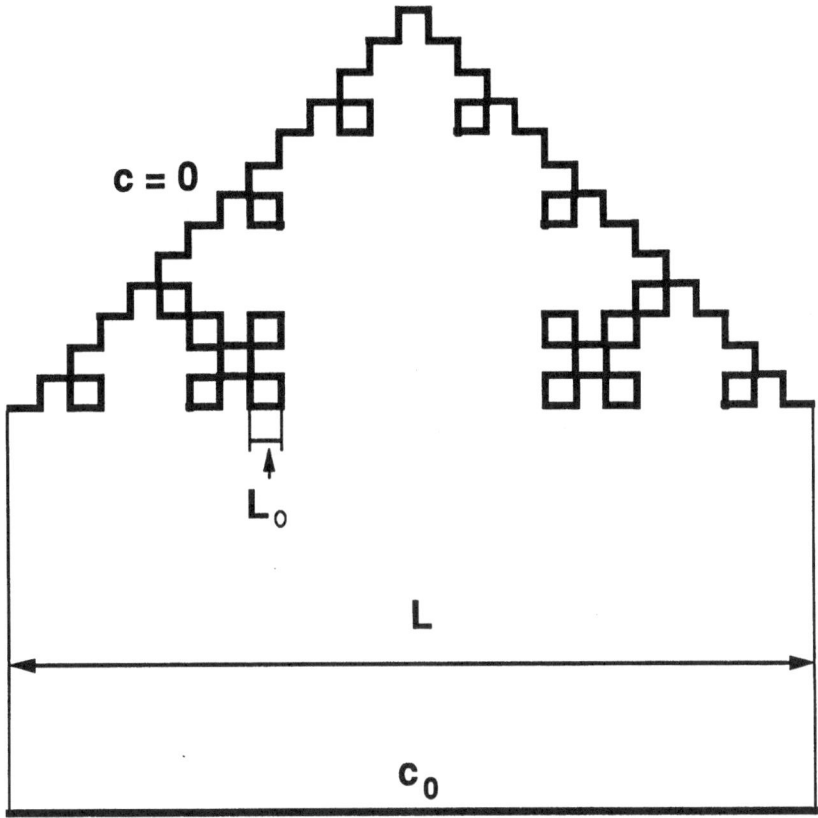

Fig. 1 Schematic representation of a fractal membrane system in $d = 2$. Here the dimension is $D_f = \log 5 / \log 3$. The membrane has a smaller scale L_0 below which it can be considered as smooth. The size of the system is L. We assume that an external mechanism maintains a constant concentration c_0 at the diffusion source. The diffusion source is situated at a distance of the order of L. In our discussion of the lung the diffusion source is supposedly situated near the entrance of the acinus.

electrochemical cell, ions have to be transported through the electrolyte and then have to undergo a redox reaction on the active electrode to finally contribute to the current. The total system then presents a macroscopic diffusion admittance Y_D defined by

$$\Phi = Y_D\, c_0 \tag{2}$$

where Φ is the total flux and c_0 is the concentration at the source. The admittance must then be written, in analogy with Eq.(1) as

$$Y_D = (R_B + Y_M^{-1})^{-1} \tag{3}$$

where R_B is the resistance of the bulk (the gas or the liquid region that diffusing particles have to cross) and Y_M the admittance of the membrane. The two transport parameters that characterise the problem are the diffusion coefficient D and the coefficient of transport across the surface or permeability W. (If c is the concentration at the surface of a linear membrane of length L the net flux across the membrane is cWL). The membrane here is supposed to be passive and we neglect back diffusion coming from the other side of the membrane where the concentration is supposed to be zero.

It is useful to keep in mind a lattice picture of the diffusion process. If diffusion results from random walks on a square lattice with lattice spacing a and jump rate per unit time $(1/\tau)$ we have $D = a^2/4\tau$ and $W = a\sigma/4\tau$. Note that nature gives D and W but not σ. The absorption probability σ is not determined uniquely from D and W but depends on the lattice spacing a in this picture of diffusion. The true physical variable in that problem is the length [8]

$$\Lambda = D/W = a/\sigma. \tag{4}$$

which is defined without a necessary lattice representation.

There exists two regimes for the diffusional transfer. In the small permeability regime the resistance to the transfer is limited by the membrane resistance itself and not by the difficulty for the particles to reach the membrane or bulk resistance. The diffusion admittance is then the admittance of the developed fractal line

$$Y_D = Y_M = L_0(L/L_0)^{D_f} W. \tag{5}$$

where L is the macroscopic size of the electrode, L_0 is the size of the smaller linear part of the electrode and D_f the fractal dimension. This is the classical regime. For higher values of the permeability one reaches a fractal regime in which the membrane admittance is

$$Y_M = L(D/L_0)^{(D_f-1)/D_f} W^{1/D_f}. \tag{6}$$

The crossover between the high and low permeability regime is obtained when these expressions are equal or

$$D/WL_0 = L/L_0 = (L/L_0)^{D_f} \tag{7}$$

In the fractal (high permeability) regime it is useful to consider the euclidian distance that a random walker that hits the membrane with absorption probability s travels in space before being finally absorbed. We call this distance $L_c(\Lambda)$. The above lattice representation of the diffusion process is useful to compute $L_c(\Lambda)$. In the picture of a walk on a lattice, the walk is terminated (by final absorption on the membrane) when the number of fractal sites within the euclidian distance $L_c(\Lambda)$ (equal to $(L_0/a)(L_c(\Lambda)/L_0)^{D_f}$) is equal to the average number of collisions needed by the walker to be absorbed. This number is of order $1/\sigma$. Writing the equality gives the relation

$$L_c(\Lambda) = L_0(\Lambda/L_0)^{1/D_f}. \tag{8}$$

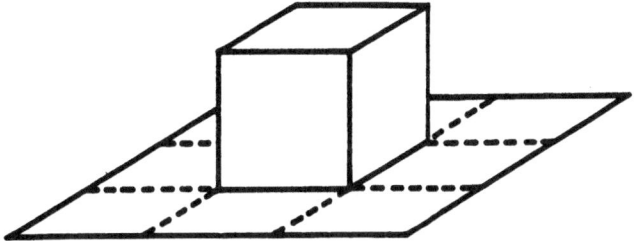

Fig. 2 The generator of an hypothetic deterministic fractal membrane in $d = 3$ with a fractal dimension $D_f = \log 13/\log 3$.

When this length is equal to the size of the system L, then the random walker has the same probability to return to the source (and then be lost for the transfer) than to be absorbed. Then we expect that when relation (7) is satisfied we have

$$R_B = Y_M^{-1} \tag{9}$$

In $d = 2$, R_B is typically the resistance to diffusion of a square of side L. Its value is then $R_B = D^{-1}$ and one can verify that this condition is satisfied by replacing (7) in (5).

3 Extension to $d = 3$ Membranes

The generator of a possible deterministic fractal membrane in $d = 3$ is shown in Fig.2. In the case of self similar membranes embedded in $d = 3$ space there exits also a low permeability regime where [9]

$$Y_D = Y_M = L_0^2 (L/L_0)^{D_f} W. \tag{10}$$

and a high permeability regime where

$$\begin{aligned} Y_M &= L^2 (D/L_0)^{(D_f-2)/(D_f-1)} W^{1/(D_f-1)} \\ &= (\Lambda/L_0)^{(D_f-2)/(D_f-1)} L^2 W \end{aligned} \tag{11}$$

The crossover is obtained in that case when

$$L = L_0(\Lambda/L_0)^{1/(D_f-1)}. \tag{12}$$

For this crossover value the value of the membrane admittance is equal to the value of the admittance to reach the surface, which, in $d = 3$ is typically the admittance of a cube of size $L : Y_B = DL$. Note that the statement that, at the crossover point, the value of the classical admittance of the fractal membrane is equal to the access resistance is very simple. It may then be of more general value and may apply to

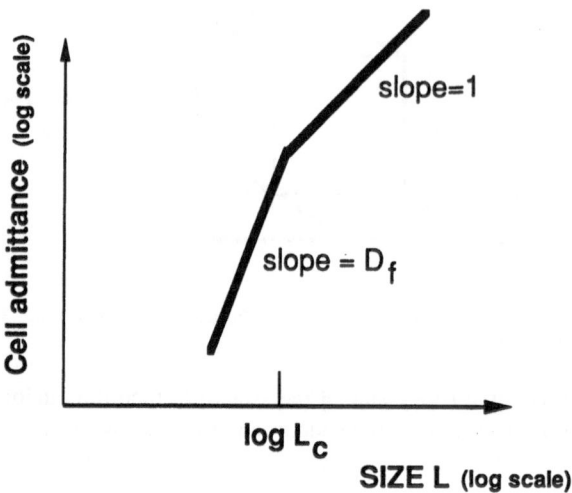

Fig. 3   Schematic behavior of the admittance of the total cell as a function of the size of the fractal membrane. For small sizes the admittance is proportional to L^{D_f} while for larger size the admittance is limited by the access resistance inversely proportional to the size. The optimum for the acinus for which $D_f = 3$ should be found at the crossover between the two regimes because the specific admittance Y_D/L^3 in that case would decrease above the crossover.

many irregular geometries. This is probably why the above considerations seem to apply to the acinus of several animals, as discussed below, although their real geometry is not a simple fractal [11].

　　If one considers systems of small sizes (with $L > L_0$ but not too large) the resistance of the system is dominated by the membrane admittance which varies as L^{D_f}. Considering larger and larger systems the admittance decreases with this power law of the size. This is true up to the crossover point after which the fractal membrane admittance varies as L^2 but the access resistance, which dominates now the total admittance, varies only with the power one of the size L. This is shown schematically in Fig. 3. The cross over condition, inverting relation (12) gives $\Lambda = L_0(L/L_0)^{(D_f-1)}$. This corresponds to the fact that the length of the perimeter of a cut of the fractal membrane by a plane, which is a fractal with a fractal dimension equal to $(D_f - 1)$, has a length equal to Λ.

4   Comparison between fractal Membrane Geometry and ordinary Geometry

Before comparing fractal and non fractal exchangers it is useful to realize that $d = 2$ and $d = 3$ fractal systems behave differently. In $d = 2$ the system admittance is $Y_D = (D^{-1} + Y_M)^{-1}$. If one considers larger and larger exchangers, the admittance increases with the size up to the situation where the resistance of the system is dominated by the access resistance to diffusion D^{-1} which do not depend on the size. In consequence in $d = 2$ a system with a size larger than $L_c(\Lambda)$ is not a better exchanger than a system of size L_c.

In contradistinction, in $d = 3$ the bulk admittance is equal to LD and the admittance is $Y_D = ([LD]^{-1} + Y_M^{-1})^{-1}$ which increases for larger and larger fractal systems (see Fig. 3). One can compare the admittances given by relations (10) and (11) with that of a flat membrane of same area L^2 which is simply equal to $Y_{fl.} = WL^2$. The admittance ratio $Y_M/Y_{fl.}$ is equal to $(\Lambda/L_0)^{D_f-2}$ in the low permeability regime and to $(\Lambda/L_0)^{(D_f-2)/(D_f-1)}$ in the high permeability regime. In both cases there is a net increase of the admittance due to the fractal geometry. Note that the smaller L_0 and the larger the fractal dimension D_f, the larger the fractal admittance. These results give some insight to what extent fractal geometry increases pratically the effectiveness of transfer through irregular membranes.

The robustness or adaptability of fractal exchangers is related to the fact that the membrane admittance in the fractal regime is a power law of the transport coefficients with exponent $(D_f - 2)/(D_f - 1)$. This number is smaller than one. This means that the effective flux through fractal membranes is less sensitive to the physical conditions or that a same membrane can absorb different species which have different permeabilities and/or diffusivities.

Also, in the fractal regime, the regions of the fractal surface which are really active for the transport are different for different species. This is discussed in [10]. For a multiple species transfer there could exist a space selection or filtering along the fractal surface. (In contrast, in the low permeabilty regime the transfer is uniform.)

5   Application to Airways

The above discussion indicates that the admittance per unit volume or specific admittance defined as $Y_D/L^3 = ([LD]^{-1} + Y_M^{-1})^{-1}/L^3$ varies as L^{D_f-3} for small L and L^{-2} for large L. If the structure is dense $D_f = 3$ and the specific admittance does not depend on the size of the fractal up to the critical size discussed above where it decreases with the power L^2. From this discussion a geometry with a dense ($D_f = 3$) arrangement of cells of smaller size L_0, up to a scale L satisfing condition (12) should be optimum because lower losses in the higher airways ask for the larger size compatible with a large specific admittance. In that sense the «best possible acinus» size should be that for which the perimeter of a cut of the acinus by a plane should have a length of the order of Λ.

The data that was obtained by E. Weibel [11] allows to compare the above considerations with what is observed in natural systems. The transport coefficients governing diffusion of oxygen in air and its capture by the alveolar membrane are $D = 0.2$ cm^2 sec^{-1} and $W = 10^{-2}$ cm sec^{-1} leading to $L = 20$ cm. In the lung the transport process is due to hydrodynamic flow in the superior

Animal	Mouse	Rat	Rabbit	Human
V_a $(10^{-3} cm^3)$	0.4	1.76	3.46	187
S_a (cm^2)	0.42	1.36	1.55	69
Computed sac diameter $(10^{-4} cm)$	9.5	13	22	27
Computed perimeter $S_a/V_a^{1/3}$ (cm)	5.7	11.2	10.3	120

Table 1 Dense model parameters of the acinus computed from the data of reference [11–14] listed on the two top lines. The third line gives the diameter of a sac and the last line gives the perimeter deduced in the dense model.

airways and to diffusion within the alveoli. The acinus region is the region where there is a transition between these processes so that it is not sure that the above calculations which takes into account only diffusion can really apply. Nevertheless in a simplified picture we consider the acinus as a dense [11] arrangement of alveoli of size L_0 up to a size L. In this simplified picture L_0 is of the order of the diameter of an alveolar sac while L is of the order of the macroscopic size of the acinus. The length of the perimeter of a cut is then of the order of $L_0(L/L_0)^2$.

Following these guidelines one can make a comparison for several animals: the mouse, the rat, the rabbit and the human lung using the data of references [11–14] In these papers, the quantity of interest which have been determined are the acinus volume V_a together with the alveolar surface per acinus S_a. In terms of the above quantities $V_a = L^3$ and $S_a = (L/L_0)^3 L_0^2 = L^3/L_0$. The quantity L_0 can then be found as V_a/S_a and the perimeter $L_0(L/L_0)^2$ is equal to $S_a/V_a^{1/3}$. If our interpretation is correct the length V_a/S_a should be found of the same order of magnitude as the typical small alveola or sac diameter and the quantity $S_a/V_a^{1/3}$ should be found of the same order of magnitude as Λ that is of the order of 20 cm. The data that we compute from V_a and S_a is given in Table I.

The values of L_0 found from these calculations are a little smaller but have the same order of magnitude as the sizes of the sacs observed in the pictures of references [11–14]. Concerning the perimeters, although there exist discrepancies, the general agreement can be considered as satisfactory if one takes into account the oversimplified model that we use and the fact that the value $\Lambda = 20$ cm is found from transport measurements only and not from geometrical determinations.

It seems that these considerations can be extended to the gills of fishes in which the geometry is regular. Because the diffusion coefficient of oxygen in water is of the order of 10^{-5} cm^2 sec^{-1} the value of Λ should be of the order of 20 micrometers if the permeabilty keeps the same value. No conclusive results exist in this case but the size of the gills is comparable to that value and the gills have a regular structure which corresponds to $D_f = 2$ and not to $D_f = 3$ [15]. In that case, equation (12) indicates $\Lambda = L$ as approximately observed for several fishes.

6 Conclusion

The overall agreement between the anatomical and physiological data and the expressions that we predict for the «best possible acinus» seems to indicate that the design for the oxygen exchanger in the lung of several animals is not far to be the best: The morphometry seems to be in satisfactory agreement with an optimized geometry which takes into account diffusion only in the acinus region. A better model should take care of the possible remaining effect of air flow in this region. In our restricted situation, we have shown that the best fractal membrane should have a fractal dimension equal to 3 and a size such that the perimeter of a cut is of the order of $\Lambda = D/W$. As quoted above, the statement that at the crossover point which represents an optimized situation, the value of the classical admittance of the fractal membrane is equal to the access resistance is very simple. It may then apply to the acinus although the real geometry of the acinus is not «simply» fractal.

One should note that the lung is a simple gas exchanger in the sense that only two gases are exchanged. If one considers the transfer of several species with very different transport parameters, the simultaneous optimisation of the transfer should imply that different Λ's should correspond to the same morphometry. This cannot be realized by an homogeneous membrane. In an optimized multi-species filter or exchanger membrane system this could be nevertheless realized by a suitable distribution on the membrane surface of cells permeable to specific species. In that sense it is possible that the micro geometry of the distribution of specific cells (or of cells with active transfer) in inhomogeneous membranes permits to optimize the transfer of very different species in the same organ. That could be the case of the filtration by the kidney. In such a case the above simplified picture of the transfer can help to understand the relation between anatomy, morphology and physiological function.

Acknowledgement:
The author gratefully acknowledges helpful discussions with R. Gutfraind and E.R. Weibel. Laboratoire de Physique de la Matière Condensée is U.R.A. D1254 of C.N.R.S.

References

[1] B. Mandelbrot, The fractal geometry of nature (W.H.Freeman and company, San Francisco, 1982)

[2] B. Sapoval, Fractals (Aditech, Paris,1990)

[3] G.H. Bell, D. Emslie-Smith and C.R. Paterson, Physiology and Biochemistry, (9th Ed.Churchill Livinstone:Edimburgh, 1976)

[4] B. Sapoval, Acta Stereologica, 6/III, 785 (1987).

[5] B. Sapoval, «Fractal electrodes, fractal membranes and fractal catalysts» in Fractals and disordered systems, (Ed. A. Bunde and S. Havlin, Springer-Verlag, Heidelberg,1991) p.207.

[6] T.C. Halsey and M. Liebig, Europhys. Lett.,14,815 (1991) and Phys. Rev.A 43, 7087 (1991).

[7] R. Ball, in «Surface Disordering, Growth, Roughening and Phase Transitions» ed. by R. Julien, P. Meakin and D. Wolf (Nova Science Publisher,1993)

[8] B. Sapoval, J. Electrochem. Soc.137,144C (1990) and Extended Abstracts, Spring Meeting of the Electrochemical Society, Montreal, Canada, 90-1, 772, (1990) and P. Meakin and B. Sapoval, Phys. Rev.A,43,2993 (1991).

[9] B. Sapoval, R. Gutfraind, P. Meakin, M. Keddam and H. Takenouti, Phys. Rev.E,(1993)

[10] R. Gutfraind and B. Sapoval, «Scaling and active surface of fractal membranes», this volume and Journal de Physique I,1993.

[11] E.R. Weibel in «Respiratory Physiology, an Analytical Approach» ed. by H.K. Chang and M. Paiva. (M. Dekker, Inc., 1989)p.1 and E.R. Weibel, private communication.

[12] E.R. Weibel, «Design of biological organisms and fractal geometry», this volume.

[13] M. Rodriguez, S. Bur, A. Favre and E.R. Weibel, Amer. Journ. Anat.,155,143 (1987).

[14] B. Haefeli-Bleuer and E.R.Weibel, Anatom. Rec.,220,401, (1988).

[15] A. Beaumont and P. Cassier, Biologie animale (Dunod Université, Paris 1978) p.443.

Scaling and Active Surface of Fractal Membranes

Ricardo Gutfraind[1]) and Bernard Sapoval
Laboratoire de Physique de la Matière Condensée, C.N.R.S.
Ecole Polytechnique
91128 Palaiseau, France

1)   present address: Groupe Matière Condensée et Materiaux, Université de Rennes I, Bât. 11B, Campus de Beaulieu, 35042 Rennes Cedex, France.

Abstract. We study the properties of a Laplacian potential around an irregular object of finite surface resistance. This describes the concentration in a problem of diffusion towards an irregular membrane of finite permeability. We show that using a simple fractal generator one can approximately predict the localization of the active zones of a fractal membrane of infinite permeability. When the the permeability of the membrane is finite there exists a crossover length L_c: In pores of size smaller than L_c the flux is homogeneously distributed. In pores of size larger than L_c the same behavior as in the case of infinite permeability is observed, namely the flux concentrates at the entrance of the pore. From this consideration one can predict the active surface localization in the case of finite permeability. We then show that a coarse-graining procedure, which maps the problem of finite permeability into that of infinite permeability, permits to obtain the dependence of the resistance and of the active surface on the surface and bulk properties. Finally, we show that the fractal geometry can be the most efficient for a membrane that has to work under very variable conditions.

1   Introduction

Many natural as well as industrial processes take place in the environment of surfaces or at the interface between two media. The surfaces that one encounters in various of these processes are complex and irregular. Examples of natural processes are the exchange of water and inorganic salts between the roots of a tree and its surrounding environment [1] or the transport of oxygen to the blood flow through the surface of the pulmonary alveoli [2]. Any process that is limited by transport across a surface or interface can be enhanced using large-surface objects. This is probably the reason why so many natural systems are found that have ramified structures. Many of these surfaces can also be approximately described using fractal geometry [3–6]. Therefore, using the tools provided by the fractal approach can help in analyzing such processes.

In the simplest physical situation the mathematical problem is to find the solution of the Laplace equation with mixed boundary conditions on the irregular surface. This problem has been first considered in the case of electrochemistry where the Laplace equation describes the electric potential in the electrolyte. In electrochemistry, frequency dependent transport experiments can be performed which have shown that in the presence of an irregular electrode many electrolytic cells have an impedance which behaves as [7–9]:

$$Z_{cell} = R_b + k(1/r_s + j\gamma\omega)^{-\eta} \tag{1}$$

where R_b can be considered as the bulk or electrolyte resistance, k is a constant, r_s is the Faradaic resistance which describes the finite rate of the electrochemical reaction, γ is the specific capacitance describing the charge accumulation at the surface and η is an exponent which satisfies $0 < \eta < 1$. This behavior is known as constant phase angle (CPA). For a smooth electrode $\eta = 1$ and η decreases with the degree of roughness of the electrode. In direct current conditions the surface contribution to the admittance will be of the form $Y_{surf.} = k^{-1}(r_s)^{-\eta}$ [the inverse of the second term in Eq. (1) under direct current conditions]. It can be easily shown that this problem can be mapped into that of the response of an irregular membrane [10–11]. This is the case we study here. In the problem of diffusion towards and across a membrane the Laplace equation describes the concentration field which satisfies the steady-state diffusion equation:

$$\Delta C = 0 \tag{2a}$$

with the boundary condition $C = C_o$ on a source of molecules at certain distance of the irregular membrane and

$$-\mathcal{D}\nabla C = r_s^{-1}C \tag{2b}$$

on the irregular membrane (see Fig. 1a), where $\mathcal{D}$ is the bulk diffusivity and the term r_s is the surface resistance which represents the resistance due to the finite permeability (W) of the membrane itself. Eq. (2b) arises from conservation of matter which imposes that transport to the membrane is equal to transport across the membrane.

In the case of self-similar fractals like DLA clusters, diffusion fronts or deterministic fractals as the one shown in Fig.(1b) the following result can be written [12,13]:

$$R_T = R_b + k(\rho/Lb)(r_s/\rho)^{1/D}L_0^{1-(1/D)} \tag{3}$$

where R_T is the total resistance to transport, ρ is the bulk resistivity (which is equivalent to $1/\mathcal{D}$), L is the size of the system, L_0 is the smallest feature size of the self-similar membrane and b is the thickness of the system. Thus $\eta = 1/D$, where D is the fractal dimension. This result has been obtained by four different methods. First a dimensional analysis with a scaling argument [14], second an iteration method [12,13], third a scaling argument based on a simplified picture of the working regions of a self-similar interface [15], and extensive numerical simulations based on the exact analogy between electrical and diffusion Laplacian fields [13,14].

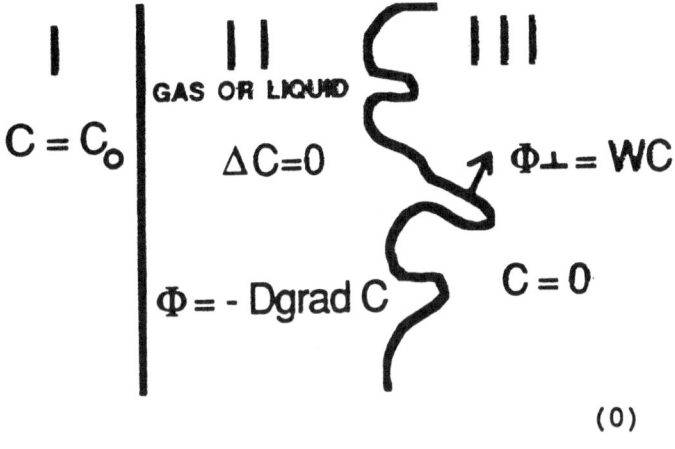

$$C = C_0 \quad \Delta C = 0 \quad \Phi_\perp = WC$$

$$\Phi = -Dgrad\ C \quad C = 0$$

GAS OR LIQUID

(0)

(b)

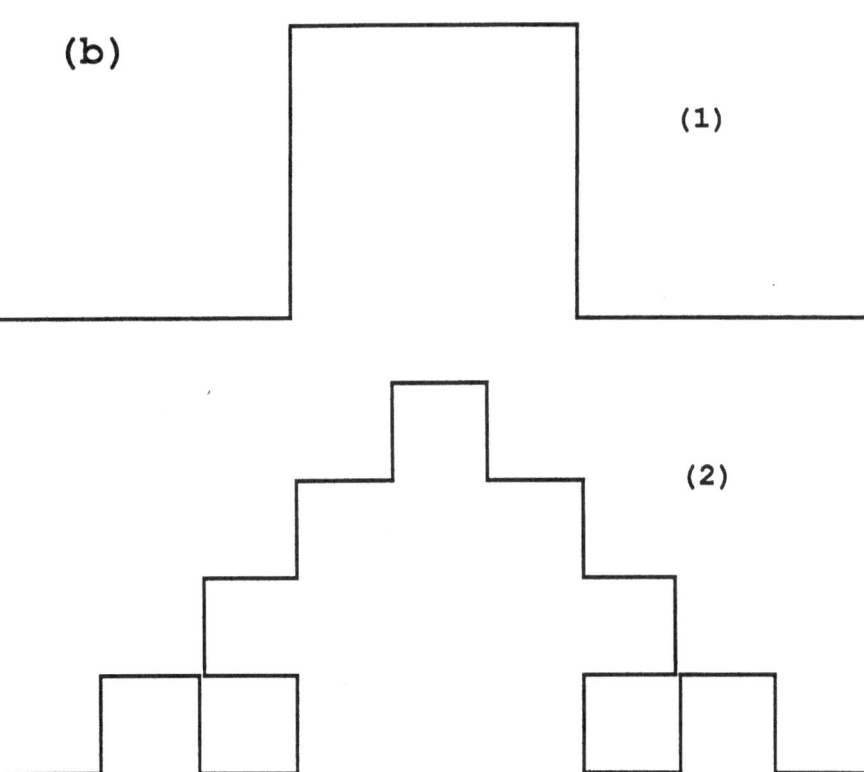

(1)

(2)

Fig. 1 a) Schematic representation of the system. The rough surface is a fractal generator. b) Generation of a two-recursion level fractal. At each generation the middle third of each segment is substituted by a square; the fractal dimension is $D = \log(5)/\log(3)$.

In this paper we discuss the concept of active surface which is the subset of an irregular interface where the flux concentrates [12,15]. Although this notion is an approximation, it provides a simple way to visualize the active zones of fractal membranes. This concept also permits to obtain equation (3) through a coarse-graining which maps the problem of $r_s \neq 0$ into that of $r_s = 0$ (Section 2). We also show that an important property arises from the physical picture of the studied process: this is how the fractal surface adapts to environment conditions (Section 3). In Section 4 we compare the numerical calculations of the flux distribution with the geometrical interpretation given in Section 3.

2   Active zone

In this section we show a practical way of visualizing the regions of a membrane that can be considered as the active zones for the case $r_s = 0$ (in this case the boundary condition $C = 0$ substitutes the boundary condition 2b on the fractal membrane). In this situation we search for the regions of the membrane which receive the flux of molecules coming from the bulk (called the information set in Ref. [12]).

Consider first the problem of the flux distribution onto the fractal generator shown in figure 1a. If the source of diffusing molecules is not too close to the fractal membrane the concentration map for a single pore is expected to behave as shown in figure 2 [16]. In the representation of the picture the boundary between the stripes are lines of equal concentration and the concentration varies with a factor two from one line to the next one. The flux is perpendicular to the lines of equal concentration and inversely proportional to their separation. One can see that at the entrance of the pore, the density of equipotential lines next to the pore walls is much larger than the density in the central part of the pore. This shows that there is little penetration through the central part of the pore. Therefore, one can see qualitatively that there is a zone of high flux (active zone) and a zone of small flux.

Based on this and according to the scaling properties that the active zone must have (it has to scale as L^1 [17]), we construct a generator as shown in figure 3 (top). From this generator we build the active zone by iteration (figure 3) but to be consistent with the above requirement, it is necessary for the size of the generator to be exactly equal to L. This is the only possibility if one wants to build the set from a single generator. If the generator would be equal to $(1 + \alpha)L$, α being any number different from zero, the active zone will not scale as L^1 in the iterative process used to build the object. Thus the dashed line in the generator must extend to exactly half of the depth of the well in figure 3 (top).

In summary, if we use the (approximate) notion of an active zone, for this notion to be consistent with the known properties of the Laplacian field, the size of its generator has to be taken equal to L. This idea is approximative in two aspects: first not all the molecules coming from the bulk arrive in this fraction of

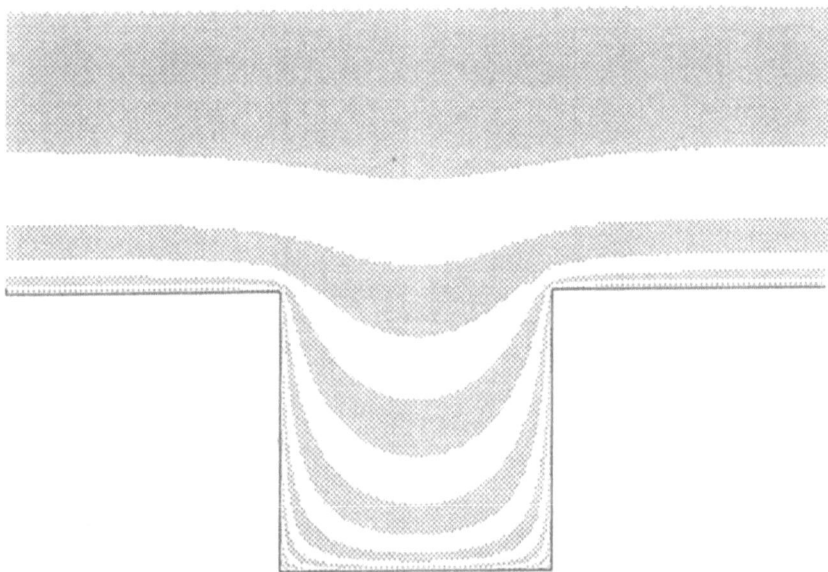

Fig. 2 Laplacian field around the fractal generator in the case $r_s = 0$ [16]. Each stripe represents a drop in the concentration by a factor of 2. The concentration C=1 at a distance of the fractal object equal to the width of the central pore.

the surface and second the flux is not homogeneously distributed on it. However, it is a good approximation when one calculates the exponents that describe the behavior of the system as it is shown in Section 3 and 4.

3 Scaling argument

3.1 Scaling form of the surface contribution to the total resistance

In this section we present the scaling argument which permits to obtain Eq. (3). This argument is based on a coarse-graining procedure which maps the problem with boundary condition (2b) on the above situation with $r_s = 0$.

For a real system with finite ρ and r_s the first problem is to compare the surface resistance $R_{surf.}$ with the resistance to access the surface $R_{acc.}$. For very large values of r_s, the resistance in the bulk (determined by the value of the resistivity ρ) becomes negligible in comparison to the resistance of the surface. In this case, the entire surface which can be considered as exposed to the same concentration. The total resistance, which is determined only by the surface contribution, is then proportional to r_s and inversely proportional to the total area of the fractal surface:

$$R_{surf.} = \frac{r_s}{b(L/L_0)^D L_0} \qquad (4)$$

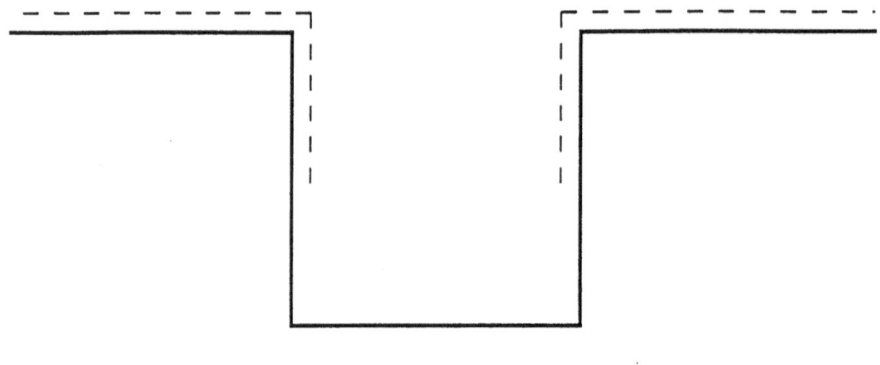

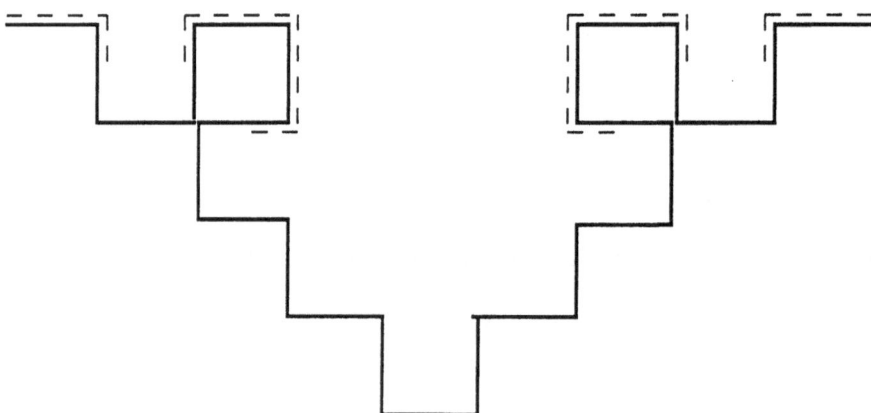

Fig. 3   Iterative process for the generation of the active surface. The solid line is the fractal object, the dashed line is its associated active surface. At each recursion level each segment is substituted by a version of the generator (top figure) rescaled by a factor $(1/3)^n$, where n is the recursion level. This process assures that independently of the surface irregularities the total length of the active surface is the linear size of the object.

For a two-dimensional system, independently of the system size, the access resistance is of the order:

$$R_{acc.} \sim \rho/b \tag{5}$$

For given values of ρ and r_s and when $r_s/L_0 > \rho/b$, namely the resistance of the smallest feature of the fractal is larger than the access resistance, the geometrical features can be separated depending on their size: Small parts have a surface resistance larger than the resistance to access them, whereas large features have a negligible surface resistance. Consequently, there exists a crossover length, $L_c(r_s/\rho)$, which separates two different geometrical behaviors: a) In pores smaller

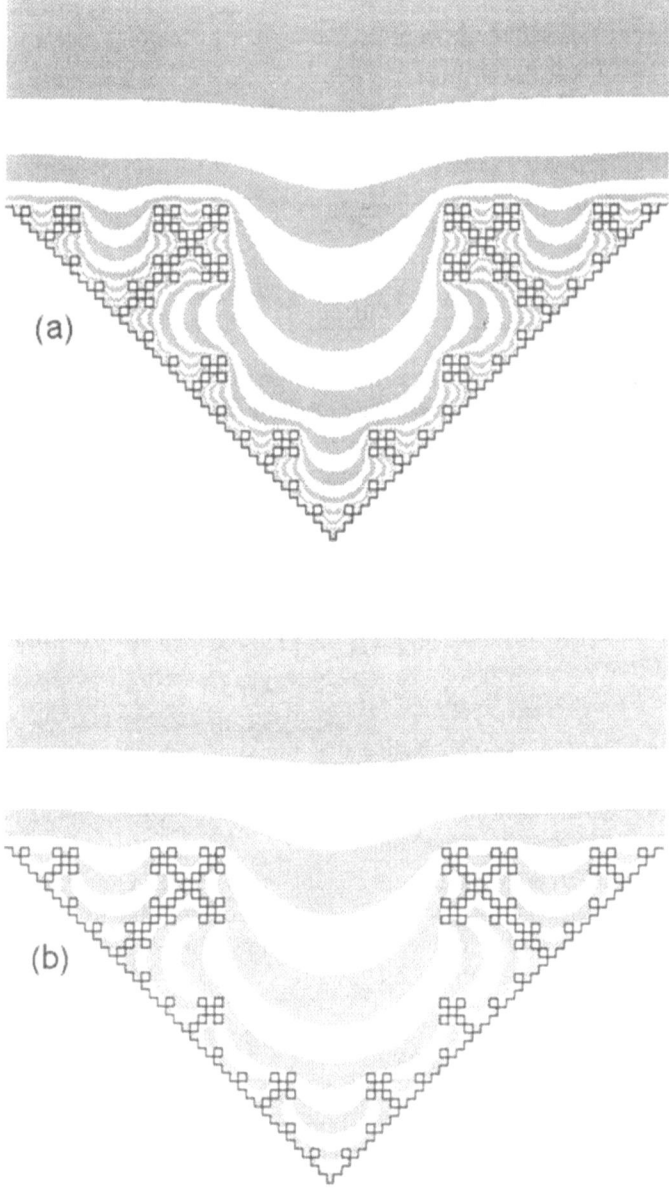

Fig. 4 The Laplacian potential around a four-iteration object: a) $r_s = 0$ and b) for $r_s/(\rho L_0) = 3$. The smallest pores are covered only by one stripe. In these pores the surface resistance is larger than the resistance to access to the pore walls. They are approximated as exposed to a constant concentration. In the largest pores the access resistance is much larger than the wall resistance and the concentration map practically recovers the behavior of $r_s = 0$ [see (a)], the only difference is a local effect near the walls.

than L_c, the resistance is dominated by the surface and the pore walls can be approximated as exposed to the same concentration. b) In pores larger than L_c the system recovers the behavior of $r_s = 0$ because the surface impedance is smaller than the access resistance in the bulk (see figure 4). Then the crossover length is determined by the equality between the surface resistance [Eq. (4)] and the access resistance [Eq. (5)]:

$$\frac{r_s}{b(L_c/L_0)^D L_0} \sim \frac{\rho}{b} \tag{6a}$$

or,

$$L_c \sim L_0 \left(\frac{r_s}{\rho L_0}\right)^{1/D} \tag{6b}$$

This means that if one considers a coarse graining of the initial membrane to the size L_c (see figure 5), any larger feature will behave with a surface resistance small as compared to the access resistance ρ/b. Once the object is coarse grained to a new one where the size of the elementary unit is L_c, the resistance of the generated macrosites is small as compared to the access resistance and then the Laplacian field recovers the behavior of $r_s = 0$. This implies the existence of an active surface of the linear size of the object L (Section 2). The difference now is that the bulk resistance R_b (Eq. 1) is in series with a contribution of the surface. The surface contribution to the resistance is then equivalent to (L/L_c) elements of order (ρ/b) or

$$R_{sur.} \sim (\rho/b)(L_c/L) \tag{7}$$

which is equivalent to Eq. (3). Notice that the coarse graining does not affect the geometry of the cell (the height-length relation), so the bulk resistance [R_b in Eq. (1)], being independent of the surface irregularities, will be the same in both the original and the coarse-grained object.

We would like to point out that the scaling argument developed in this section is general, it is based on the existence of an active zone of dimension one, but this does not have to be exactly located as it was shown in Section 2. However, the iterative procedure that we proposed allows to visualize this set in a very simple manner.

3.2 Active surface and adaptability

The scaling form dependence of the active surface on the resistance, r_s, can also be estimated using the preceeding argument. For a given value of r_s/ρ, the surface which corresponds to each macrosite is r_s/ρ itself and the number of these macrosites is of the order L/L_c. The total surface of the electrode is $S_T = L_0(L/L_0)^D$ so that the fraction of active surface S_A/S_T is:

$$S_A/S_T \sim (L/L_0)^{1-D}(r_s/\rho L_0)^{(D-1)/D} \tag{8}$$

An important property arises from the physical picture developed in this section: This is how the fractal surface adapts to environment conditions. One can easily

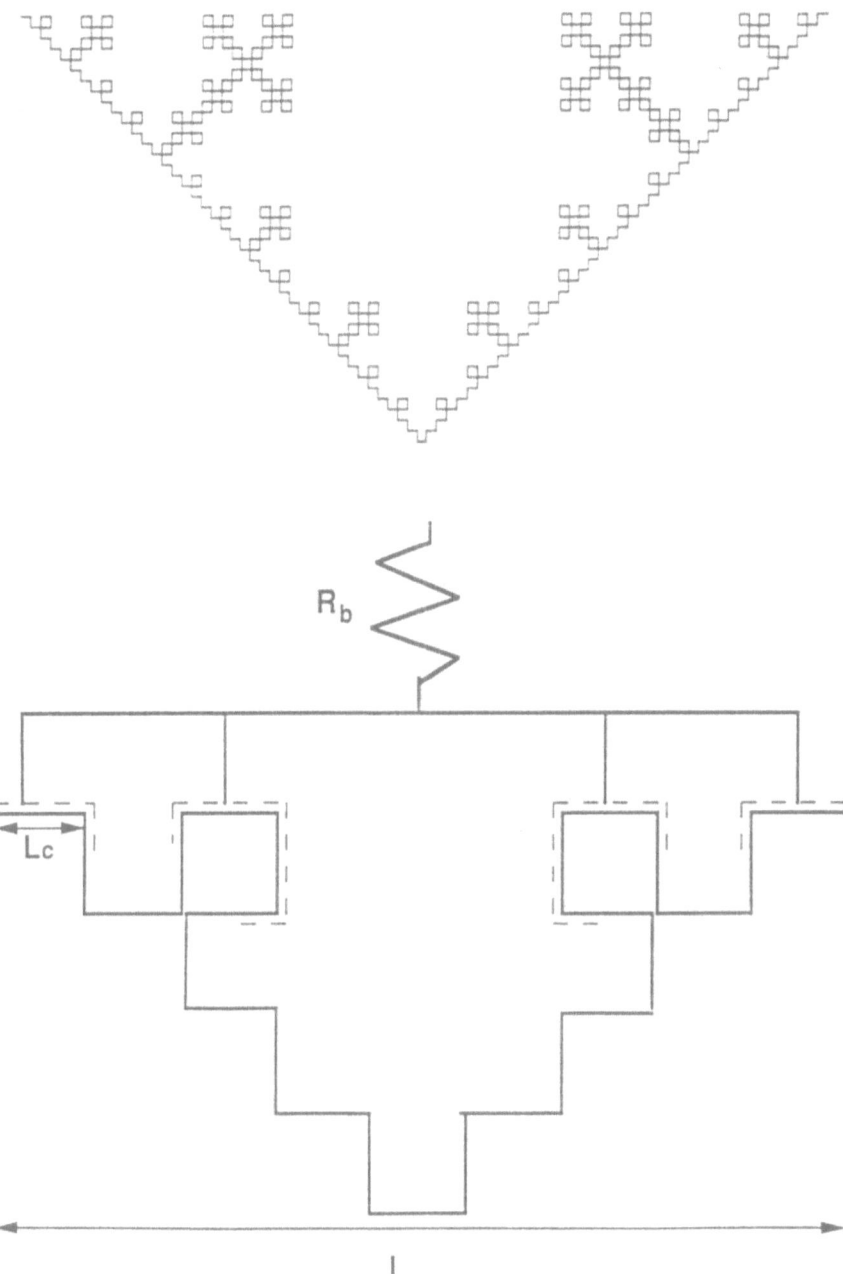

Fig. 5 A three-iteration object (top) is coarse-grained using the characteristic length shown in the figure. In the coarse-grained object (bottom) the active surface recovers the behavior of $r_s = 0$, namely its total length is L.

see that this is because the fractal symmetry provides new active surface when r_s increases and this part compensates for the larger value of r_s. Then a fractal membrane has the capability of providing new active surface whenever it has to absorb molecules having a lower permeability constant, avoiding in part a reduction in the transfer rate. This property can be crucial in the life of many natural systems, where a rough membrane can provide a solution to keep a suitable supply rate for nutrients being in a broad range of permeability values.

4   Numerical Results

In this section numerical results are compared to the theoretical predictions of the previous sections. To verify that the response of the studied object behaves according to Eq. (3), we used a relaxation method to numerically compute the response of a five-iteration object of the type shown in figure 1b. In this calculation one obtains the total resistance, R_T which is the sum of the bulk and membrane resistances. The resistance of the bulk R_b is the resistance $R_T(r_s = 0)$. The exponent η is then obtained from the slope of the log-log plot of $[(R_T - R_T(r_s = 0)]$ against r_s. The value $\eta = 0.64$ was obtained in good agreement with the value $\eta = 1/D = 0.68$ as shown in figure 6 [15]; the same result was obtained for several other fractal objects [14].

The idea of the theoretical active surface presented in Section 2 for the case $r_s = 0$ can be tested by comparing its theoretical localization with actual numerical observations. The first thing one can ask is what is the fraction of the total flux that indeed reaches it. When $r_s = 0$ the S_A/S_T ratio is equal to $(L/L_0)^{1-D}$. In the case of a five-iteration object of the type shown in figures 1–3 $S_A/S_T = 7.8\%$. From the solution of Laplace equation the flux that arrives in each surface site (j_o) is computed. For comparison purposes one can look at the 7.8% of the surface associated with the largest j_o values. First, one finds that 86% of the flux arrives in this fraction of the surface. Second, the localization of these sites, as shown by the black points in figure 7, is close to that predicted from the theoretical active surface as suggested in figure 3 [15]. Notice then that the simple procedure suggested in Section 2 gives a good approximation of the localization of the most active zones of the fractal membrane.

We present now the visualization of the working regions of the membrane for different values of the surface resistance (r_s). To build the active surface we take the sites associated to the largest j_o values up to the accumulation of a given percentage of the total flux. The case of 86% of the flux for $r_s/\rho L_0 = 5$ is shown in figure 8a. One can distinguish two classes of pores, those that behave homogeneously (a continuous line of active zone extends to the bottom of the pores) and those in which the behavior characteristic of $r_s = 0$ is recovered (the active zone penetrates one half of the pore width). It is on this fact that is based the idea of the coarse-graining procedure: The existence of a crossover length upon which one can coarse-grain the object (figure 8b) and recover the behavior of $r_s = 0$.

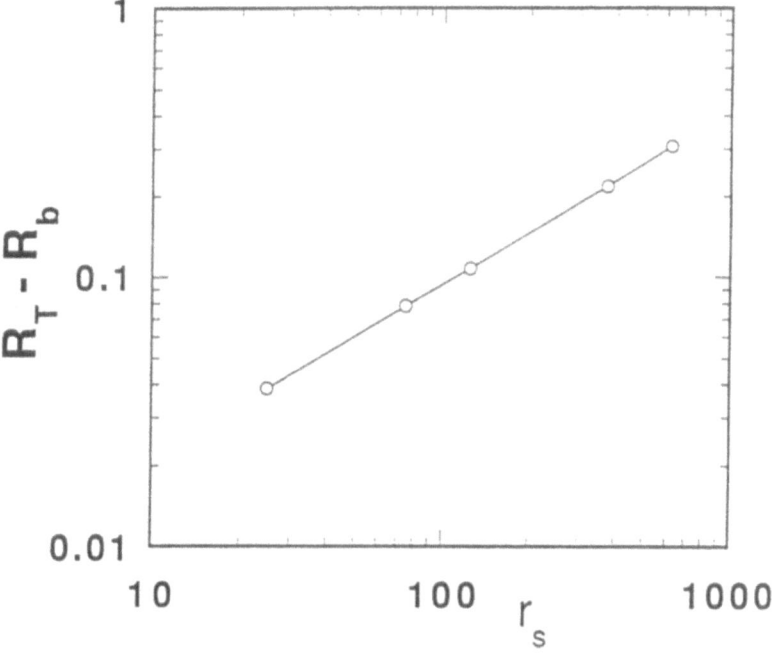

Fig. 6 Dependence of $(R_T - R_b)$, in computer units, on r_s for a five-iteration fractal object of the type shown in Fig. (1b). $\eta = 0.64$ is obtained from the slope in good agreement with $\eta = 1/D = 0.68$.

Fig. 7 Numerical estimation of the active surface ($r_s = 0$) for a five-iteration object (the active sites are in black). The linear size of the object is $L = 729$ lattice units. The active surface are the $L = 729$ boundary sites associated with the largest values of flux per site; 86% of the total flux really arrives in this zone. Notice that the part of the surface considered as active is approximately one half of the pore width at each pore wall.

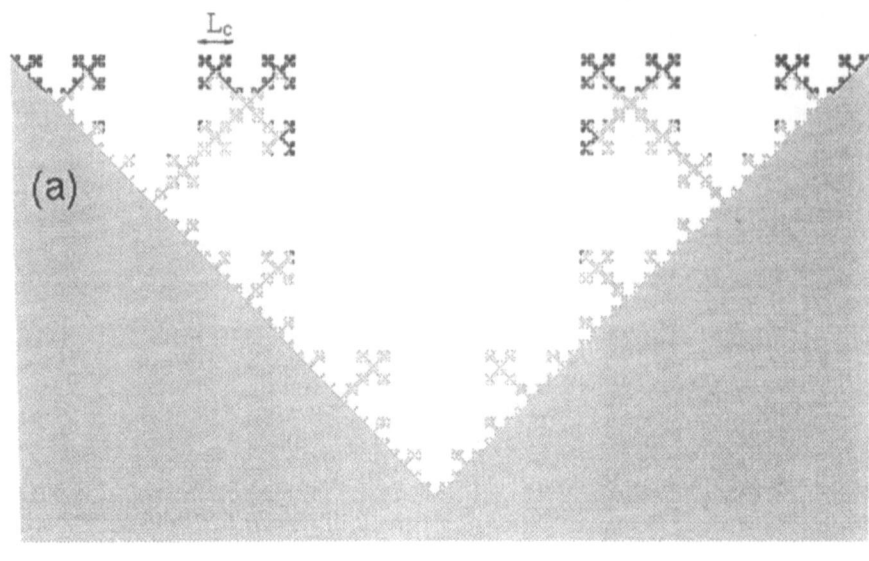

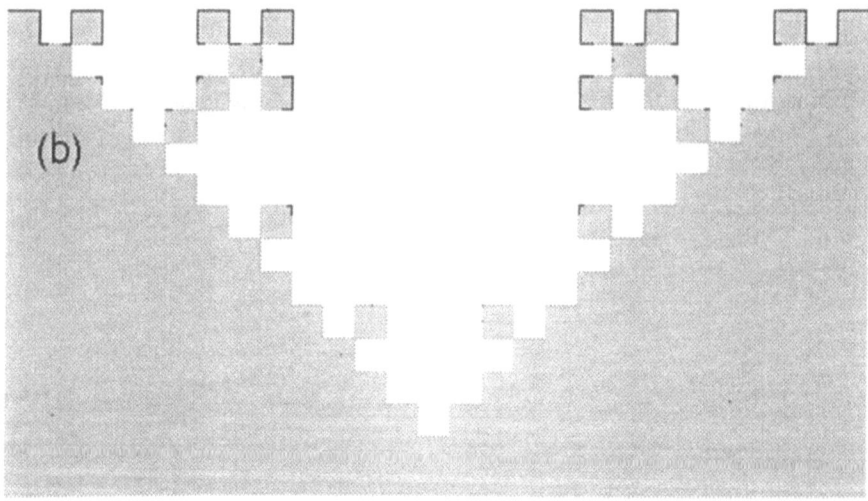

Fig. 8   Numerical estimations of the active surface for a five-iteration object when $r_s/(\rho L_0) = 5$. a) In the original object and b) in the coarse-grained object. Notice that the localization of the active surface in the coarse-grained object is close to its theoretical localization.

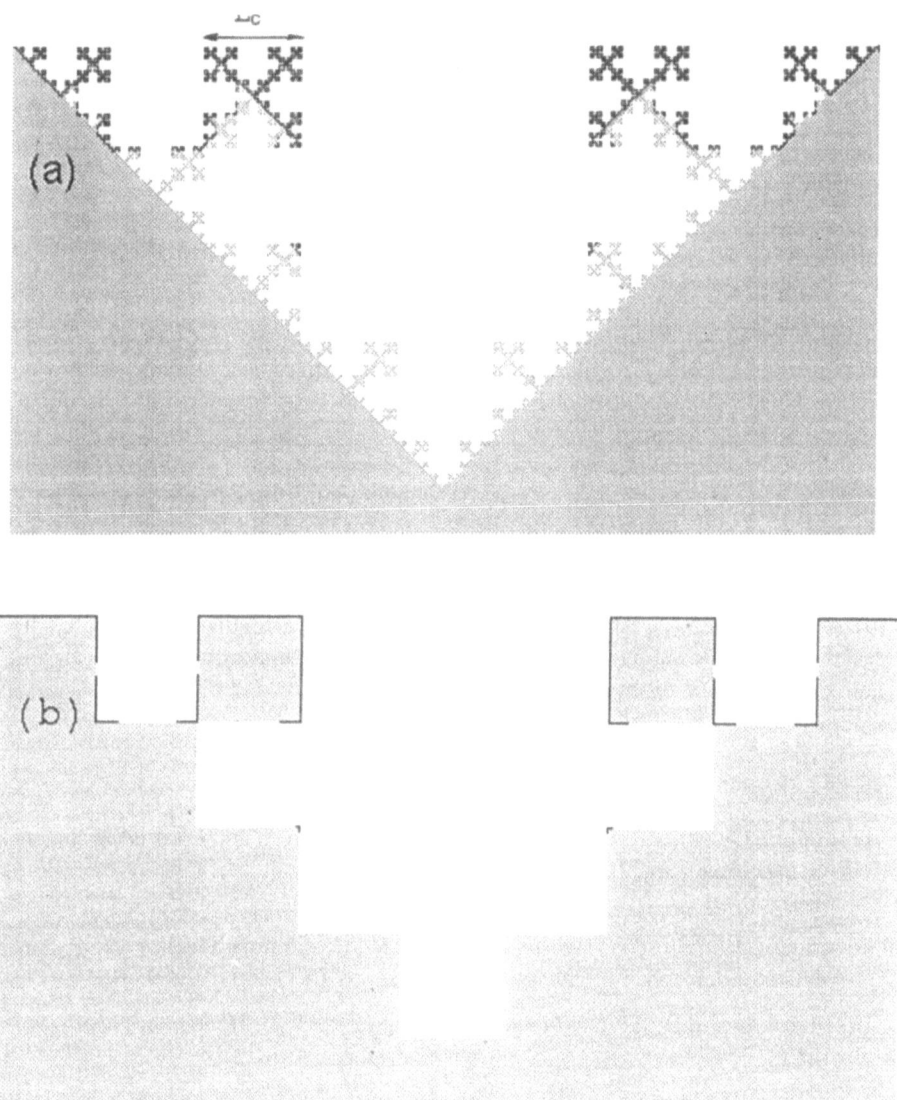

Fig. 9 Numerical estimations of the active surface for a five-iteration object when $r_s/(\rho L_0 = 25)$. a) In the original object and b) in the coarse-grained object. Notice that there is a shift in the value of L_c by approximately a factor of three with respect to that of figure 8, in agreement with Eq. (6).

The case of $(r_s/\rho L_0) = 25$ is shown in figure 9. The picture shows again where arrives 86% of the total flux associated with the largest j_o values. The same type of behavior as in the case $(r_s/\rho L_0) = 5$ is observed, but with an increase of the crossover length. Comparing figure 8 and figure 9, one can see that the visual crossover length has been shifted approximately by the factor $(L_1/L_2)^{1/D} = (25/5)^{1/D} = 3$ as predicted by Eq. (6).

5 Conclusions

The concept of active surface can be a powerful tool in the study of the macroscopic behavior of an irregular membrane. It gives first a very simple picture of the working regions of a fractal membrane in the case $r_s = 0$. Then a simple coarse-graining procedure permits to find the active surface when $r_s \neq 0$. This procedure provides a simple visualization of the process and a simple explanation of the transfer equation (3). We have also shown that a fractal membrane is self-adapted to variable working conditions. We think that a fractal membrane may be the optimal solution to keep a suitable rate of transfer for absorbing simultaneously molecules having different permeability values. This property can be crucial in the life of many natural systems, where a surface showing irregularities at all length scales can provide sufficient surface for the transfer at an adequate rate of all the nutrients necessary for the metabolism.

References

[1] Hall F., Oldeman R.A.A., Tomlinson P., Tropical Trees and Forests (Springer-Verlag, New York, 1978).

[2] a) Weibel E.R, The Pathway for Oxygen (Harvard University Press, Cambridge Massachusetts, 1984). b) Weibel E.R., Respiratory Physiology, an Analytical Approach, ed: Chang H.K. and Paiva M. (Dekker, Basel, 1989), p.1.

[3] Mandelbrot B.B., Fractal Geometry of Nature, 3rd ed. (Freeman, New York, 1983).

[4] Sapoval B., Fractals (Aditech, Paris, 1990).

[5] a)The Fractal Approach to Heterogeneous Chemistry: Surfaces, Colloids, Polymers, ed: Avnir D. (Wiley, Chichester, 1989).

[6] Feder J., Fractals (Plenum, New York, 1988).

[7] Scheider W., J. Phys. Chem. **79** (1975) 127.

[8] Armstrong D. and Burnham R.A., J. Electroanal. Chem. **72** (1976) 257.

[9] Rammelt U. and Reinhard G., Electrochimica Acta **35** (1990) 1045.

[10] Sapoval B., Acta Stereologica, **6/III** (1987) 785.

[11] Sapoval B., Fractal electrodes, fractal membranes and fractal catalysts in Fractals and disordered systems, ed: Bunde A. and Havlin S. (Springer-Verlag, Heidelberg, 1991) p. 207.

[12] a) Sapoval B. and Gutfraind R., Workshop on Surface Disordering, Growth, Roughening and Phase Transitions, Les Houches April 1992. ed: Julien R., Kertesz J., Meakin P. and Wolf D. (Nova Science Publisher, 1993, page 285). b) Sapoval B., Spring meeting of the Electrochemical Society St. Louis, Missouri 92-1 (1992) 514.

[13] Sapoval B., Gutfraind R., Meakin P., Keddam M. and Takenouti H. (Phys. Rev. E, accepted)

[14] Meakin P. and Sapoval B., Phys. Rev. A **43** (1991) 2993.

[15] R. Gutfraind and B. Sapoval (Journal de Physique I, to be published)

[16] Evertsz C., Mandelbrot B.B. and Normant F., J. Phys. A: Math Gen. (1992) 1781.

[17] Jones P. and Wolff T., Acta Math., **161** (1988) 131.

Structure Formation in Excitable Media

Martin Lüneburg
Bielefeld University,
Fakultät für Mathematik
Universitätsstraße
D-33615 Bielefeld

Excitable media are widespread in nature. They play important roles in physics and chemistry as well as in biology and medicine. Applications comprise phenomena as diverse as the pigmentation patterns of vertebrate skins or of shells of molluscs, cardiac arrythmia, formation of galaxies, energy metabolism, aggregation of slime mold amoebae, spatio-temporal EEG-patterns and circadian rhythms in physiology and biological populations (cf. [6], [10], [11]). The most prominent feature of an excitable medium, of course, is its ability to receive and distribute excitation. The main phenomenon observed in such media, therefore, is that of waves formed by the propagation of excitation gradients. As the examples above might suggest, the supporting medium need not be a continuum but can also have discrete structure. The former, as a rule, is modeled by means of partial differential equations, the latter more often by cellular automata. Obviously, many reaction-diffusion and aggregation processes are intimately connected with the phenomena encountered in excitable media.

Starting point for our own investigations in Bielefeld on excitable media was the self-similar curve depicted in figure 1. It shows the varying degree of overall activity of certain catalytic metal surfaces which has been measured by N. Jaeger et al. at Bremen university (cf. [7] and the literature quoted there). In their experiments, a flow of carbon monoxide was sent from below through a heap of granular pieces of catalysts. Theses catalysts were supposed to oxidize continuously at a constant rate the carbon monoxide to carbon dioxide. Instead, measurements showed perplexing patterns of spontaneous activity reduction. Now, compare figure 1 with figure 2 which has been gained from a completely different situation. The latter shows for $0 \leq n \leq 120$ the number $K(n)$ of binomial coefficients $\binom{n}{k}$ which are odd. If you visualize these coefficients in the Pascal triangle modulo 2, where black stands for odd and white for even (fig. 3), $K(n)$ is simply the number of black spots in the nth column. As is well known, the Pascal triangle modulo 2 for $n \longrightarrow \infty$ is equivalent to the Sierpinski gasket, so that the self-similarity of figure 2 is no surprise. Furthermore, a fact which will be of importance later on, this triangle can be interpreted as the temporal record of a 1-dimensional cellular automaton, Pascal's parity automaton: starting with exactly one nonzero entry in an infinite vertical line of cells (the leftmost column) it develops by the rule that the state of a cell at time $t + 1$ is defined to be the sum of its own state and that of its lower neighbour at time t.

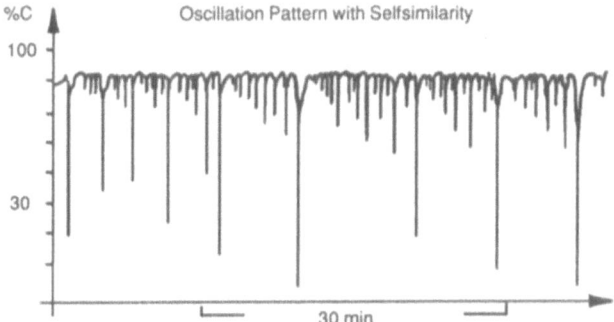

Fig. 1

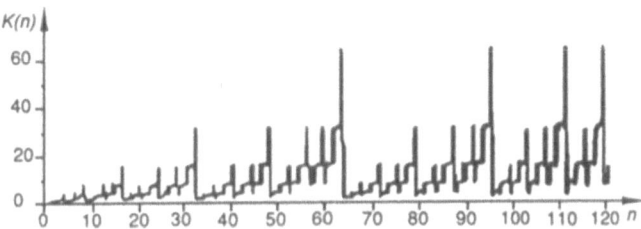

Fig. 2

Is there any reason for the resemblance of the activity record and the Pascal histogram? There is, if you are willing to believe in the following first approximation model [1]: Below a certain temperature threshold each single catalyst oscillates periodically between its active and its passive state, while above this threshold it exhibits bistability. Bistability means that it remains active when active and passive when passive. Moreover, we may assume that activity of a catalyst raises the temperature of the catalyst above it, so the upper one will remain or become bistable, preserving its state of activity or passivity, respectively. In contrast, passivity may cool the catalyst just above — so in this case the upper one remains or becomes oscillating and immediately starts changing its state of activity. If you look more closely to these rules, you see that finally you can forget about bistability and oscillations since only activity and passivity are really relevant for deciding how the process will continue. Namely, imagine two catalysts lying above each other. Then the upper one will be active after the next time step if and only if it and its lower neighbour are both passive or both active. If you identify active with 0 and passive with 1 this coincides with addition modulo 2. Hence, the catalyzing rules correspond precisely to the rules of Pascal's parity automaton. Thus, the model would explain the experimental observations and provide a rather simple description of a seemingly highly complex phenomenon.

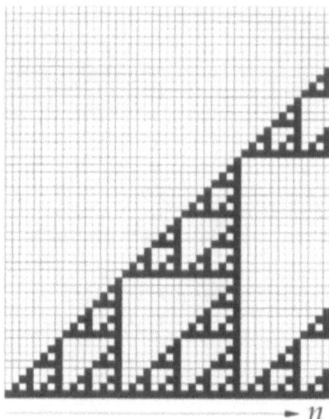

Fig. 3

Elaborating on Pascal's parity automaton in order to develop a chemically more realistic 2-dimensional variant, Martin Gerhardt and Heike Schuster from Bremen and Bielefeld university designed what they later came to call the hodge-podge machine (cf. [2], [4]): This is a 2-dimensional cellular automaton consisting of cells all of which would perform simple sawtooth oscillations if left alone. By diffusion coupling of neighbouring cells the speed at which they run through their cycles is either enhanced or reduced, depending on the phase of their neighbours. These simple rules produce moving wavefronts which organize themselves into an intricate scrollwork of spiral and concentric patterns (fig. 4) resembling to an amazing degree those known from the Belousov-Zhabotinsky reaction. Moreover, recently a group of chemists in Berlin has developed new experimental methods which enabled them to observe, in fact, such patterns on a real catalytic metal surface [8].

Of course, coupled sawtooth oscillations as in the hodge-podge machine cannot explain but in a roughly qualitative way what happens on a catalyst's surface. Therefore, A. Dress in close cooperation with the aforementioned Bremen chemists group developed a more realistic model, called «ideal storage», for the behaviour of a single catalyst [3]. The discrete time steps it employs are only for the sake of convenience and simplicity. The following are the rules for a single catalytic crystallite:

$$Q_{n+1} := Q_n + S - R_n \cdot Q_n$$
$$R_{n+1} := F(R_n \cdot Q_n)$$

Thereby, Q is the quantity of reactive substance, R is the reaction rate, S is the supply of reactive substance sent «continuously» to the catalyst and F is

Fig. 4

a monotonously increasing function defined on the non-negative real axis with values in the intervall $(0, 1]$.

An analysis of this discrete dynamical system and its unique equilibrium $(R_0, Q_0) = (F(S), S/F(S))$ showed that the system can indeed exhibit oscillatory behaviour. More precisely, for standard choices of F and with increasing values of the control parameter S, the system undergoes two discrete Hopf bifurcations, first from a low level stable equilibrium to an intermediate periodic attractor K_S and then back to another stable equilibrium, now of high reactivity. Since this coincided well enough with what was known from experiments, the question was: What behaviour do *coupled* catalysts exhibit, do they, for instance, produce spiral waves?

More precisely, we asked for the evolution of a certain cellular automaton. Its cells were arranged like the junctions of a regular rectangular grid, the neighbours of each cell being its 3, 5 or 8 surrounding companions depending on whether the cell lied on a corner, on the boundary or in the inner part of the rectangle. After random initialization, the (Q, R)-values of the cells were simultaneously updated by applying the above mentioned equations and thereafter averaging the reaction rates of neighbouring cells as follows:

$$r_{n+1} := w \cdot r_n + (1 - w) \cdot a_n$$

Thereby, r_n and r_{n+1} are the reaction rates of a single cell at time n or $n + 1$, respectively, a_n is the arithmetic mean of the reaction rates of its neighbouring cells at time n and w is a weight with $0 \leq w \leq 1$. Hence, for $w = 1$ no coupling is present and every cell evolves without being influenced by others, and for $w = 1/9 = 0.\overline{1}$ all reaction rates of a cell's neighbourhood play the same role for the dynamics of the automaton. A coupling *via* the reaction rates is reasonable, since the reaction rate of a single catalyst depends monotonously on the amount of heat produced during the reaction process (cf. [3]) and heat gradients move and dissolve by conduction and diffusion. An interpretation of the coupling parameter w will be given later.

Before presenting our results (for details and references see [9]) we have to introduce some notions: Let $I(F, S)$ be an «ideal storage» built on the monotonous function F and with supply S, and $A(F, S, w)$ a 2-dimensional cellular automaton the cells of which are such ideal storages coupled as described above with coupling parameter w. Typically, $A(F, S, w)$ consists of at least 100 times 100 cells. Furthermore, let

$$\Omega(F) := \{S \mid 0 < S \in \mathbb{R}, \ I(F, S) \text{ is a global oscillator}\},$$
$$\Sigma(F, w) := \{S \in \mathbb{R} \mid A(F, S, w) \text{ permits spirals stable in time}\}$$

and

$$\sigma(F, w) := \{S \in \mathbb{R} \mid A(F, S, w) \text{ permits spontaneous formation of spirals}\}.$$

«Stable in time» means that such a spiral does not vanish as long as the parameters F and S are not changed. The fact that after random initialization the automaton produces spiral waves is called «spontaneous formation of spirals». If $w = 0.\overline{1}$, which will be the case for most of the time, we simply omit it, i.e. $A(F, S) = A(F, S, 0.\overline{1})$, etc.

The occurrence of spontaneous formation of spirals and the adaptability of spiral waves to variations of the supply S are the two principal phenomena we have studied. For this purpose the following experiments were carried out:

E1) After initialization at random, $A(F, S, w)$ was run with $S \in \Omega(F)$

E2) After spirals stable in time had evolved during E1 with $w = 0.\overline{1}$, the supply was changed.

E1 proved that spontaneous formation of spirals is possible with suitable choices of F, w and S. The combination of E1 and E2 showed a strong discrepancy between the attractivity and the stability of spirals: $\sigma(F)$ is a relatively small subintervall of $\Sigma(F) \cap \Omega(F)$. It consists of those S in whose neighbourhood the size of the cycle K_S changes drastically. In contrast, $\Sigma(F)$ is an intervall of similar size as $\Omega(F)$ but not a subinterval of $\Omega(F)$; even for values S not in $\Omega(F)$ spirals can remain stable. Hence, coupling of cells which would tend to the same equilibrium, if left alone, can give rise to highly non-trivial forms of symbiosis, manifesting itself by endlessly turning spiral waves.

Starting with spirals stable in time, variation of the supply in general changes the «spectrum» of the spirals which adapt to the new conditions, this spectrum essentially being the states every cell belonging to the spiral runs through cyclically.

Increase or decrease of S in general induces an increase or decrease, respectively, of the spirals' curvature. Figures 5–7 show snapshots of a typical example (the two variables Q and R of each cell are depicted on the left and the right, respectively): $A(F, S, 0.\overline{3})$ was started with F defined as a function of x by

$$F(x) = \frac{\left(\frac{x}{D}\right)^2 + 1}{\left(\frac{x}{D}\right)^2 + 1 + D}$$

and $D = 86$, and with $S = 91 \in \sigma(F)$. After spirals had fully developed (fig. 5), S was set first to 140 and then to 200, the process after every increase running long enough for complete adaptation of the spirals (fig. 6–7).

Variation of the coupling parameter w between $0.\overline{1}$ and 1 showed that even quite weak a coupling induces strong correlations between the phases of neighbouring cells. With weaker coupling only the size of the single spirals decreases while their total number increases. This to a certain degree can be interpreted as a scaling phenomenon: Consider the catalyst as an ensemble of coupled subcatalysts by subdividing it into parts of equal size. Then, of course, the coupling intensity must depend on the size of the boundaries these parts have in common. In a coarser subdivision these boundaries in comparison to the area of the subcatalysts are smaller than in a finer one. Since a weaker coupling of the automaton's cells, as stated above, corresponds to a smaller degree of heat exchange, such a coupling could be interpreted as a coarser subdivision of the catalyst in question, i.e. as a change of scale implying that the same number of cells represents a larger catalyst, though of similar character, than before. The scaling problem we have touched hereby is one of the major problems if not the main one which has to be solved for a quantitative application of cellular automata.

Spontaneous formation of spirals stable in time is a non-trivial example of self-organization. The temporal stability of these spirals is due to replication: a situation develops which is non-stationary itself but the characteristics of which are invariant in time as they replicate themselves in any time step.

Another important property of these spirals is their tendency to spread by growth over all the space not yet occupied by other spirals. It implies that after some time the whole automaton's surface is tiled by spirals. This final situation coincides with minimal temporal variance, though not constance, of the histogram representing the number of cells in a certain state. Hence, spiral waves make the automaton run as steadily as possible without resting. This could be most desirable for a catalyst but maybe fatal for the heart since the working rhythm of the latter, in contrast, must be fractal in case of health (cf. [5]).

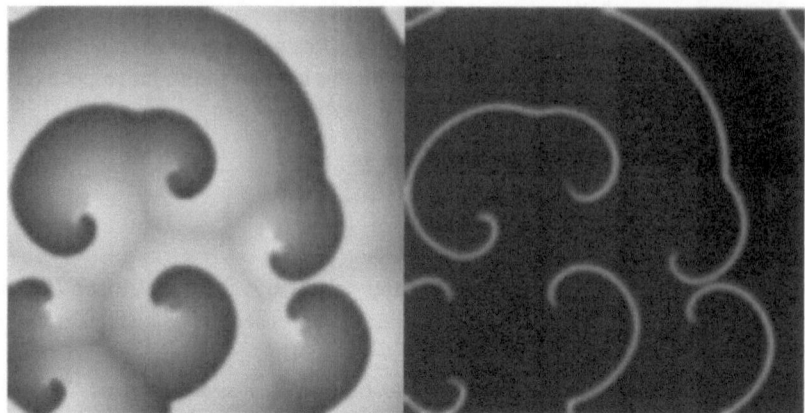

Fig. 5

Fig. 6

Fig. 7

References

[1] A. Dress: Cellular automata: a simplistic approach to complexity. To appear in: Neurobionics (eds. Bothe et al.), Elsevier Science Publ. 1993

[2] A. Dress, M. Gerhardt, H. Schuster: Cellular automata simulating the evolution of structure through the synchronization of oscillators. In: From Chemical to Biological Order (eds. L. Rensing and N.I. Jaeger), Springer Series in Synergetics **39** (1988), 134–145

[3] A. Dress, N.I. Jaeger, P.J. Plath: Zur Dynamik idealer Speicher. Ein einfaches mathematisches Modell. Theoret. Chim. Acta **61** (1982), 437–460

[4] M. Gerhardt, H. Schuster: A cellular automaton describing the formation of spatially ordered structures in chemical systems. Physica D **36** (1989), 209–221

[5] A.L. Goldberger: Fractal mechanisms in the electrophysiology of the heart. IEEE Engineering in Medicine and Biology, June 1992, 47–52

[6] A.V. Holden, M. Markus, H.G. Othmer (eds.): Nonlinear Wave Processes in Excitable Media, Proceedings of a NATO Advanced Research Workshop. Plenum Press, New York and London 1991

[7] N.I. Jaeger, K. Möller, P.J. Plath: On the classification of local disorder in globally regular spatial patterns. In: Temporal Order (eds. L. Rensing and N.I. Jaeger), Springer Series in Synergetics **29** (1985), 96–100

[8] S. Jakubith, H.H. Rotermund, W. Engel, A. von Oertzen, G. Ertl: Spatiotemporal concentration patterns in a surface reaction: propagating and standing waves, rotating spirals, and turbulances. Phys. Rev. Lett. **65** (1990), 3013–3016

[9] M. Lüneburg, in preparation

[10] A.T. Winfree: The Geometry of Biological Time. Springer-Verlag Berlin Heidelberg New York 1990

[11] D.A. Young, A local activator-inhibitor model of vertebrate skin patterns, Math. Biosciences **72** (1984), 51–58

Colony Morphology of the Fungus Aspergillus Oryzae

Shu Matsuura and Sasuke Miyazima[1]
School of High-Technology for Human Welfare
Tokai University
Nishino 317, Numazu
Shizuoka 410-03, Japan

1) Department of Engineering Physics, Chubu University, Kasugai, Aichi 487, Japan

Abstract. Thick and homogeneous colonies of fungus *Aspergillus oryzae* with connected growth fronts have been found to grow on solid agar media or on the media with high nutrient concentration, and the roughness of their growth front have been characterized by self-affine fractals. On the other hand, fairly ramified colonies with low growth rate have been seen on the media of low stiffness with very low glucose concentrations.

1   Introduction

Most filamentous fungi are found in the soil, where the mineral matters, bacterial colonies, waste products of living organisms, gaseous pores, water, etc., form random crumb structures. The branched structure of filamentous fungi is advantageous as a strategy to grow in such inhomogeneous substrates.

Mycelium is one giant cell consisting of the basal hypha, the aerial hypha, and other organs for reproduction such as conidia. Leading hypha can continue to extend even through the unfavorable field using nutrients transported from other part of the mycelium, and branch out when they reach the nutritious areas. Different physiological tasks are assigned to each part of the thallus through the active transport of materials within the thallus. Such adaptation to the inhomogeneous environments will contribute to the pattern formation of the fungal colony.

On the other hand, growth of thin filaments of fungi must be largely subject to the environmental conditions, such as the nutrient content, pH, water, temperature, and gaseous conditions. Also, several physical conditions such as the attachment to substrate, osmotic pressure on the cell wall, the degree of complexity of the medium, etc., might influence the regulation of colony formation.

In this study, we present the morphology of the 2-dimensional colonies of fungus *Aspergillus oryzae* cultivated at 24°C from line inoculum on the synthetic agar media with various glucose and agar contents. The strain forms thick colony on the solid agar medium. Dependence of the roughness of growth fronts of these colonies on the glucose concentrations is analyzed in terms of self-affine fractal.

The roughness of the growth front of 1-dimensionally growing colony is determined in the following way. Let us choose a point on the growth front and indicate as the position i. Then, the height h_i of the point is defined by the distance from the inoculation line to the front point i. Next, let us define a horizontal range of l parallel to the inoculation line. When there are N digitized front points in this range, the mean height $h(l)$ of the colony within l is written as,

$$h(l) = \sum h_i/N. \tag{1}$$

Now, the roughness of front within l is estimated by,

$$\sigma(h,l) = \sqrt{\left(\sum (h_i - h)^2/N\right)}. \tag{2}$$

Generally, σ increases with increasing h in the growth process. When the horizontal range l is broadened, σ includes longer wave length fluctuations of front.

As for the random growth processes, it is known that σ is scaled with l and h, in the form of so called self-affinity relations as[1],

$$\sigma(h,l) \sim l^\alpha \quad \text{for} \quad l \ll h, \tag{3}$$
$$\sigma(h,l) \sim h^\beta \quad \text{for} \quad l \gg h.$$

Here, α is within the range of $0 < \alpha < 1$. A typical random growth is represented by Eden model[2] generated from a line seed at initial. In this model, every surface point grows with an equivalent probability, and α and β values have been determined as $\alpha = 1/2$ and $\beta = 1/3$.

Several examples are known to satisfy the above self-affinity relations such as the fungal colonies[3], bacterial colonies[4], burning fronts[5], and rupture lines in paper sheets[6]. In this paper, we present the dependence of h, σ, and α on the glucose concentrations for the thick colonies cultivated on the solid agar media.

2 Experimental methods

The strain used is *Aspergillus oryzae* (supplied by the Institute for Fermentation Osaka), cultivated on the Czapek synthetic agar medium containing $NaNO_3$ (0.3wt%), K_2HPO_4 (0.1wt%), $MgSO_4 \cdot 7H_2O$ (0.05wt%), KCl (0.05wt%), $FeSO_4 \cdot 7H_2O$ (0.001wt%), glucose (varied from 0.01wt% to 5wt%), and Difco Bact-Agar (1.5wt%, 0.3wt%, and 0.15wt% for «solid», «soft», and «liquid-like» medium, respectively). Stiffness of the medium is controlled by the amount of agar. 25 ml of sterile medium is poured into each petri plate of 9 cm in diameter. Line inoculation is carried out by the method previously described in reference 3, and the strain is cultivated at 24°C.

Photographs of colonies are taken with a 35mm camera. Colonies are illuminated from the bottom side by a halogen lamp. The black-and-white prints of colony images are digitized using an image-scanner. The final resolution of images is approximately 600 pixels in length of the total inoculation line.

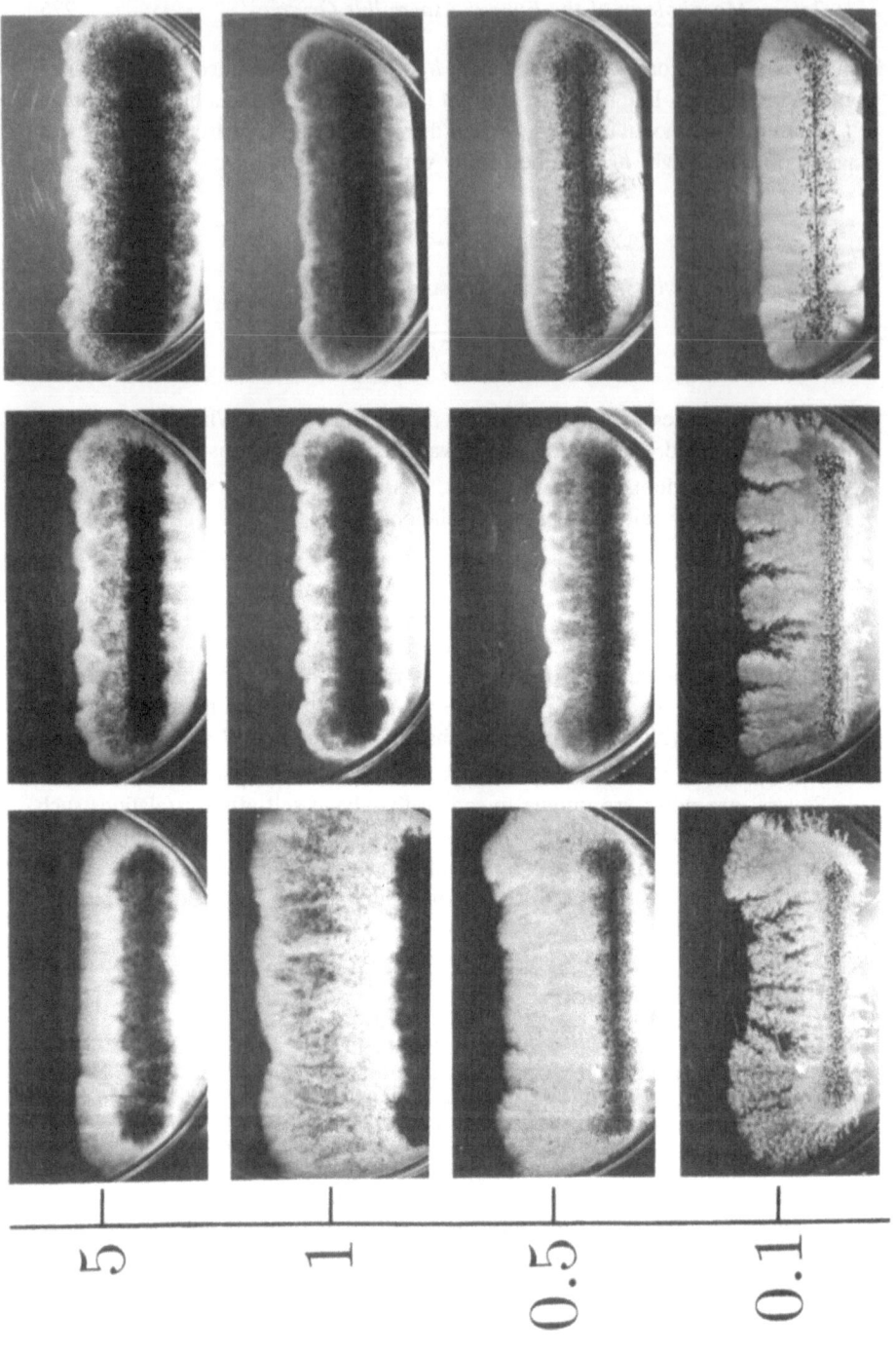

glucose concentration (wt%)

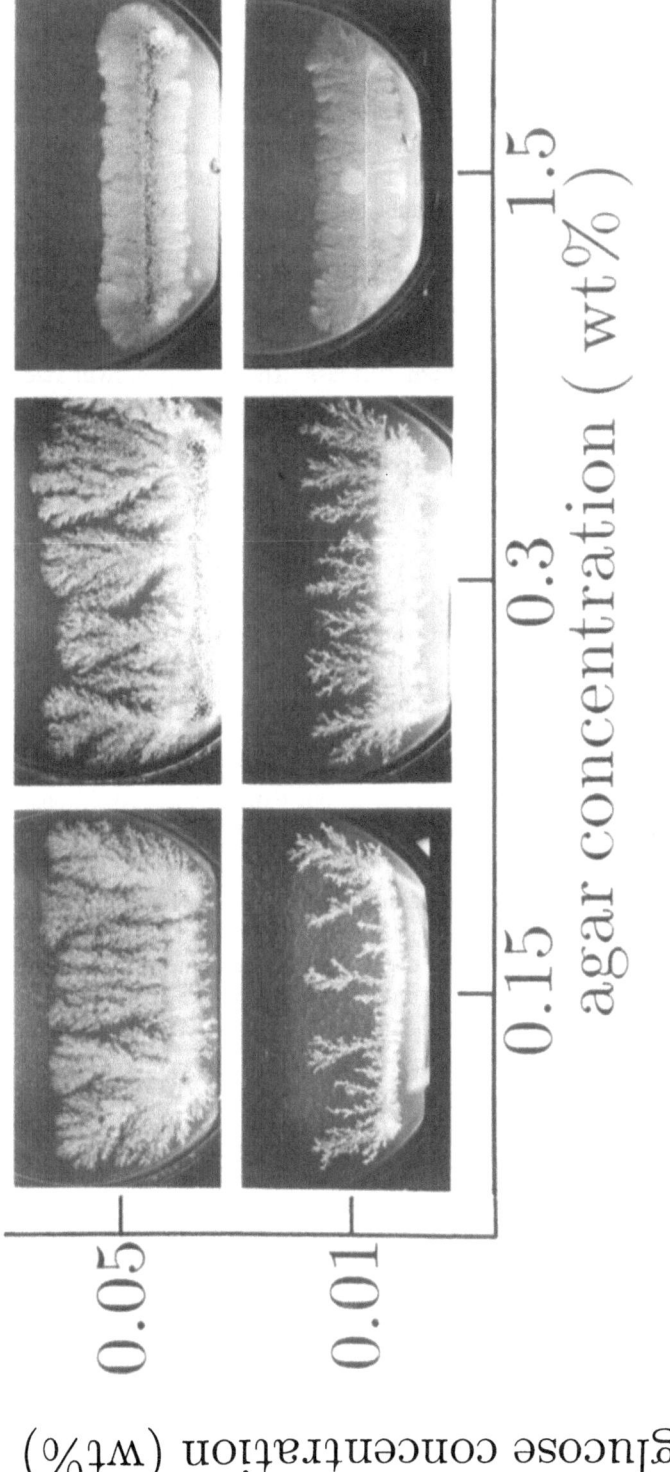

glucose concentration (wt%)

agar concentration (wt%)

Fig. 1 *Aspergillus oryzae* colonies cultivated on the media with various glucose and agar concentrations. The cultivation periods for colonies on liquid-like media (0.15wt% agar) are 10 days for 5wt% glucose, 20 days for 1wt% glu., 20 days for 0.5wt% glu., 20 days for 0.1wt% glu., 30 days for 0.05wt% glu., and 30 days for 0.01wt% glu., for colonies on soft media (0.3wt% agar) are 10 days for 5wt% glu., 10 days for 1wt% glu., 10 days for 0.5wt% glu., 15 days for 0.1wt% glu., 30 days for 0.05wt% glu., and 30 days for 0.01wt% glu., and for colonies on solid media (1.5wt% agar) are 8 days for 5wt% glu., 8 days for 1wt% glu., 8 days for 0.5wt% glu., 8 days for 0.1wt% glu., 8 days for 0.05wt% glu., and 10 days for 0.01wt% glu., respectively. The length of the inoculation line is approximately 55 mm for soft and solid media.

3   Results and discussion

3.1 Shapes of colonies

Photographs of *A. oryzae* colonies grown with various glucose and agar concentrations are shown in Fig.1. On the solid agar media, colonies become thick and compact. The band shaped colony is an assembly of sub-colonies as clearly seen in the photos of low glucose cases. The front shape of the whole colony is the connection of the front lines of the sub-colonies. These sub-colonies have the most advancing portion in the middle of their fronts, exhibiting smooth arch shapes.

At high glucose concentration ($\geq$ 1wt%), thick colonies cover the medium homogeneously, independent on the medium stiffness. The colony growth rate is, however, higher, and the front is rougher for the colonies on solid medium, indicating active growth at the front.

Stiffness of the agar medium causes remarkable morphological change of the colony. Roughening of front and ramification of whole colonies appear on the soft and liquid-like media (Fig.1). On the liquid-like media, localization of growth points occurs with decreasing glucose concentration. At very low glucose condition of 0.01wt%, only strong leading hyphae continue to extend, and most secondary hyphae seem to cease growing at some physiological age. The inactive hyphae are seen to form the chains of conidia at their apical portions.

Hypha secretes various enzymes to decompose and absorb nutrient materials. If the nutrients are distributed uniformly in the substrates, the diffusion of nutrient to the mycelium or the individual hypha must be one of the important factors which determine the mycelial pattern formation. In our experimental conditions, diffusion of glucose must be easier both in the medium of low agar and high glucose concentration.

However, at a fixed glucose concentration, low stiffness of the substrate seems to be unfavorable condition for mycelial growth. Selection of hyphal filaments proceeds, and the active regions are localized at the top of a few leading branches. At present, there is no concrete explanation for this phenomenon. The attachment of vegetative hypha to the substrate might be a significant factor for branch creation.

The box-counting fractal dimensions D_{box} of individual branch colonies formed on the liquid-like medium with 0.01wt% glucose take the values between $\sim$ 1.6 and $\sim$ 1.8 at 20 days since inoculation. Then, D_{box}'s are between $\sim$ 1.55 and $\sim$ 1.75 at 40 days since inoculation. In many cases, D_{box} decreases with time. This suggests that the growth process of branch colony is not perfectly self-similar, and then there is a high singularity of growth probability only at the apical portion.

Many fungi are known to produce peculiar root-like aggregation of hyphae, called rhizomorph, in the soil. In this kind of structure, the strand hyphae are thought to take part mainly in the transport of nutrients etc. Our branched colony might be an example of the prototype of such differentiated structures, and might be a positive activity of exploiting an unfavorable substrate.

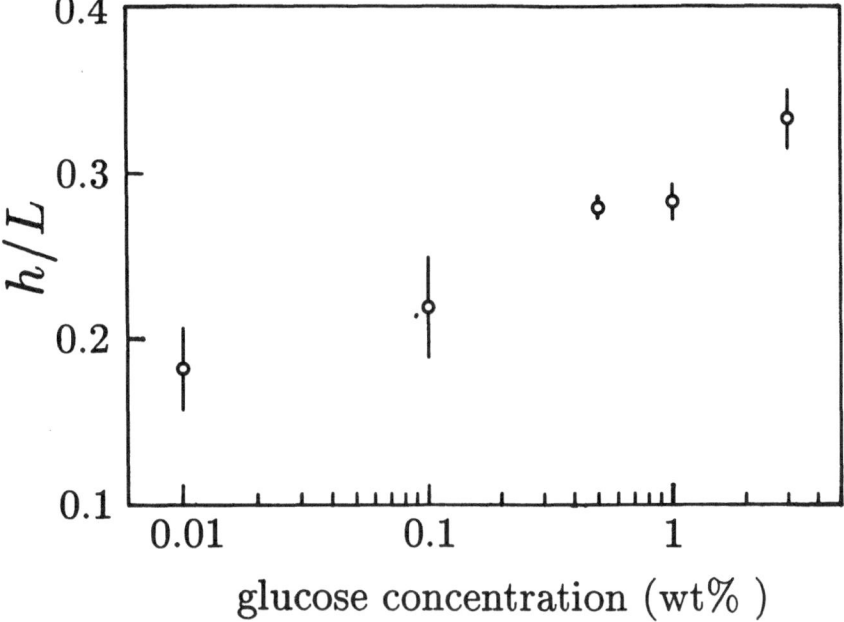

Fig. 2 Dependence of the mean height of colonies h/L at 8 days since inoculation on the glucose concentrations. Points indicate average values over $10 \sim 20$ samples. Bars indicate the standard deviations. The colony images for 0.1 and 0.01 wt% glucose were processed by a slightly modified method (see ref. 3).

3.2 Self-affine growth fronts formed on solid media

Let us now turn to the dependence of the roughness of growth front on the glucose concentration for colonies formed on the solid media. Figure 2 shows the mean height of colonies $h(L)$, calculated over the total length of the inoculation line L. As seen in the figure, h increases with the glucose concentration. The roughness $\sigma(L)$ calculated over the total length of the inoculation line L is plotted against the glucose concentration in Fig.3. $\sigma(L)$ seems to take minimum values at around 0.5wt% glucose and then increases with the glucose concentration.

By visual inspection of colony images in Fig.1, it is clearly seen that colony is sparse rough at low nutrient (0.01wt% glucose) and become homogeneous with flat front at 0.05wt% glucose. Then, as the glucose concentration is increased, colony produces dense hyphal filaments and forms an uneven growth front again. The roughened surface is the connection of the growth fronts of sub-colonies described in the previous section.

Formation processes of the sub-colony fronts seem not to be identical for various nutrient conditions. At low nutrient, competition of colonies will be controlled by the limiting nutrient. Low hyphal density interface between neighboring sub-colonies seen in the case of 0.01wt% glucose is thought to be exhibiting an

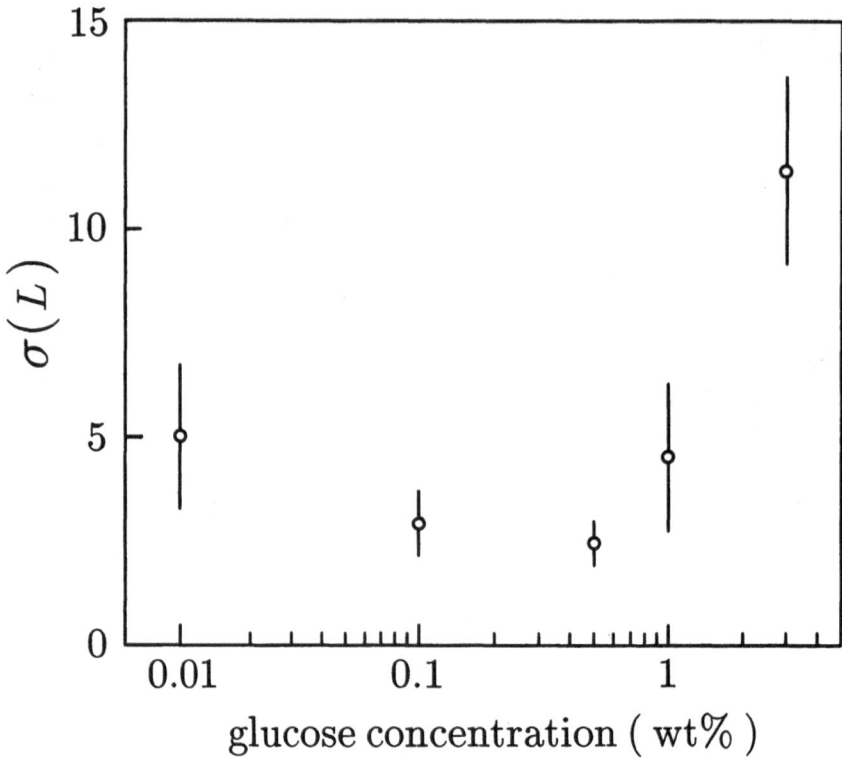

Fig. 3 Dependence of the roughness $\sigma(L)$ of growth fronts at 8 days since inoculation on the glucose
concentrations.

inhibitory interactions between colonies. The branches formed in the interface
tend to cease growing. At high nutrient, mycelium can get sufficient resources to
produce hyphae. Then, the colony growth may be controlled mainly by the inherent growth ability of the individual hypha. In this case, very strong hyphae will
branch out vigorously to the outer fresh medium space, so that the front roughness is thought to exhibit a positive exploitation. At 0.5wt% glucose, colony front
becomes remarkably flat and smooth. This is probably a balanced state between
2 limiting states mentioned above. The front hyphae have enough resources to
sustain an approximately equivalent growth rate to form a flat front.

Figure 4 shows the self-affinity exponents α for colonies on the solid media
plotted against the glucose concentration. α is larger than or near to 0.5, except
for the flat colonies formed on the medium with 0.5wt% glucose.

Although the biological meaning of α has not been established as yet, an
analogical conceiving with the case of the fractional Brownian motion (fBm)[1]
may be possible. The fractional Brownian function $V_H(t)$ is a function of time t,

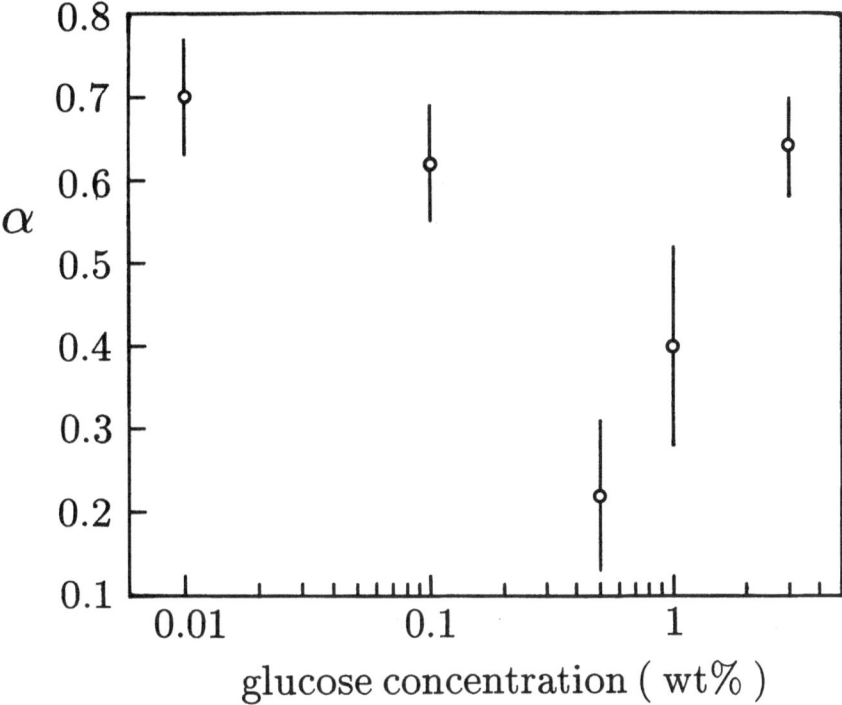

Fig. 4 Dependence of the self-affinity exponent α of growth fronts on the glucose concentrations. The data for 0.1 and 0.01wt% glucose are of colonies at 10 days, and others are those at 8 days since inoculation.

with its increments $V_H(t_2) - V_H(t_1)$ obeying a Gaussian distribution, as,

$$\langle\{V_H(t_2) - V_H(t_1)\}^2\rangle \sim (t_2 - t_1)^{2H}, \tag{4}$$

where the brackets denote ensemble averages and H is $0 < H < 1$. The case $H = 1/2$ is the Brownian motion and the increments are uncorrelated. Also, the increments are positively correlated for $H > 1/2$, and they are negatively correlated for $H < 1/2$. If we plot the increments against the time, then a self-affine fluctuation is obtained, and H corresponds to α.

An arch-shaped growth front of each sub-colony may corresponds to the case of positively correlated fBm, where the increase (or the decrease) of increments often persists relatively long time period. In this way, large α-value suggests larger sized arches formed at the front.

At 0.5wt% glucose, α is found to be lower than 1/2. The long wave length components which produce the arch-shaped contours are reduced, and the short wave length components which are due to the fluctuation in the local hyphal distribution remain in the front roughness. This corresponds to the negatively correlated increments, where the long wave length fluctuations are relatively reduced and increments change very frequently.

Finally, it seems worth notable that we find a fairly wide variation of σ and α for colonies under the equivalent conditions. Range of the variations of $\sigma(L)$ seems to be broadened as the glucose concentration is increased (Fig.3). This suggests that much dominant branch systems with high growth rates can be created from time to time, and the fluctuations of front shape are amplified. Further, α varies significantly from sample to sample, and this fact is found in other experimental works[5),6)]. Variation of α is found even for the flat colonies with 0.5wt% glucose, where $\sigma(L)$'s are relatively close to each other (Fig.3). Thus, the self-affinity relation (3) is considered to be very sensitive to the geometry of the individual colony, in comparison with the parameters such as the colony size.

4   Concluding remarks

Morphology of the fungal colony is thought to be a direct appearance of the flexible adaptation and, sometimes, positive strategies to exploit the environment. The mycelial controls governing the complex hyphal growth may gain an enough efficiency when the formation of whole colony becomes appropriate to the complex environment. Even on artificially prepared homogeneous media, the mycelial growth will bring about complex conditions to their environment through physiological activities. In this sense, it seems to us intriguing to consider the mycelial control from the colony shapes formed and the relationships between them.

References

[1] F. Family and T. Vicsek, *Dynamics of Fractal Surfaces*, (World Scientific, Singapore, 1991).

[2] M. Eden, *Proc. 4-th Berkeley Symp. on Math. Statistics and Probability*, vol.4, ed. F. Neyman, (University of California Press, Berkeley, 1961).

[3] S. Matsuura and S. Miyazima, *Fractals*, 1 (1993) 11.

[4] T. Vicsek, M. Cserzo, and V. K. Horvath, *Physica A*, 167 (1990) 315.

[5] J. Zhang, Y.-C. Zhang, P. Alstrom, and M. T. Levinsen, in press.

[6] J. Kertesz, V. K. Horvath and F. Weber, *Fractals*, 1 (1993) 67.

Estimation of the Correlation Dimension of All-Night Sleep EEG Data with a Personal Super Computer

Peter Achermann, Rolf Hartmann[1], Anton Gunzinger[1],
Walter Guggenbühl[1], and Alexander A. Borbély
Institute of Pharmacology
University of Zürich

1)   Electronics Laboratory, Swiss Federal Institute of Technology (ETH), Zürich, Switzerland

Abstract. The correlation dimension (CD) is a gauge of the complexity of the recorded signal. An algorithm proposed by Takens (1985) was implemented on the personal super computer MUSIC (multi processor system with intelligent communication) which allowed to analyze for the first time the data of entire sleep episodes. The all-night sleep EEG of a subject recorded under three different conditions (baseline (BL), sleep in a sitting condition after intake of placebo (PL) or triazolam (TR)) served as the data base. The CD of the EEG was modulated by the sleep cycle. Median values decreased from nonREM (non rapid eye movement) sleep stage 1 to 4. For REM (rapid eye movement) sleep the values were between those of stage 1 and 2. For sleep in a sitting position, the CD in REM sleep (PL and TR) and stage 2 (PL) was increased compared with BL. In stage 4 the CD was lower for TR than for BL and PL, and higher for PL than BL.

1   Introduction

In addition to the conventional scoring of sleep stages (Rechtschaffen and Kales 1968), methods based on quantitative EEG analysis are increasingly used (i.e. spectral analysis e.g. Borbély et al. 1981; period-amplitude analysis, e.g. Feinberg et al. 1978). Various mathematical models of processes underlying the regulation of sleep and circadian rhythms have been established since the early 1980s (for an overview see Borbély and Achermann 1992). EEG slow-wave activity (SWA; power of the sleep EEG in the 0.75 to 4.5 Hz range) in particular has been shown to be a useful measure for describing sleep homeostasis, since it is a function of the duration of prior waking and sleep. SWA exhibits a global declining trend as well as an ultradian variation reflecting the nonREM-REM sleep cycle (Borbély et al. 1981, Achermann and Borbély 1987, Dijk et al. 1990, Achermann et al. 1993).

    The study of the dynamics of non-linear systems opened new ways for characterizing signal properties of biological processes. Simple and deterministic nonlinear systems can display a very regular behavior as well as a highly disordered, turbulent (chaotic) one. Within chaos order is observed. Sensitive dependence on initial conditions is a further property of such systems. Various procedures have been developed to investigate the dynamics of a nonlinear system. Usually only a few variables can be measured and little is known about the complex mechanisms generating their temporal evolution (see e.g. Schuster 1988).

The correlation dimension (CD) is a gauge of the complexity of a recorded signal. In recent years the CD was estimated for various sleep states. The results of the early studies were summarized by Mayer-Kress and Layne (1987). The recent (Ehlers et al. 1991, Röschke and Aldenhoff 1991, 1992, 1993) studies provided statistics and were based on a sample of 8 to 12 subjects. However, only one, short, artifact-free segment of the EEG (5 to 164 s) was selected and analyzed for each sleep stage. The CD of the EEG was reported to decrease from nonREM sleep stage 1 to 4, while for REM sleep the values were between stage 1 and 2. The complexity parameter has not yet been evaluated for the data of an entire sleep episode and its changes as a function of time have not been analyzed. The present preliminary report is the first to provide such data. Computation of the CD is very time consuming. By using the personal super computer MUSIC it was possible to perform the analysis almost in real time.

2   Methods

2.1 Recording and data processing

The present analysis is based on the data of one individual who participated in a study in which 8 healthy male subjects were recorded for a baseline night (BL), a night in a sitting position after intake of placebo (PL) or triazolam (TR 0.25 mg), and a recovery night in bed (Aeschbach et al., in prep.). The recordings were preceded by an adaptation night. The data were subjected to a dimension analysis and a spectral analysis for the three conditions BL, PL, and TR. The C3-A2 EEG derivation was amplified by a Grass 7P511 amplifier (time constant: 0.9 s). The combined action of the amplifier's 50-Hz notch filter and an analog low-pass filter served to attenuate high frequency components (-3 dB at 27 Hz). After AD-conversion (sampling rate 128 Hz, resolution 12 bit) and digital low-pass filtering (4th order Butterworth with 3-dB attenuation at 25 Hz) the data were stored on optical disk for further processing. Sleep stages were scored for 20-s epochs according to Rechtschaffen and Kales (1968). Power spectra were on-line calculated for consecutive 4-s epochs. Artifacts were eliminated on a 4-s basis. SWA was averaged for 1 min using a moving window in 20-s steps (BL) or 40-s steps (PL and TR).

2.2 Algorithm for the estimation of the correlation dimension

Usually, the Grassberger-Procaccia (GP) algorithm (Grassberger and Procaccia 1983) or modifications thereof had been used in previous studies to estimate the CD. Since the GP algorithm has certain shortcomings (Gershenfeld 1992, Theiler 1990), we applied a more elaborated algorithm that has been proposed by Takens (1985), and further examined by Cawley and Licht (1987). This algorithm requires a much larger number of computations than the GP algorithm, and has therefore been little used. We implemented the algorithm on the personal super computer MUSIC (multi processor system with intelligent communication) (Gunzinger et al. 1992a,b) designed and built by the Electronics Laboratory of the ETH Zürich.

Up to 63 processing elements operate in parallel on information partitioned in the data space, resulting in a peak performance of $3.7 \cdot 10^9$ floating point operations per second 3.7 GFlops. The MUSIC achieves super computer performance in a desktop unit with low cost and low power dissipation.

The state vector was reconstructed on the basis of time-delay embedding (Takens 1981). The estimator is based on Fischer's maximum likelihood rule using the interpoint distances in the calculation of the CD d_2 (Takens 1985). The method uses all distances less than an upper boundary ϵ_0:

$$d_2 = -1/ < \log(\epsilon/\epsilon_0) > .$$

The upper boundary ϵ_0 was calculated as

$$\epsilon_0 = \epsilon_{min} + \alpha(< \epsilon > -\epsilon_{min}).$$

Angle brackets indicate mean values. For calculating distances the L^1-norm (Manhattan-norm) was used. Since a fit of a straight line in a log-log plot was not required, a user interaction was not needed. The CD was estimated for embedding dimensions of 25, 30, and 35. The other parameters used in the algorithm were: Time delay (τ) 39.06 ms (5 sample points); number of data points 7680 (per analysis sequence of 1 min); $\alpha = 0.55$. The algorithm was applied to 1-min epochs of sleep EEG data. A 1-min window was moved through the data of an entire night by shifting the window in 20-s (BL) or 40-s (PL, TR) steps. For the further analysis only the dimension values with an embedding dimension 30 were used. Furthermore, only 1-min epochs were included which (1) contained three identical 20-s sleep scores; (2) were without artifacts; and (3) showed a rise rate of the CD lower than 0.1 per embedding dimension for the embedding range of 25–35. The chosen time delay ($\tau \approx 40ms$), the number of data points (7680), and the sampling rate (128 Hz) are in accordance with the literature (e.g. Mayer-Kress and Layne 1987, Ehlers et al. 1991, Röschke and Aldenhoff 1991). Since a robust algorithm for the estimation of τ does not exist, and to avoid additional variability, we have kept τ constant throughout the analysis. Two recent papers (Gershenfeld 1992, Theiler 1990) give an excellent overview of the theoretical background and the problems encountered in the numerical estimation of the CD.

The calculation speed depended on the number of processors used; the MUSIC can be extended in steps of three processors (i.e. one board). For a 1-min epoch the algorithm as specified (see above) needed 374 s of computer time with 3 processors, 58 s with 21 processors, and 32 s with 45 processors. To perform the analysis in real time, 21 processors (7 boards) are sufficient if the analyzed 1-min epochs do not overlap.

2.3 Statistics
Non-parametric statistics (notched box plot; Velleman and Hoaglin, 1981) were used to establish significant differences ($p < 0.05$) between stages and between conditions.

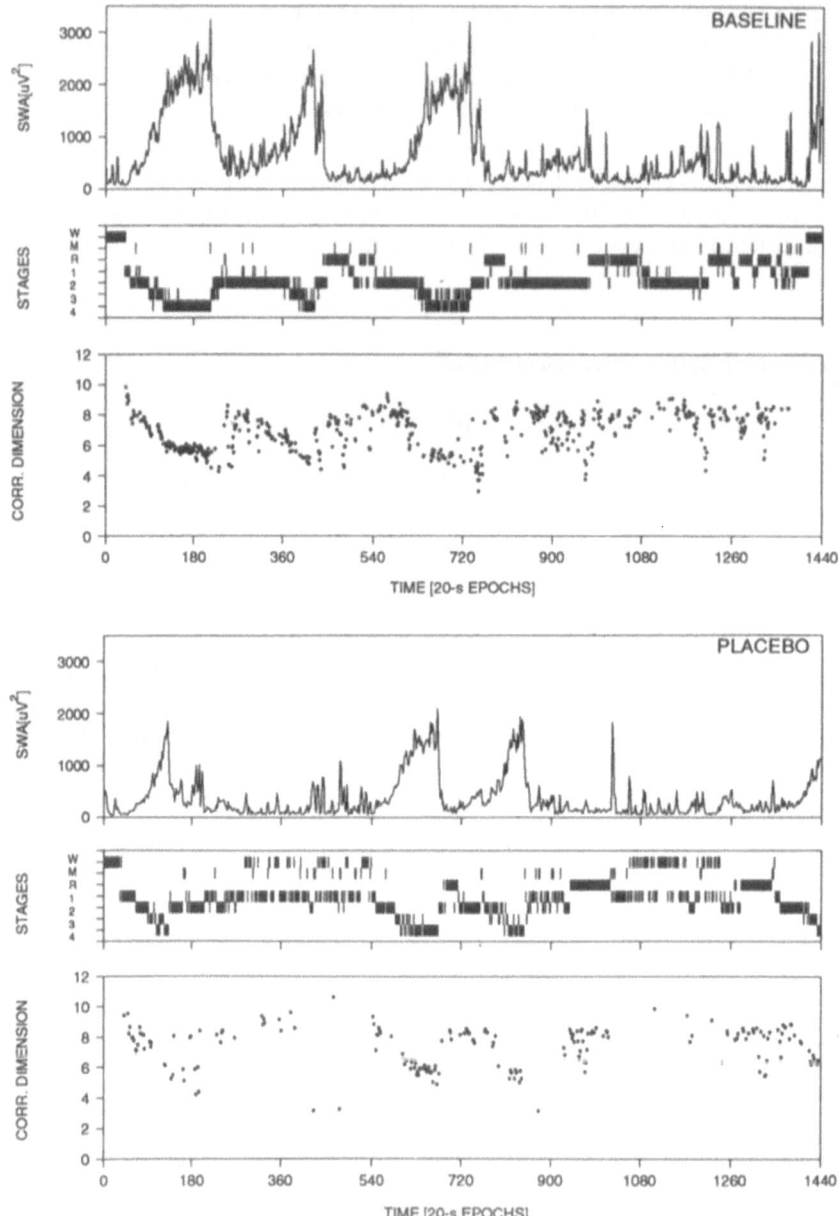

Fig. 1   Slow-wave activity (SWA; power in the 0.75 to 4.5 Hz range), vigilance states (W: waking, M: movement time, R: REM sleep, 1–4: nonREM sleep stages) and correlation dimension for a subject under baseline (BL) condition (sleep in bed) and placebo (PL) condition (sleep in a sitting position). Sleep stages are plotted on a 20-s basis (overlaps occur due to plotting resolution). SWA and dimension values were calculated for 1-min epochs by shifting a 1-min window by 20 s (BL) or 40 s (PL).

3   Results

The algorithm was first tested with a calibration signal and a theoretical test signal. For the calibration signal (10-Hz sine wave) the mean dimension estimate was 1.03 (sd 0.00086, $n = 8$). The analysis of the test signal (seven sine waves with incommensurate frequencies) resulted in a CD of 6.77 (sd 0.017, $n = 8$) with a rise rate of the dimension of 0.011 per embedding dimension (sd 0.002, $n = 8$). For both signals the estimated dimension was close to the theoretically expected value of 1 and 7, respectively.

Figure 1 illustrates the time course of SWA, of the vigilance states, and of the CD for the baseline condition and for sleep in a sitting position after placebo intake. For the drug condition (TR) the data (not shown) were very similar to those of the PL-condition. In the PL-condition the number of plotted dimension values is low because moving the 1-min window by 40 s and not by 20 s increased the relative number of state transitions and reduced the relative number of artifact free 1-min epochs with identical sleep scores. Moreover, due to the sitting position, more state changes and artifacts occurred in PL than in BL. In all conditions the CD was clearly modulated by the sleep cycles. Low values with low variability were observed during consolidated episodes of slow wave sleep. Sleep in a sitting position (in both the PL- and the TR-condition) was impaired in comparison to the BL-condition (sleep in bed). Sleep efficiency was decreased, waking after sleep onset was increased, and SWA was decreased (Aeschbach et al., in prep.).

Median and quartile values of the CD and SWA are summarized in Table 1. In all conditions, SWA increased with the deepening of sleep whereas the CD decreased. The CD was significantly different between the sleep stages, except between stage 2 and REM sleep. SWA was significantly different between all sleep stages. Computed for all 8 subjects, SWA in the TR-condition was significantly reduced in comparison to the PL-condition and both conditions differed significantly from BL (Aeschbach et al., in prep.). This is also reflected in the median values of SWA of the single subject analyzed in this study (Table 1). On the other hand, the CD in stage 4 was lower for TR than for PL and BL, and higher for PL than BL. The PL and TR values in REM sleep and the PL value in stage 2, were higher than the corresponding values in BL.

4   Discussion

The present study demonstrates the feasibility of computing the CD for an entire sleep episode with a personal super computer. The analysis showed that the sleep cycles are reflected in changes of the CD, particularly when SWA peaks were present. The CD showed a decline from stage 1 to stage 4 with REM sleep values being intermediate between stage 1 and 2. These results correspond to those of previous authors who had based their analysis on short EEG segments (e.g. Mayer-Kress and Layne 1987, Ehlers et al. 1991, Röschke and Aldenhoff 1991). However, it is also clear that a considerable range of values occur within a single sleep stage

Stage	Parameter	Corr. Dimension			SWA		
		BL	PL	TR	Bl	PL	TR
1	Q1	8.34	8.48	8.70	108.5	70.4	68.4
	MED	8.85	9.00	8.75	122.0	81.9*	89.7
	Q3	9.24	9.39	8.86	145.1	99.8	132.5
2	Q1	6.39	7.49	6.40	310.2	220.3	215.2
	MED	7.48	8.06*	7.71	451.1	268.6*	324.4*
	Q3	8.10	8.25	8.27	640.7	356.9	480.1
3	Q1	5.57	6.44	5.92	1085.8	694.3	732.4
	MED	6.25	6.84	6.37	1244.5	762.2*	851.2*
	Q3	6.83	7.35	6.82	1478.7	884.4	1149.1
4	Q1	5.36	5.63	5.03	1880.5	1352.3	1269.8
	MED	5.63	5.81*	5.36*$	2070.6	1470.5*	1389.9*
	Q3	5.86	6.02	5.68	2250.4	1592.3	1580.8
REMS	Q1	7.17	7.70	7.81	149.7	111.7	123.1
	MED	7.70	8.15*	8.17*	176.0	128.0*	139.5*
	Q3	7.99	8.38	8.38	222.9	172.7	167.0

Table 1 Correlation dimension and slow-wave activity (SWA; Q1: first quartile, MED: median, Q3: third quartile) for the different sleep stages of one subject. BL: baseline, PL: placebo, and TR: triazolam. SWA is in μV^2. * indicates a significant difference with respect to BL; $ indicates a significant difference between TR and PL ($p < 0.05$; see text for method). The number of data points per stage ranged from 7 to 273.

(e.g. see stage 2 baseline record in Fig. 1). The selection of single short segments for analysis is therefore unlikely to provide a reliable estimate of the CD. This conclusion is supported by the differences between the values for sleep in bed (BL) and sleep in a sitting position (PL, TR). The more restless sleep in the sitting position was evident from the frequent sleep disruptions and the reduced consolidation of stage 2 sleep (Fig. 1). The increased CD in stage 2 in PL as well as the reduction of SWA (Tab. 1) indicate that a more superficial type of stage 2 prevailed under this condition. The increase in the REM sleep value in both PL and TR is a comparable finding. Of course, the data from a single subject cannot be generalized and analysis of more nights are required to substantiate these observations. It will be also interesting to see whether the CD in REM sleep is more sensitive to pharmacology and pathology than the CD in other sleep stages as some studies seem to indicate (Röschke and Aldenhoff 1992, 1993).

The present study demonstrates that the use of a personal super computer opens new possibilities for performing complex EEG analyses in real time.

Acknowledgement:
We thank Daniel Aeschbach and Christian Cajochen for providing the data (recording and scoring), Markus Hürzeler (Institute of Mathematics, ETH Zürich) for calculating the boxplots, Bernhard Bäumle, Peter Kohler, Urs Müller, Walter Scott, Hans-Ruedi vonder Mühll for their support (MUSIC group), and Dr. Irene Tobler for comments on the manuscript. The study was supported by the Swiss National Science Foundation grant nr. 31.32574.91.

References

[1] Achermann, P. and Borbély A.A. Dynamics of EEG slow wave activity during physiological sleep and after administration of benzodiazepine hypnotics. *Hum. Neurobiol.,* 1987, 6: 203–210.

[2] Achermann, P., Dijk, D.J., Brunner, D.P. and Borbély A.A. A model of sleep homeostasis based on EEG slow-wave activity: quantitative comparisons of data and simulations. *Brain Res. Bull.,* 1993, 31: 97–113.

[3] Borbély, A.A. and Achermann, P. Concepts and models of sleep regulation: an overview. *J. Sleep Res.,* 1992, 1: 63–79.

[4] Borbély, A.A., Baumann, F., Brandeis, D., Strauch, I. and Lehmann, D. Sleep deprivation: effect on sleep stages and EEG power density in man. *Electroencephalogr. Clin. Neurophysiol.,* 1981, 51: 483–493.

[5] Cawley, R. and Licht, A.L. Maximum likelihood method for evaluating correlation dimension. *Lect. Notes in Phys.,* 1987, 278: 90–103.

[6] Dijk, D.J., Brunner, D.P. and Borbély, A.A. Time course of EEG power density during long sleep in humans. *Am. J. Physiol.,* 1990, 258: R650–R661.

[7] Ehlers, C.L., Havstad, J.W., Garfinkel, A. and Kupfer, D.J. Nonlinear analysis of EEG sleep states. *Neuropsychopharmacology,* 1991, 5: 167–176.

[8] Feinberg, I., March, J.D., Fein, G., Floyd, T.C., Walker, J.M. and Price, L. Period and amplitude analysis of 0.5–3 Hz activity in NREM sleep of young adults. *Electroencephalogr. Clin. Neurophysiol.,* 1978, 44: 202–213.

[9] Gershenfeld, N.A. Dimension measurement on high-dimensional systems. *Physica D,* 1992, 55: 135–154.

[10] Grassberger, P. and Procaccia, I. Measuring the strangeness of strange attractors. *Physica D,* 1983, 9: 189–208.

[11] Gunzinger, A., Müller, U., Scott, W., Bäumle, B., Kohler, P. and Guggenbühl, W. Architecture and realization of a multi signalprocessor system. In: J. Fortes, E. Lee and T. Meng (Eds) *Proc. Int. Conf. on application specific array processors.* IEEE Computer Society Press, Los Alamitos, CA, 1992a: 327–340.

[12] Gunzinger, A., Müller, U.A., Scott, W., Bäumle, B., Kohler, P., Vonder Mühll, H.R. Müller-Plathe, F., Van Gunsteren, W.F. and Guggenbühl, W. Achieving super computer performance with a DSP array processor. In: R. Werner (Ed) *Supercomputing '92*. IEEE Computer Society Press, Los Alamitos, CA, 1992b: 543–550.

[13] Mayer-Kress, G. and Layne, S.P. Dimensionality of the human electroencephalogram. In: A.S. Mandell and S. Koslow (Eds) *Perspectives in biological dynamics and theoretical medicine*. Annals of the New York Academy of Sciences, New York, 1987, 504: 62–87.

[14] Rechtschaffen, A. and Kales, A. A manual of standardized terminology, techniques and scoring system for sleep stages of human subjects. Bethesda, MD, U.S. Department of Health, Education, and Welfare, 1968.

[15] Röschke, J. and Aldenhoff, J. The dimensionality of human's electroencephalogram during sleep. *Biol. Cybern.*, 1991, 64: 307–313.

[16] Röschke, J. and Aldenhoff, J.B. A nonlinear approach to brain function: deterministic chaos and sleep EEG. *Sleep*, 1992, 15: 95–101.

[17] Röschke, J. and Aldenhoff, J.B. Estimation of the dimensionality of sleep EEG data in schizophrenics. *Eur. Arch. Psychiatry Clin. Neurosci.*, 1993, 242: 191–196.

[18] Schuster, H.G. Deterministic chaos. An introduction. VCH Verlagsgesellschaft mbH, Weinheim, 1988.

[19] Takens, F. Detecting strange attractors in turbulence. *Lect. Notes in Math.*, 1981, 898: 366–381.

[20] Takens, F. On the numerical determination of the dimension of an attractor. *Lect. Notes in Math.*, 1985, 1125: 99–106.

[21] Theiler, J. Estimating fractal dimension. *J. Opt. Soc. Am. A*, 1990, 7: 1055–1073.

[22] Velleman P.F. and Hoaglin D.C. Applications, basics, and computing of exploratory data analysis. Duxbury Press, Boston MA, 1981.

Fractals
in Pathology

Changes in Fractal Dimension of Trabecular Bone in Osteoporosis: A Preliminary Study

C. L. Benhamou[1], R. Harba[2], E. Lespessailles[1],
G. Jacquet[2], D. Tourliere[1] and R. Jennane[2]

1) Pole d'Activité Rhumatologie, CHR Orléans La Source, France.
2) Laboratoire d'Electronique Signaux Images, Université d'Orléans, France.

1 Introduction

The trabecular bone network represents a complex structure, either in a macroscopic examination, in a microscopic histological study, or in a radiological projection [1, 2]. Particularly for radiological images, euclidean geometry does not lend itself to the description of such a complex structure. Fractal geometry is much better suited to this analysis, allowing a characterization of this type of rough irregular texture, or of its projection [3,4].

The choice of the fractal model is dependant on the characteristics of the images. Some models necessitate a binarisation of the image [5]. This binarisation may be simple for some textures. On radiological images of bones, which represent a 2 D projection of a 3 D structure, the binarisation implies a definition of a threshold between trabeculae and intertrabecular spaces. Unfortunately, this definition of a threshold remains arbitrary, and the threshold may vary in different parts of the image due to the variations of bone thickness. So it seems better for our study to use a fractal model taking into account the grey level variations. The Fractional Brownian Motion (FBM), and its increment the Fractional Gaussian Noise (FGN) respond to this condition [6]. It has been used to undergo an analysis of radiological bone images [3,4]. FBM, noted BH, is characterized by the H parameter ($0 < H < 1$). H is related to the fractal dimension D: for a one dimensional signal $D = 2 - H$. The smaller is H, the bigger is the roughness of the texture.

The trabecular bone microarchitecture organization is nowadays considered as a very important factor in biomechanical competence of bone [7,8,9] accounting in a great part for the properties of resistance and elasticity of this tissue. But bone is not only an amorphous framework, its trabecular microarchitecture plays a great role in the calcium homeostasis function: this structure offers a very large surface for ions exchanges between trabecular bone itself and the highly vascular bone marrow which fills the intertrabecular spaces.

Bone ageing and osteoporosis may alter cortical or trabecular bone. In trabecular bone, the main changes consist in a decrease of bone mass [10]. In addition to this loss of bone tissue, qualitative changes are considered as a very important component of the biomechanical alterations of bone [11]. They are mainly linked to abnormalities outcoming in the trabecular network, for instance perforations, dysconnections, loss of trabeculae [12,13]. These abnormalities would be much more deleterious in terms of biomechanical competence than a simple thinning of trabeculae [14]. Their analysis requires at the present time an histological study, implying an invasive procedure of bone biopsy [15].

We report the results of a new method of trabecular microarchitecture fractal analysis on radiological images of calcaneus, in a study comparing osteoporosis cases and control cases.

2 Material and Methods

2.1 Patients

We have studied 31 X ray-views from 17 osteoporotic patients and 24 X ray-views from 12 control subjects.

Osteoporosis was defined by fractures occurring after a fall from a standing position or spontaneus fractures: mainly hip fracture or vertebral crush fracture except in two cases (rib and sacral recurrent fractures in one, ankle and clavicle for the other one). The 17 osteoporotic patients, 3 men and 14 women, were all Caucasian (age 74.4 ± 14.3 years). They had no history of metabolic bone diseases except osteoporosis.

In the control group, 5 men and 7 women, all subjects were Caucasian (age 77.3 ± 12.1 years), they had no history of fracture, of diseases interfering with bone metabolism and they were not taking medication known to affect bone metabolism (e.g., fluoride, calcitonin, estrogen, bisphosphonates, thiazide diuretics, corticosteroids).

2.2 Acquisition of images

Os calcis X-ray images have been performed with the same standardized conditions. Focal-calcaneus distance was fixed at one meter. Gridless cassettes, Plastilix with a thickness of 1 cm were used. Calcaneus was placed in contact with the film. The same X-ray tube, voltage (48 KV), and exposure conditions (18 mA.s for 0.08s) were used. Radiographs were taken on a Kodak min R film (18 $\times$ 24 cm) usually devoted to mammography. Development was made at $34°$ Celsius in a Kodak developing bath. Tested in the Kodak laboratory conditions, this film has given a resolution of fifteen pairs of lines a millimeter.

Radiographs were put on a lighting table and images were acquired with a CCD camera. The lens aperture was always at its optimum: 8. The numerisation process is performed with a matrox frame grabber at a format of 512×512 pixels of 8 bits. The gain and offset of the frame grabber were set in order to have a correct histogram with the greatest dynamic without saturation. The acquisition field was kept constant which led to a pixel size of 105μm. A specific acquisition software

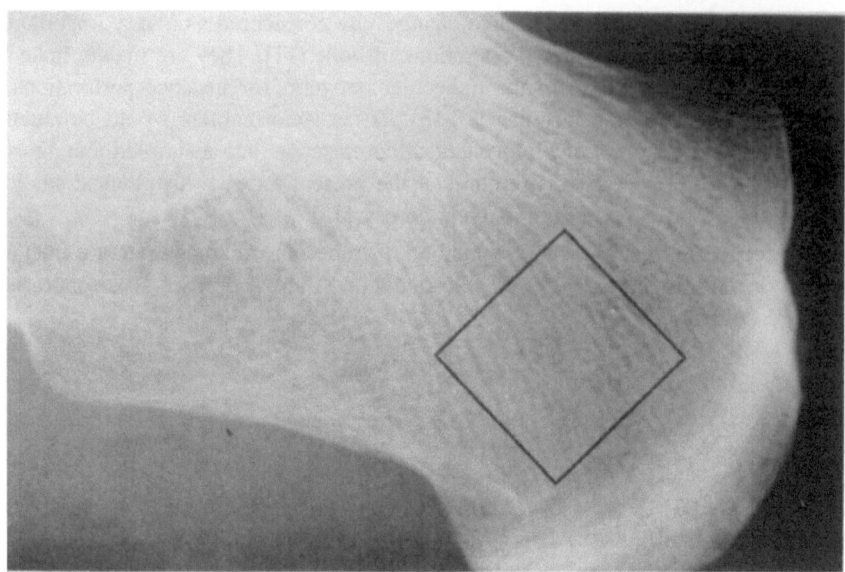

Fig. 1 A typical radiograph of calcaneus. The drawn square delimitates the region of interest.

ran on an IBM PC-AT computer which allowed to select an area of 256×256 pixels corresponding to the central part of the trabecular bone. Figure 1 shows a typical radiograph of a calcaneum and the selected region of analysis. Then, data files were sent through an ethernet network to IBM RS/6000 workstations for numerical computations.

2.3 Estimation of the fractal dimension

Because of the anisotropy of the image, a one dimensional analysis is performed. 400 lines of 100 samples each are taken from images in a given direction as seen in figure 2. The H value is computed on each line and the final value for a given direction is the average on these 400 lines. 36 different angles are computed. A polar diagram represents the H value as a function of the direction as seen on figure 3. The anisotropy of the images can be seen with this kind of representation. For this study, we only use the global value for the whole image which is the average on the 36 H values for the different angles. The H value has to be estimated for each line. Among the numerous methods, we choose the Maximum Likelihood Estimation method because it gives the smallest theoretical variance for the estimation of H. This method consists in maximizing the conditional probability density function $p(x; H)$. x is a FGN vector, increment of the data set modelized by the FBM BH:

$$x[i] = B_H[i] - B_H[i-1] \tag{1}$$

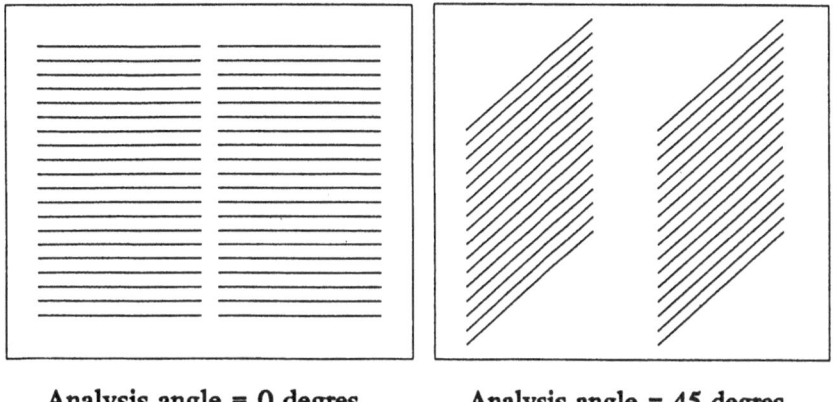

Analysis angle = 0 degres   **Analysis angle = 45 degres**

Fig. 2   Lines selection for the oriented analysis.

x is stationary, zero mean and has Gaussian distributed samples as:

$$p(x; H) = \frac{1}{2\pi)^{N/2}|R|^{1/2}} \exp\left(-x^T R_x^{-1} x / 2\right). \tag{2}$$

N is the length of the data set, R_x is the covariance matrix which elements depends on the autocorrelation function of the FGN such as $[R_x]_{ij} = r_x[i-j]$ with:

$$r_x = \frac{\sigma^2}{2}\left(|K+1|^{2H} - 2|K|^{2H} + |K-1|^{2H}\right). \tag{3}$$

σ^2 is the variance of the FGN vector.

No analytical solution is available to find the maximum of $p(x; H)$. The shape of this function with H from 0 to 1 is unimodal wich allows to implement numerical methods. Since the logarithm is a monotonic function, the log of $p(x; H)$ is maximized:

$$\log p(x; H) = -\frac{N}{2}\log 2\pi - \frac{1}{2}\log|R| - \frac{1}{2}x^T R^{-1} x. \tag{4}$$

Two parameters have to be estimated: H and σ^2. The R matrix can be decomposed as:

$$R = \sigma^2 R' \tag{5}$$

Using this in the log likelihood function and letting the derivative go to zero with respect to σ^2 gives the final function to be maximized:

$$\log p(x; H) = -\frac{N}{2}\log 2\pi - \frac{1}{2}\log|R'| - \frac{N}{2}\log\frac{x^T R'^{-1} x}{N} - \frac{N}{2}. \tag{6}$$

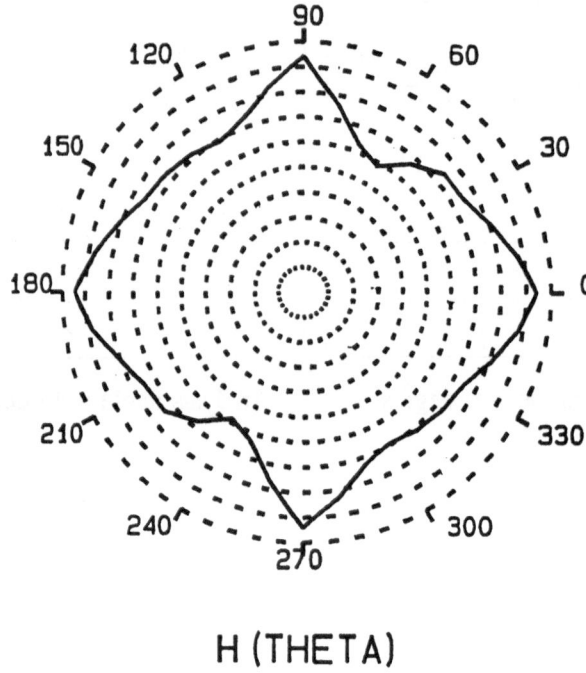

H (THETA)

Fig. 3   Polar representation of the H value for an image.

R' has to be inverted and its determinant computed. R' is definite positive and can be decomposed using the Levinson algorithm as

$$R'^{-1} = LD^{-1}L^T \qquad (7)$$

where L is a lower triangular matrix with ones on the diagonal and D is diagonal. Then, the determinant of R' is equal to the determinant of D. Both determinant and inverse are computed in one step. The golden search technique has been implemented to find the maximum of the log likelihood function. The total time to compute the global H value for an image is about 40 minutes on an IBM RS/6000 workstation.

3   Results

Mean results of H mean value were 0.401 ± 0.065 in the osteoporotic group and 0.500 ± 0.052 in the control group. Student's paired t test showed a statistical significant difference $p < 0.0001$ between these two populations as shown in table 1. There was no statistical difference between these two groups for the mean age ($p = 0.42$). We have found no statistical significant correlation between age and H mean. Figure 4 represents the histogram of the two populations and allows a visual appreciation of the overlap.

	H	age
Osteoporotic cases	0.401 ± 0.065	74.4 ± 14.3
Control cases	0.500 ± 0.052	77.3 ± 12.1
Statistical significance	$p < 0.0001$	$p = 0.42$

Table 1 Mean result for the H parameter for the two groups

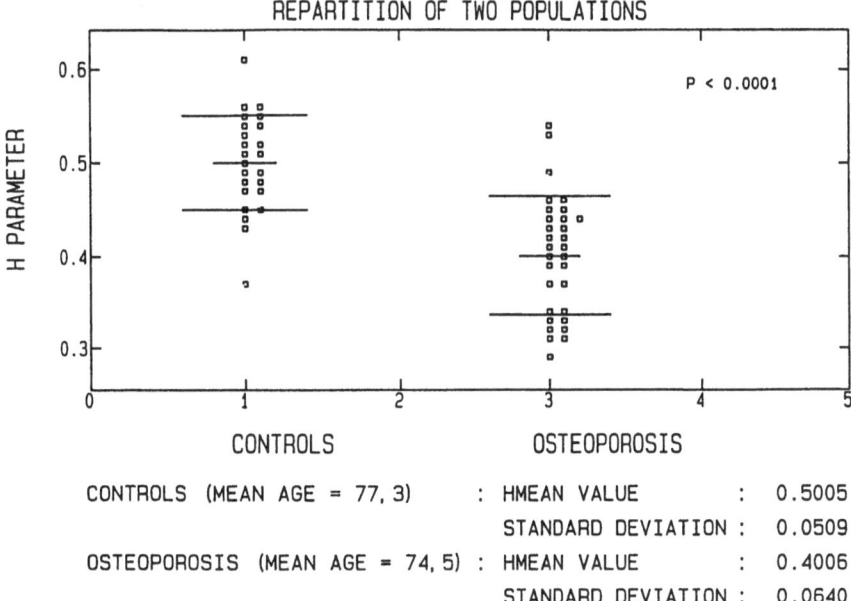

Fig. 4 The histogram of H mean in osteoporosis cases and control cases.

4   Discussion

These data indicate that our fractal evaluation allows to discriminate an osteo-porotic elderly population from an age-matched control group. We have to under-line the high statistical significant difference between these two populations, in spite of a limited number of subjects. This fractal model has been previously used in preliminary experiences. In one immobilized patient, the H value has shown the same trend to a progressive lowering [3]. Acid dissolution of bone samples also results in a reduction of H Value [4]. Other models have been tested by other groups in such preliminary studies: Van der Stelt and Geraets also report a decrease of H value following menopausis [5]. Ageing is generally considered to result in a loss of complexity, implying a lowering of the fractal dimension [16], particularly in dynamic processes. Our decrease of the H value corresponds to an increase of the fractal dimension, probably accounting for the disorganization of the trabecular network.

One of the advantages of our method consists in the ability to describe an image with a various range of grey. This method allows to analyze a lot of informations, coming from a 2 D projection of a 3 D structure.

The study of calcaneus radiological images offers the advantages of a rapid, simple and safe non invasive procedure. It is not expensive. The irradiation is limited, allowing epidemiological studies, and successive evaluations in the same subject. Spine, hip and wrist are the preferential sites of osteoporotic fractures [17]. However, spine and hip are surrounded by large soft tissues, so that the results may differ widely with morphological variations. We have tested some evaluations on the radius, but this site offers a limited area of analysis. Our calcaneus study does not involve these difficulties. Furthermore, in terms of bone mineral density, it has been shown to be the best site for the determination of the relative risk of osteoporotic fractures [18].

5   Conclusion

In our experience, this fractal model based on the FBM, has been very interesting for trabecular bone microarchitecture characterization on X-Ray images. Before to start large epidemiological studies, we have searched to assess its discriminant power between controls and osteoporosis cases. Our preliminary study has shown that this model could valuably separate these two populations, with a limited degree of overlap. We have not found a statistical correlation whith age, but this was not the aim of the study, so that both groups were mainly constituted of elderly subjects, with limited variations in age.

Bone mineral density measurements are not able to fully explain the mechanisms of osteoporosis. We think that our model is able to account for structural qualitative changes.

Acknowledgements:
We thank Rhone-Poulenc Roser Int¹. for its support, Kodak Pathé Division Santé for its technical advice, and Mrs. Nadège Marchant for her secretarial assistance.

References

[1] Rockoff S.D., Scandrett J. and R. Zacher: Quantitation of relevant image information: automated radiographic bone trabecular characterization. Radiology, 101, 435–439 (1971).

[2] Aggarwal N.D., Singh G.D., Aggarwal R. et al.: A survey of osteoporosis using the calcaneum as an index. Int. Orthop., 10, 147–153 (1986).

[3] Lundahl T., Ohley W.J., Kay S.M., Siffert R.: Fractional Brownian motion: a maximum likelihood estimation and its application to image texture. IEEE Trans. med. imaging, MI-5, 152–160 (1986).

[4] Jacquet G., Ohley W.J., Mont M.A., Siffert R., Schmukler R.: Measurement of bone structure by use of fractal dimension. Proc. Annual Conf. of IEEE/EMBS, 1402–1403 (1990).

[5] Van der Stelt P.F., Geraets: The fractal dimension of the trabecular pattern in patients with increased risk of alveolar ridge resorption. Proc. Annual Conf. of IEEE/EMBS, 12, 2071–2072 (1990).

[6] Mandelbrot B.B., van Ness J.: Fractional brownian motion, fractional noise and applications. SIAM 10, 422–438 (1968).

[7] Kleerekoper M., Villanueva A.R., Stanciu J., Sudhaker Rao D., Parfitt A.M.: The role of three dimensional trabecular microstructure in the pathogenesis of vertebral compression fractures. Calcif. Tissue Int., 37, 594–597 (1985).

[8] Parfitt A.M.: Trabecular bone architecture in the pathogenesis and prevention of fracture. Am. J. Med., 82 (suppl 1B), 68–72 (1987).

[9] Mosekilde Li.: Age-related changes in vertebral trabecular bone architecture. Assessed by a new method. Bone, 9, 247–250 (1988).

[10] Dempster D.W., Lindsay R.: Pathogenesis of osteoporosis. LANCET, 341, 797–801 (1993).

[11] Parfitt A.M.: Implications of architecture for the pathogenesis and prevention of vertebral fracture. Bone, 13, S41–47 (1992).

[12] Vernon-Roberts B., Pirie C.J.: Healing trabecular micro-fractures in the bodies of lumbar vertebrae. Ann. Rheum.Dis. 32, 406 (1973).

[13] Parfitt A.M.: Age related structural changes in trabecular and cortical bone: cellular mechanisms and biomechanical consequences. Calcif. Tissue Int., 36, S123–S128 (1984).

[14] Weinstein A.S., Hutson M.S.: Decreased trabecular width and increased trabecular spacing contribute to bone loss with aging. Bone, 8, 137–142 (1987).

[15] Birkenhager-Freukel D.H., Courpron P., Hupscher E.A., Clermonts E., Coutinho M.F., Schmitz P.I.P.M. and Meunier P.J.: Age related changes in cancellous bone structure. Bone and Mineral 4, 197–216 (1988).

[16] Lipsitz L.A., Goldberger A.L.: Loss of complexity and aging. JAMA, 267, 1806–1809 (1992).

[17] Nordin B.E.C.: The definition and diagnosis of osteoporosis. Calcif. Tissue Int. 40, 57–58 (1987).

[18] Black D.M., Cummings S.R., Genant A.K. et al.: Axial and appendicular bone density predict fractures in older women. J. Bone Miner. Res., 7, 633–638 (1992).

Use of the Fractal Dimension to Characterize the Structure of Cancellous Bone in Radiographs of the Proximal Femur

Curtis B. Caldwell[1], John Rosson[2],
James Surowiak[2] and Trevor Hearn[2]

1) Departments of Medical Biophysics and Radiology, University of Toronto, and Department of Radiological Sciences, Sunnybrook Health Science Centre, 2075 Bayview Avenue, North York, Ontario, Canada M4N 3M5

2) Orthopaedic Biomechanics Research Laboratory, Sunnybrook Health Science Centre, 2075 Bayview Avenue, North York, Ontario, Canada M4N 3M5

Abstract. Preliminary work on the development of a quantitative method of characterizing the radiographic appearance of cancellous bone is described. The method is based upon calculation of the «fractal dimension» in regions of interest of digitized radiographs. The method was applied to standard pre-operative radiographs of the hip and proximal femur of fifteen patients who were to undergo total hip arthroplasty. Cancellous bone samples were retained at arthroplasty and their compressive strength and ash density assessed. The fractal dimension was found to correlate more strongly with compressive strength than did a conventional method of assessing bone quality (the Singh Index). Among the potential benefits of the new technique is the possibility of providing guidance in the pre-operative selection of the optimal method of implant fixation for an individual patient.

1 Introduction

Fracture of the hip is one of the major clinical risks associated with osteoarthritis. If a hip replacement becomes necessary, the possibility of implant failure poses an additional risk to the patient. Although these risks are well known, reliable, non-invasive diagnostic methods which accurately discriminate among patients at risk for hip fracture or implant failure are not currently available. The majority of current methods attempt to measure gross bone density, while ignoring the influence of bone structure. While bone density is clearly an important parameter, assessing risk by density alone neglects any possible influence of the geometrical distribution or structure of cortical and cancellous bone. Some would contend that bone density alone is sufficient to predict fracture risk (Hui et al. 1988). This is equivalent to assuming that the only change in diseased or damaged bone is loss of mass, with no change in structure. In addition, there appears to be an age-related bone loss of about 1 % a year in normal postmenopausal women. This implies that if bone mass measurements are to be meaningful for individual patient follow-up, they must have very good precision. In a study of four methods of bone mass measurement in 49 osteoporotic women obtained at time zero and at one year of treatment, it was found that none of the methods could be used to predict changes as measured by any of the other techniques (Ott et al., 1986). This result occurred despite changes in some patients which were as high as 5 % by one or other of the techniques.

The most common current diagnostic techniques for assessing the bone quality of the proximal femur include quantitative bone densitometry (e.g., using the dual energy x-ray technique — DEXA or the dual photon radioisotope technique — DPA) and qualitative evaluation of plane film radiographs. Quantitative x-ray computed tomography (Lotz et al. 1990) has been suggested as an alternative method of assessing bone quality, but is a relatively expensive technique which is often not available. Singh et al. (1970) developed a method of subjectively evaluating changes in the trabecular pattern of the femur (as seen in plane film radiographs of the hip), as an aid to distinguishing the degree of osteoporosis. While Singh's work addresses the problem of assessing trabecular bone structure, the Singh Index is, unfortunately, subject to high inter- and intra-observer variability and has not proved to correlate well with fracture risk.

We have been working on the development of a method of quantitatively evaluating bone structure from digitized radiographs of the hip. The potential benefit of such a method is that it may provide structural information complementary to the bone density information already available. Since we propose to derive the information from conventional radiographs of the hip, the patient is exposed to no more invasive procedure or to any further radiation dose. Together, information on structure and density should provide a better means of risk stratification, when considering treatment for degenerative bone disease. The combination of structural and density information may provide a more precise method of assessing longitudinal changes in patients undergoing drug or dietary treatment for osteoporosis. A method of evaluating the structure of cancellous bone may be particularly important when deciding upon the best method of fixing an implant in place in an area such as the femoral head.

Fractals are a powerful tool for modelling biological objects. While no universal definition of a «fractal» has been agreed upon, fractal shapes and patterns are often identified by their properties of deterministic or statistical self-similarity and independence to scaling. The use of the fractal concept is based on the work of Mandelbrot (1983), who showed that many physical systems in nature are well described by sets of numbers which show elements of self-similarity and randomness over some range of scale. Such a set may be considered to have a particular «fractal dimension» characteristic of the physical system or biological structure in question. Fractals are especially useful in providing a means of quantitatively characterizing complex natural structures which are not well described by classical geometry (i.e., the Euclidean geometry of straight lines, circles, spheres, etc which describes most man-made objects very well).

Image analysis using fractal mathematics has been investigated by a number of researchers. It has been shown that image textures which are subjectively rough or highly disorganized have higher fractal dimensions than images that are subjectively more smooth or more regularly organized (Pentland 1984). Fractals have been used in medical imaging to discriminate between normal and abnormal Nuclear Medicine liver scans (Cargill et al. 1989), mammograms (Caldwell et al. 1990), and images of retinal vessels (Mainster MA 1990).

While it is known that fractals are useful in describing many complex natural structures, why would one propose that the structure of cancellous bone in particular would be «Fractal»? We were lead to this hypothesis through the work of Kaandorf (1991) who applied fractal methods to modelling the structure of sea sponges. Cancellous or «spongy» bone has a subjectively similar structure to that of sea sponges. Moreover, other researchers have found the fractal dimension to be a useful descriptor of radiographic bone quality in the knee (Lynch et al. 1991) and in the peridental alveolar bone (Ruttimann et al. 1992). Ruttimann contends that a type of fractal scaling may be seen in bone as the texture of bone tends to change from coarse to fine as one moves across the boundary from cortex to spongiosa.

2   Methods

Fifteen patients undergoing total hip arthroplasty were studied. The underlying pathology was osteoarthritis in all patients. Standard pre-operative radiographs of the hip and proximal femur were taken with the proximal femur rotated fifteen degrees to the plane of the x-ray beam. These radiographs were digitized using a Konica laser scanning microdensitometer with a pixel size of 0.175 mm and a ten-bit grey-scale resolution. Digital images were subsequently analyzed using computer programs written in «C» on a Sun workstation (Sun Microsystems, California, USA).

Each of the patients was assigned a Singh Index value (one to six in accordance with the original method) by an experienced radiologist who had no knowledge of the patients' age or sex. Note that the radiologist used the original radiographs (i.e., not the digitized images of these radiographs).

At surgery, the femoral neck resection specimens were retained and stored at -20 degrees Celsius. Cubes of cancellous bone measuring one centimetre on each side were cut from the specimens while still frozen using a water cooled saw equipped with a diamond tipped blade and a micrometer. The anatomical orientation of the cubes was recorded and compression was applied in a consistent direction in all samples during mechanical testing.

Prior to testing the specimens were warmed to 37 degrees Celsius and the dimensions of each cube were measured to within 0.1 mm. The specimens were then compressed in a servo-hydraulic materials-testing machine (MTS, Minnesota, USA) at a rate of 0.1 mm per second (strain rate 1 % per second). The specimens were not otherwise constrained. The force displacement history was recorded digitally for 75 seconds at a sampling rate of 20 Hz.

Following mechanical testing the specimens were ashed in a box furnace for 24 hours at 580 degrees Celsius (S 1800 model, Lindberg Unit of General Signal,

Wisconsin, USA). Ash mass of each specimen was then determined allowing calculation of ash density from the previously measured, pre-compression volume.

Irregular regions of interest (ROIs) corresponding to the projected area of the femoral head were traced on the digital images displayed on a computer screen. The pixels within these ROIs were used in the subsequent digital analysis. To obtain an estimate of the «fractal dimension» of the ROI, a method similar to that of Lundahl et al. (1985) was applied. Conceptionally, the radiograph was considered as a topologically three dimensional object, with the optical density at each point in the radiograph (x, y) plane treated as a distance measure representing the third topological dimension. Thus, the digital image was taken as representing a «mountainous» surface. The problem of assessing the fractal dimension of a mountainous surface can be solved by fitting an equation of form:

$$A(\epsilon) = \lambda \epsilon^{2-D} \tag{1}$$

where

$\quad (A\epsilon)$ is the area of the surface measured with a square of side ϵ

$\quad \lambda$ is a scaling constant

$\quad D$ is a constant characteristic of the surface (i.e., the «fractal dimension»)

The surface area of the radiograph was calculated using different «square sides». In computing the area as a function of square size, the surface of the image was considered to be a collection of adjacent «skyscrapers», that is, a set of rectangular columns of differing heights but with square «roofs» having side length ϵ. The area of the surface is given by the sum of the area of the «roofs» plus the sum of the areas of the exposed sides of the columns:

$$A(\epsilon) = \sum_{x,y} \epsilon^2 + \sum_{x,y} \epsilon \mid I(x,y) - I(x+1,y) \mid + \mid I(x,y) - I(x,y+1) \mid$$

where $I(x, y)$ is the height of a column, found for a particular value of ϵ by averaging the values stored in adjacent image array elements to produce pixels with a side length of ϵ.

$A(\epsilon)$ was first calculated with ϵ equal to single pixel width (0.175 mm). Subsequently, the grey levels in squares of 4, 9, 16, 25, and 36 contiguous pixels were averaged to form new image arrays. $A(\epsilon)$ was re-calculated for each ϵ.

As expected for a fractal surface, the measured area was dependent on the ruler size. From Equation 1, the fractal dimension for the «mountainous» surface is directly related to the slope of a plot of $\log(A(\epsilon))$ versus $\log(\epsilon)$. A more rugged or complex surface results in a higher fractal dimension. A gentler or smoother surface results in a lower fractal dimension.

In addition to the fractal dimension, mean pixel grey level and grey level standard deviation were derived for the pixels in each ROI.

3 Results

The data derived from the 15 patients are presented in Table 1. Ash density, Singh Index, fractal dimension, mean pixel grey level, pixel standard deviation and compression strength are listed. The value of the correlation coefficient relating to the degree of linear correlation between each parameter and compressive strength is found in Table 2. Note that the highest correlation is between ash density and compressive strength. Of the remaining parameters, only the fractal dimension displayed a high degree of correlation with compressive strength.

Compress Strength (MPa)	Density (mg/cm3)	Singh Index	Fractal D	Mean Grey Level	Standard Deviation in Grey Level
0.57	83	2	2.64	639	54
0.75	89	3	2.64	678	55
0.80	78	4	2.69	586	57
0.86	109	5	2.66	633	31
0.96	95	2	2.65	773	57
1.0	124	3	2.61	792	27
1.3	103	6	2.66	648	50
1.6	143	6	2.74	667	60
1.6	108	2	2.65	669	47
1.7	128	3	2.86	773	42
1.7	116	5	2.85	709	44
1.7	103	5	2.77	741	60
1.9	128	3	2.81	813	41
2.3	144	4	2.79	772	54
3.5	183	5	2.78	676	54

Table 1 Measurements for 15 Subjects

Parameter Tested versus Compressive Strength	Linear Correlation Coefficient	p-value
Density	0.89	< 0.001
Fractal Dimension	0.65	< 0.01
Singh Index	0.33	> 0.1
Mean Grey Level	0.27	> 0.1
Standard Deviation in Grey Level	0.15	> 0.1

Table 2 Correlations with Compressive Strength

4 Discussion

In this preliminary study, we found that patients who had relatively strong cancellous bone also had radiographs with higher fractal dimension than did patients who had relatively weak cancellous bone. While the number of patients studied was small ($N = 15$) this may indicate that normal cancellous bone has a sharply defined structure of high fractal dimension which may be «smoothed out» to a lower fractal dimension by the action of osteoarthritic disease. Two simpler measures (the mean and standard deviation in pixel grey level) did not appear to be useful predictors of bone strength. The Singh Index was also not strongly correlated with bone strength. Bone density, as assessed by ash mass measurements, was highly correlated with bone strength, as expected. However, conventional methods of measuring bone density *in vivo* are unlikely to correlate with strength to this degree. Bone density measurements made by conventional (i.e., non-CT) methods *in vivo* are, in effect, averages of the density of both cortical and trabecular bone. Variations in the amount of fat in tissue surrounding the bone can also disturb the estimate of bone density by these techniques. In addition, even if knowledge of cancellous bone density were sufficient to predict cancellous bone strength, cancellous bone density is not measured directly by either DPA or DEXA. It may be that the quality of cancellous bone is of great importance in some clinical applications. For example, the role of cancellous bone in the support of an uncemented femoral component remains unknown. It has been postulated that the quality of cancellous bone is important in resisting rotational forces at the prothesis bone interface (Spotorno and Rognoli 1988). A parameter based on a combination of bone density and structural measurements (possibly using the fractal dimension) may prove to be useful in the evaluation of cancellous bone using only digitized conventional plane film radiographs. This would be preferable to the current status, where neither bone density measurements nor subjective evaluations of plane films provide adequate information to predict the quality of cancellous bone. It may also be preferable to extensive use of quantitative x-ray CT, a relatively expensive and not widely available procedure.

Acknowledgements:
The authors are grateful to Dr. Brian Howard for his help in assigning a Singh Index value and to Dr. Keith Willett for helpful discussions.

References

[1] Browne MA, Gaydecki PA, Gough RF, Grennan DM Khalil SI, Mamtora H Radiographic image analysis in the study of bone morphology Clin Phys Physiol Meas 1987 8: 105–121.

[2] Cargill EB, Donohoe K, Kolodny G, Parker JA Zimmerman RE Analysis of lung scans using fractals Proc SPIE 1989 1092: 2–9.

[3] Hui SL, Slemenda C, Johnston CC Age and bone mass as predictors of fracture in a prospective study J Clin Invest 1988 81: 1804–1809.

[4] Kaandorp JA Modelling growth forms of sponges with fractal techniques in: Fractals and Chaos New York: Springer Verlag 1991.

[5] Lotz JC, Gerhart TN, Hayes WC Mechanical properties of trabecular bone from the proximal femur: a quantitative CT study J Comput Assist Tomogr 1990, 14: 107–114.

[6] Lotz JC, Hayes WC The use of quantitative computed tomography to estimate risk of fracture of the hip from falls J Bone Joint Surgery 1990 72-A: 689–700.

[7] Lundahl T, Ohley WJ, Kuklinski WS, Williams DO, Gerwitz H, Most AS Analysis and interpolation of angiographic images by use of fractals Comput in Cardiol 1985 24: 355–358.

[8] Lynch JA, Hawkes DJ, Buckland-Wright JC Analysis of texture in macroradiographs of osteoarthritic knees using the fractal signature Phys Med Biol 1991 36: 709–722.

[9] Mainster MA The fractal properties of retinal vessels: embryological and clinical implications. Eye 1990 4: 235–241.

[10] Mandelbrot BB The Fractal Geometry of Nature New York: W.H. Freeman and Company 1983.

[11] Martens M, Van Audekercke R, Delport P, De Meester P, Mulier JC The mechanical characteristics of cancellous bone at the upper femoral region J Biomechanics 1983 16: 971–983.

[12] Ott SM, Kilcoyne RF, Chesnut CH III Longitudinal changes in bone mass after one year as measured by different techniques in patients with osteoporosis Calcif Tissue Int 1986 39:133–138.

[13] Pentland AP Fractal-based descriptions of natural scenes IEEE Trans Pattern Anal and Machine Intelligence 1984 PAMI-6: 661–674.

[14] Rosson J, Surowiak J, Schatzker J, Hearn T The relationship between the radiographic appearances and the structural properties of proximal femoral bone in patients undergoing total hip arthroplasty (submitted for publication)

[15] Ruttimann UE, Webber RL, Hazelrig JB Fractal dimension from radiographs of peridental alveolar bone: A possible diagnostic indicator of osteoporosis Oral Surg Oral Med Oral Pathol 1992 74: 98–110.

[16] Singh M, Nagrath AR, Maini PS Changes in trabecular pattern of the upper end of the femur as an index of osteoporosis J Bone Joint Surg 1970 52-A: 457–467.

[17] Spotorno, Rogmagnoli. Indications for the CLS stem in: Cementless Total Hip Replacement System CLS. Monograph published by Protek AG, 1988.

[18] Stulberg BN, Bauer TW, Watson JT, Richmond B Roentgenographic versus histologic assessment of hip bone structure Clinical Orthopaedics and Related Research 1989 240: 200–205.

Distribution of Local-Connected Fractal Dimension and the Degree of Liver Fattiness from Ultrasound

Carl J.G. Evertsz[1], C. Zahlten[1], H.-O. Peitgen[1],
I. Zuna[2], and G. van Kaick[2]

1) Center for Complex Systems and Visualization, University of Bremen (FB III), Box 330 440, D-28334 Bremen, Germany

2) Deutsches Krebsforschungszentrum, Im Neuenheimer Feld 280, D-69120 Heidelberg, Germany

1   Introduction

Central to fractal geometry [3] is the concept of self-similarity. Fractal and multifractal [1] analysis provide tools for the quantitative analysis and description of self-similarity in natural and mathematical sets and distributions. The arsenal of tools for fractal analysis which is growing both in diversity and sophisication provides fresh new ways to analyse geometries, even those which are non-generically fractal. This contribution to *Fractals in Medicine* discusses an application of a hybrid of fractal and multifractal analysis to a problem in medical ultrasound imaging. Our method is a continuation of work by Richard Voss [7] which he applied to the X-ray detection of malignant breast tumors and the classification of chineese landscape paintings.

Ultrasound imaging of the liver is a standard part of a checkup by the specialist in internal medicine. Besides being a harmless tool for diagnosing several life threatening diseases ultrasound is know to provide some information on the degree of liver fattiness. It is also known to be pretty much imprecise and to a large degree subjective when it comes to diagnosing the latter. Quantitative image analysis could play a role in an assessment of the limitations of ultrasound as a diagnostic tool for liver fattiness.

The right unprocessed liver ultrasound in Figure 1 is from a healthy patient and the left one from a patient with a fatty liver. Our aim is to quantify the qualitative strategy an experienced medical doctor would apply when asked to diagnose the fat condition of a patient's liver given only such ultrasound images. Depending on this person's experience a varying number of visual aspects are taken into account; amongst others the shape of certain parts of the liver, the overall intensity of the image, the liver/bloodvessel contrast, and the intensity difference between the liver and kidney tissue, etc.

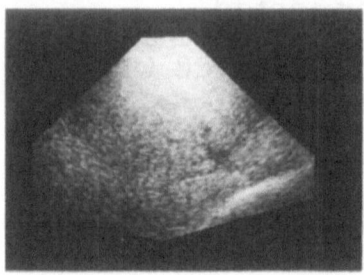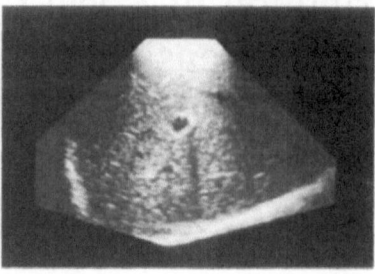

Fig. 1   Ultrasound images of a fatty (left) and a healthy (right) liver.

Ultrasound images of homogeous tissue are not smooth, but have a *speckled* stucture, which in our images have typical sizes of about 30 pixels. One quantitative approach is based on texture analysis of a small *region of interest* (ROI) in the ultrasound image containing only liver tissue [5]. The basic assumption is that the speckle structure contains information about the fattiness of the liver tissue.

The other, new approach, is much more close to the qualitative analysis of the medical specialist and tries to take as much as possible the impression conveyed by the whole image. It concentrates on the overall contrast between bloodvessels and liver tissue and that between liver tissue and liver boundary. The microscopic speckle structure is of little relevance.

We first review the relevant basic tools from fractal geometry [3][4], namely mass dimension and local mass dimension, and then move on to the specifics of the method.

2   Mass dimension

Figure 2a shows the three stages of a «growing» Sierpinski triangle. The building blocks are little triangles of size 1 and mass 1. Thus, the respective sizes L of the three stages depicted are 1, 2 and 4. The corresponding masses are $M(1) = 1$, $M(2) = 3$ and $M(4) = 9 \ldots$ and so on for increasing sizes. Therefore, the mass and length have a scaling relationship of the form $M(L) \sim L^D$ where the exponent $D = \log 3 / \log 2$ is called the mass dimension.

3   Local mass dimension

The local mass dimension D_{loc} also characterizes a mass/size relationship. Let x be a point in the set considered, and let $B_L(x)$ be a box or disc of size L centered around x. When the mass $M(B_L(x))$ of such boxes of different sizes scales like $M(B_L(x)) \sim L^{D_{\text{loc}}}$ then D_{loc} is called the local mass dimension in

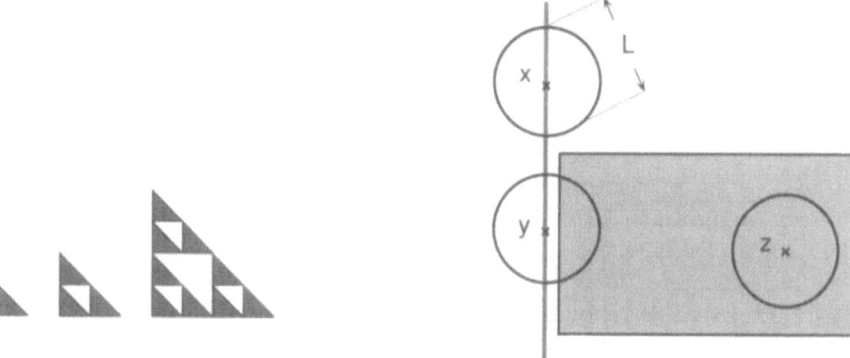

Fig. 2 *a)* Left: the first three stages of a Sierpinski triangle. *b)* Right: A black and white image containing two objects; a line and a surface.

point x. In general this dimension varies with the location x. For example in fig. 2b $D_{\mathrm{loc}}(x) = 1$ and $D_{\mathrm{loc}}(z) = 2$.

In medical images this scaling behaviour can only be tested with box-sizes between 1 and the size of the image. When the $\log M(B_L(x))$ versus $\log L$ plot yields a straight line, or oscillates around a line, then the concepts of dimension make sense and one can use linear regression to determine the exponents D.

4   Local-connected mass dimension in image analysis

Most often an image contains several objects which can be differentiated by means of preprocessing steps. When such a segementation of the image into distinct objects is possible it can be desirable to only consider the local mass dimension in those points x belonging to a particular object. However, in general boxes $B_L(x)$ will also cover parts of objects different from the one to which x belongs: $B_L(y)$ in figure 2b covers both the line and the square. When one does not want to mix the geometric properties of the object point x is part of with those of other objects, it is desirable to only count the mass contributed to the box by the object to which x belongs. For example, let the line in figure 2b be separated from the square by a distance δ, and let us take $L_{\min}$ larger than δ, say $L_{\min} = 2\delta$ and $L_{\max} = 8\delta$. The local mass dimension D_{loc} in point x will be approximately $D_{\mathrm{loc}}(x) \approx 1$ while $D_{\mathrm{loc}}(y) \approx D_{\mathrm{loc}}(z) \approx 2$. The reason for $D_{\mathrm{loc}}(y) \approx 2$ is that the boxes $B_L(y)$ will not only cover the 1-dimensional line, but also part of the 2-dimensional square. If the line and the square are recognized as two different objects then one could exclude the contribution of the square to the mass in the boxes $B_L(y)$ and only keep the contribution of the line. This would yield a local-connected mass dimension [7] $D_{\mathrm{conn}}(y) \approx 1$. So the local-connected mass dimension is an object oriented quantitative characterization of a geometry.

There are many techniques to segment images into distinct objects. The image in figure 2b consists of a white ($I = 1$) background and a black ($I = 0$) geometry. Therefore the first step in segmenting this image is simply to take all sites x with intensity $I(x) = 0$ to belong to objects and the rest of the sites to be background. This is segmentation by means of intensity thresholding. The object to which, say point x belongs, consists of all other sites with intensity 0 which can be *connected* to x by a string of nearest neighbor sites. The application to liver utrasound uses such segmentation by thresholding combined with a further segmentation in connected objects, called *clusters*. The mass M_{conn} used for the estimation of the local-connected mass dimension D_{conn} by means of the scaling relation $M_{conn}(B_L(x)) \sim L^{D_{conn}}$, only incorporates the mass of those points which are also connected to the center point x.

5   Distribution of local-connected mass dimension

The local-connected mass dimension $D_{conn}(x)$ contains quantitative information about the local geometry of the cluster to which x belongs. In general $D_{conn}(x)$ will differ with the location x. One convenient way of representing this large amount of data is to consider the probability density $p(\Delta)d\Delta$, which is the probability that a randomly picked point x on any of the segmented objects has a local-connected mass dimension with $\Delta < D_{conn}(x) < \Delta + d\Delta$. In such an analysis of figure 2b, the «line-cluster» will contribute a peak in $p(\Delta)$ at $\Delta = 1$ and the «square-cluster» a peak at $\Delta = 2$. So by only looking at the distribution $p(\Delta)d\Delta$, one can conclude that the original picture consists of linear structures (the line) and surfaces (the square).

6   Preprocessing and segmentation of the liver ultrasound image

In order to get rid of possible noise in the image due to the lower cutoff in the resolution of the ultrasound machine or due to the digitization procedure, the images are smoothed on length-scales of the order of the pixelsize. This can be done in several ways. We chose to replace the intensity $I(x)$ in each pixel x by the average intensity $I_s(x)$ over a square box of sides 7×7. For x near the boundaries of the image one only averages over the image points within the box.

Since the speckle structure, the vein/liver and liver boundaries seemed important, we decided to enhance the edges by taking the normalized gradient of the image, i.e., the gradient of the logarithm of the image

$$I_{s.g} = \nabla \log I_s(x) = \frac{|\text{grad} I_s(x)|}{I_s(x)}.$$

The new image $I_{s.g}$ has the advantages that it is independent of the setting of the intensity button on the machine or the variable layer of body fat between

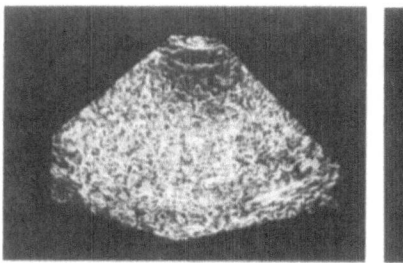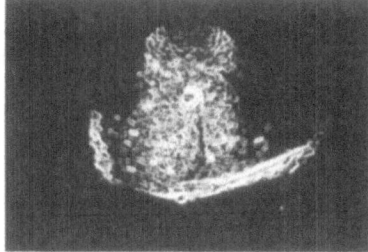

Fig. 3   The normalized gradients of the smoothed images depicted in figure 1.

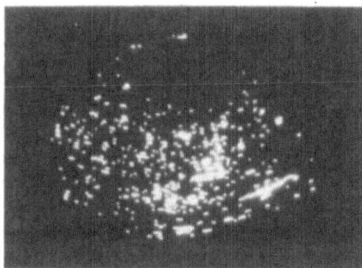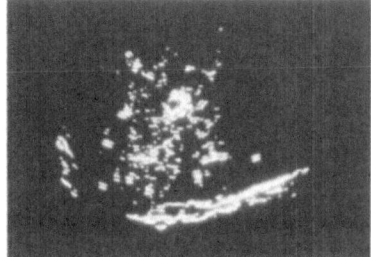

Fig. 4   The white regions have a normalized gradient larger than the threshold value T. The rest is
black. The value T used here was the one giving the largest differentiation between images
of typically fatty and healthy livers.

utrasound head and liver. Here it is assumed that the ultrasound intensity output is
linear in the relevant intensity domain. Also, the speckle structures which usually
have a weaker intensity than the vein/liver and liver boundaries, are put on a par.
The processed images $I_{s.g}$ corresponding to the images in figure 1 are shown in
figure 3. The small wormy structures are the edges of the speckles, and the large
structures are the edges of the veins and liver boundaries.

In order to segment the image $I_{s.g}$ one chooses a threshold T and makes
black all pixels with $I_{s.g}(x) < T$, and white all others. An example is shown in
figure 4. All connected objects can be found using an algorithm developed for
cluster detection in percolation theory [6]. This yields a new image I_{obj} in which
all clusters are numbered and all pixels x in a cluster have $I_{obj}(x)$ equal to the
cluster number. This is a considerable aid when computing the local-connected
mass dimensions.

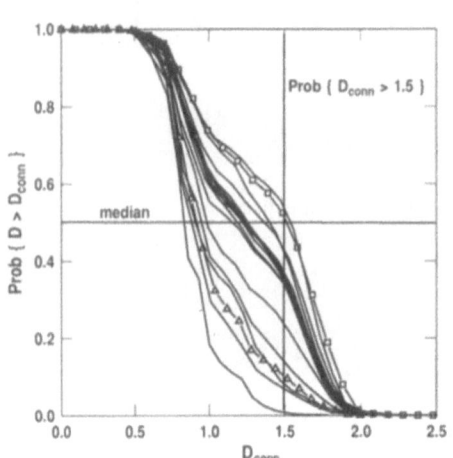

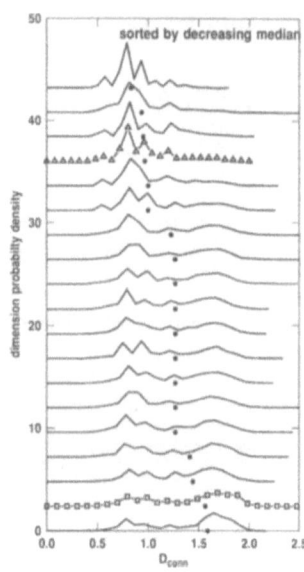

Fig. 5 Cumulative distributions of the local-connected mass dimension for twenty patients with different degrees of liver fattiness.

Fig. 6 The dimension probability densities for the 20 patients. The dots mark the median.

7 Quantitative analysis of the liver images

The analysis starts by determining the local-connected mass dimensions of all points belonging to all objects in the image I_{obj}. This is done using square boxes $B_L(x)$ of sides 2, 4 and 8 and then doing a least square fit of $\log M_{conn}(B_L(x))$ versus $\log L$ to find $D_{conn}(x)$. In order to insure that one is doing something sensible, only those points with a small error in the least square fit are considered. The probability density of the local-connected mass dimension provides a quantitative description of the geometry of the objects.

The optimal value for T is determined experimentally by doing the analysis for different values of T for both a typical ultrasound image of a healthy and fatty liver. The optimal value of T is the one yielding the largest difference among the two probability densities $p(\Delta)d\Delta$ (see the marked curves in figure 6). Such optimal threshold values T were determined for both the analysis of ROI's and the overall analysis, and are subsequently used in the analysis of liver ultrasounds of new patients.

8 Results for 20 patients

For each of twenty patients we estimated the $p(\Delta)d\Delta$, combining results from 3 ultrasound liver images slices of each. The settings of the ultrasound machine were the same for all patients.

Figure 5 shows the cumulative distribution of local-connected mass dimension for the 20 patients. Two numbers seem to be relevant: The first is the median, i.e., the value of the dimension where the different curves intersect with the horizontal line through 0.5. The second is the probability $\text{Prob}\{D_{\text{loc}} > 1.5\}$ for a local-connected mass dimension larger than 1.5, i.e., the value of the cumulative probability at which the different curves intersect with the vertical line through $D_{\text{loc}} = 1.5$. Both the curves corresponding to the typically healthy and fatty liver intersect this vertical line at very different values: For the healthy liver (squares) half of the points have dimension larger than 1.5. For the fatty (triangle) one that's only 10%. In a similar fashion also the median provides such a quantitative differentiation between healthy and fatty.

Figure 6 shows the corresponding densities $p(\Delta)d\Delta$ ordered from top to bottom by increasing median, i.e., increasing content of higher dimensional objects. Since the boundaries between the different tissues is not so clearly demarkated in fatty livers, their thresholded $I_{\text{s.g}}$ images contain less 2 dimensional structures (see figure 4) as those of healthy ones. This explains the potential relevance of both the median and $\text{Prob}\{D_{\text{loc}} > 1.5\}$ as measures of the degree of liver fattiness.

Similar plots were also investigated for ROI's containing the speckle texture of pure liver tissue. The cumulative distributions obtained from that analysis were all very much the same, making them apparently useless for quantitative diagnostic purposes.

9   Comparison with CT

As a test our results were compared with CT scan fat estimates for the same livers: the intensity of CT scans is known to correlate well with tissue's fat content. The Spearman correlation coefficient [2] between the CT ordening and the ordening according to increasing $\text{Prob}\{D_{\text{loc}} > 1.5\}$ for these 20 patients was 0.74, while that between the CT ordening and median ordening was 0.63. In absolute sense these numbers may not seem impressive. Nevertheless they are very much better than anything obtained in previous textures analysis.

Showing the ultrasound images to a specialist, asking this person to order them in 4 categories, we found remarkably close agreement with our fractal image analysis. In a sense this is not surprising, since the method of analysis discussed here was aimed at quantifying what the specialist's eye was capturing.

Two ingredients are new in this context: First the fractal analysis based on the distribution of local dimensions and second, perhaps more important, our departure from the traditional methodology of looking at small regions of interest containing only smooth liver tissue.

Acknowledgements:
We are very grateful to Richard Voss for many discussions and for his plot program VP. We also like to thank Dr. W. Hofer for his diagnoses of the unprocessed ultrasound liver images.

References

[1] Evertsz, C.J.G., Mandelbrot, B.B. *Multifractal measures* In: Ref. 4, 921–953.

[2] Hays, W.L. *Statistics* Holt, Rinehart and Winston, London, 1963.

[3] Mandelbrot, B.B. *The fractal geometry of nature* W.H.Freeman, San Francisco, 1982.

[4] Peitgen, H.-O., Jürgens, H., Saupe, D. *Chaos and Fractals* Springer-Verlag, New York, 1992.

[5] Raeth, U., Schlaps, D., et al. *Diagnostic accuracy of computerized B-scan texture analysis and conventional ultrasonography in diffuse parenchymal and malignant disease* J. Clin. Ultrasound 13, 87–99, 1985.

[6] Stauffer, D., Aharony, A. *Introduction to Percolation Theory* Taylor & Francis, London, 1985.

[7] Voss, Richard F. *The local connected fractal dimension: multifractal classification of early Chinese landscapes drawings* In: R.A.Earnshaw (ed.) Application of Chaos and Fractals, Springer-Verlag, London, 1992.

[8] Zahlten, C., Evertsz, C.J.G., et al. *Fraktale in der Analyse van Ultraschallbildern der Leber* preprint

Fractal Dimension as a Characterisation Parameter of Premalignant and Malignant Epithelial Lesions of the Floor of the Mouth

Gabriel Landini and John W. Rippin
Oral Pathology Unit, School of Dentistry
University of Birmingham
St. Chad's Queensway
Birmingham B4 6NN, England, U.K.

Abstract. Irregularity of shape of epithelial-connective tissue interfaces is a well-recognised feature of malignant and premalignant epithelial lesions, yet few attempts have been made to assess it objectively. The fractal dimension (as a measure of irregularity of shape) of the epithelial-connective tissue interface of normal, premalignant and malignant epithelial tissues of the floor of the mouth was measured using box counting and boundary trace methods.

The lowest estimated value was 0.97 (a normal mucosa using the box method), and the highest 1.61 (a carcinoma using the trace method). Analysis of the values against the histopathological diagnoses (normal, keratosis with mild dysplasia, keratosis with moderate/severe dysplasia and squamous cell carcinoma) showed no significant difference between normal epithelium and that from keratosis with mild dysplasia, but these were significantly different from the two other diagnoses, which were also significantly different from each other. The study illustrates the potential of fractal analysis for providing objective diagnostic information about irregular shapes in histopathology and objective descriptors for tumour growth.

1   Introduction

The oral cavity is covered by mucosa which is composed of two different tissues: a stratified squamous cell and underlying connective (lamina propria). Malignant transformation of the epithelial tissue into a squamous cell carcinoma may occur, and its occurrence is thought to depend on a multitude of exogenous and endogenous factors [11]. Oral cancer accounts for approximately 5% of all diagnosed tumours in western countries and as much as 60% in some South Asian regions.

The shape of the junction between the basal epithelial cells and the underlying lamina propria, the epithelial-connective tissue interface (ECTI) of the oral mucosa is regular: the shapes of rete pegs and interdigitating papillae repeat. In the (normal) floor of the mouth, which is the most susceptible area of the oral cavity to cancer transformation, the shape of this interface is relatively flat. In premalignant lesions (epithelial dysplasia), the ECTI often becomes irregular due to increased cell proliferation. In frank malignancy (squamous cell carcinoma), this effect is more marked and detached islands of cells penetrate the lamina propria. At this stage, as seen ultrastructurally, the basement membrane can be defective [8] or even absent [1]. One of the pathologist's major difficulties is classifying premalignant mucosal lesions. Most often this is done in a rather arbitrary (or at

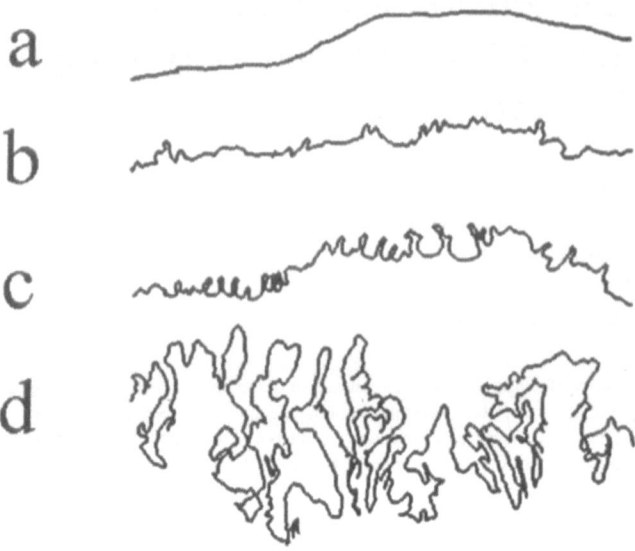

Fig. 1 Four profiles of epithelial-connective tissue interfaces form (a) normal tissues, (b) mild epithelial dysplasia, (c) moderate/severe epithelial dysplasia and (d) squamous cell carcinoma
of the floor of the mouth.

least, subjective) way based on a varying number of cytological features such as
number of mitosis, cellular and nuclear atypia and pleomorphism, nuclear hyper-
chromatism and also irregularity of the ECTI [4][5].

It has been suggested that premalignancy and malignancy progress by the
proliferation of some transformed stem cells at the expense of neighbouring normal
cells [15]. Such proliferation may be balanced by increased cell exfoliation at the
free surface of the mucosa, but it also helps to account for the irregularity of the
ECTI. Since fractal geometry can be regarded, for present purposes, as a numerical
method of describing the irregularity of shapes [7], fractal analysis may provide an
objective means of classifying the irregularity of the ECTI and eventually become
a diagnostic parameter in histopathology.

2   Materials and methods

Hematoxylin-eosin stained histological sections of lesions from the floor of the
mouth, presenting to the routine histopathology service of the University of Birmingham School of Dentistry, were selected. Sections from lesions of each of the
following were examined and re-diagnosed: 10 normal mucosae, 10 keratoses
with mild dysplasia, 10 keratoses with moderate or severe dysplasia and 10 well
differentiated squamous cell carcinomas.

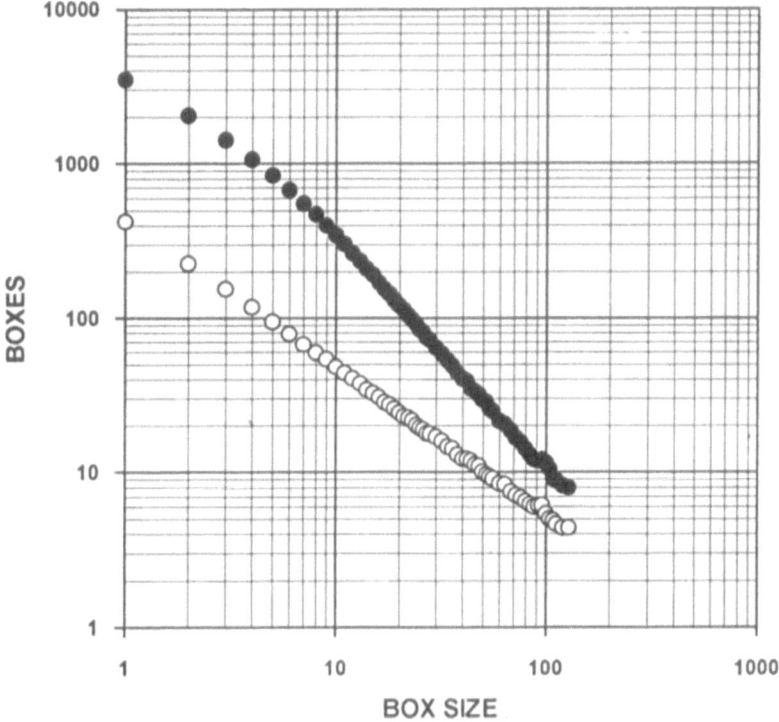

Fig. 2 The fractal dimension of the epithelial-connective tissue interface by the box method is given by the negative of the slope of the linear part of the log-log plot. Empty circles represent data from figure 1a (normal tissue, box dimension = 0.98); filled circles: data from figure 1d (squamous cell carcinoma, box dimension = 1.47). Note the tendency to flatten of the upper part of the plot. Box size is in pixel units.

The classification of dysplasia into mild, moderate and severe categories, at least as far as the oral mucosa is concerned, is largely dependent on the pathologist's subjective interpretation [11], but it is recognised practice. The dysplastic lesions were grouped into two as «mild» and «moderate/severe», since it has been found that the former are much less likely to transform into carcinoma than the latter. The distinction between moderate and severe dysplasias is less marked in this respect, and hence it is recommended that, for the floor of the mouth, moderate/severely dysplastic lesions be fully excised, while mildly dysplastic lesions may be treated conservatively [11].

Projected images of the junction between the basal epithelial cells and the adjacent lamina propria were enlarged 23.7 times, hand traced and then scanned into a personal computer (Fig. 1) achieving a final resolution of 1 pixel = 11.72 micrometres.

The fractal dimensions of the shapes were calculated using the box counting and boundary trace methods [7][2]. The box counting method measures the space

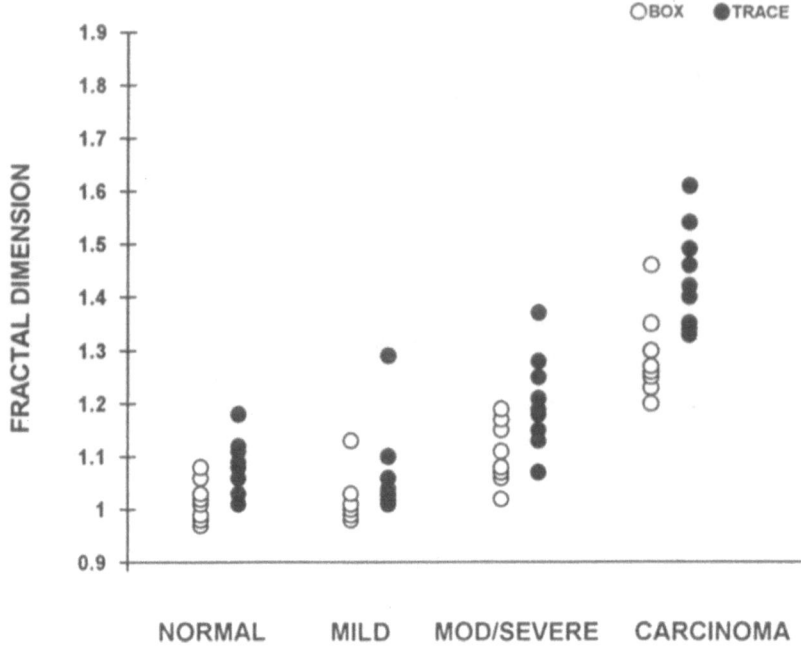

Fig. 3   Fractal dimensions of the epithelial-connective tissue interfaces of the 40 cases estimated by
the box and trace methods.

filling properties of an object at different resolutions by superimposing a grid of
increasing size on the planar space of the figure being measured, and counting the
number of «boxes» $(N(r))$ of linear size r that contain any part of the figure. If a
plot of $\log(r)$ against $\log(N(r))$ tends to a straight line in some range of r (usually
larger than one order of magnitude), then the image can be regarded as fractal,
and its fractal dimension (D) is estimated as $D = -S$, where S is the slope of
the regression line in that range. In the present study, 58 box sizes of side length
ranging from 1 to 128 were used (Fig. 2).

With the boundary trace or yardstick method [7], the length of the image (l)
is measured with, rulers of increasing size (r). The value of D is calculated as
$D = 1 - S$, where S is the slope of the linear part of the regression of $\log(r)$
on $\log(l)$. In this study the ruler sizes used were from 1 to 200 pixels (unit steps
from 1 to 10, steps of 5 from 10 to 30, steps of 10 from 30 to 200). Mean ruler
length values were used after each figure measurement; this is necessary because
of unavoidable errors associated with the square shape of the computer pixels. For
example, the possible distances in a boundary for a ruler $r = 2$ are: 2, SQR(5) and
SQR(8). The methods were validated by measuring computer generated shapes of
known fractal dimension. For the experimental data a two-way analysis of variance,
with diagnosis and method of fractal dimension estimation as the variables, was
carried out.

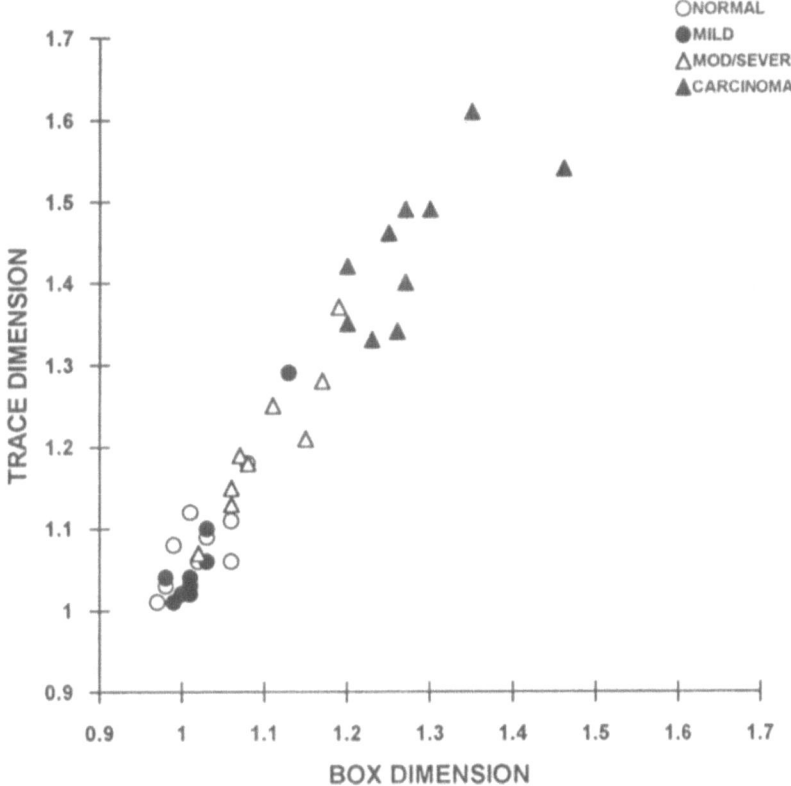

Fig. 4   Box dimension versus trace dimension. Note the clear separation of the normal from the malignant cases.

3   Results

Estimated fractal dimensions of standard figures agree very closely with their known fractal dimensions (Table 1). For the experimental data, the mean ranges of scales for which the log-log plots were linear were between 40.4 and 884.6 micrometres for the box counting method, and between 44.6 and 1134.3 micrometres for the trace method. For ranges smaller than these, the plots tended to flatten to $D = 1$ (Fig. 2). A two-way analysis of variance showed significant differences between the methods of fractal dimension estimation ($P < 0.001$) and the diagnoses ($P < 0.001$), but no significant interaction between method and diagnosis ($P > 0.4$). Further examination showed that for both methods, the data for normal mucosa was not significantly different from that for keratosis with mild dysplasia ($P > 0.7$), but that these were both significantly different from the two other diagnoses (moderate/severe dysplasia and squamous cell carcinoma), which were in turn significantly different from each other (for all, $P < 0.05$). The two methods gave significantly different results for each diagnosis ($P < 0.05$); as can be

Object	Ideal D	BOX	SD	TRACE	SD
Line	1.0000	0.99	0.004	1.00	0.000
Circle	1.0000	1.00	0.013	1.00	0.000
Square (perimeter)	1.0000	1.00	0.000	1.01	0.009
Square (filled)	2.0000	1.97	0.013	*	
Koch island coastline	1.2618	1.26	0.018	1.26	0.004
Koch island coastline	1.5000	1.49	0.018	1.48	0.018

Table 1 Ideal and mean estimated fractal dimension of known objects. The mean values were obtained from 5 measurements starting at random initial positions of the measurements. SD: standard deviation. * not applicable

seen (Figs. 3 & 4), the trace method gave fractal dimensions consistently higher than the box method by a factor of about 1.06. The minimum fractal dimension obtained was 0.97 for a normal mucosa (box method), and the maximum 1.61 for a carcinoma (trace method).

4   Discussion

Mathematical fractals show scaling at all orders of magnitude, from the infinitely small to the infinitely large, but in natural structures there are usually both upper and lower bounds. In the present study, the log-log plots showed a linear trend at relatively coarse resolutions, but asymptotic behaviour tending to $D = 1$ at relatively high ones, in accord with Rigaut's concept of an asymptotic fractal [12][13]. It may be that as the measurements approximate the highest light microscope resolutions, the size of the individual cells is reached, whence the shape of the ECTI would depend on individual cell factors (such as the variability in number of pseudopodia and hemi-desmosomes) [9][16] rather than on the lesion as a whole.

Interestingly the changes of the ECTI reveal in some way the dynamics of the tissue in its different stages and this prompts speculations about tumour growth as a self-similar process. Self-similar processes have already been proposed as being responsible for pattern formation in normal chimaeric liver of rat [3] and in simulated «cell pushing» replicative systems [6] revealing their characteristic fractal measures. However in carcinogenesis and tumour growth, it may help to understand local aggressivity, tumour infiltration and metasatatic spread.

It is not surprising that, as measured by fractal dimension, the shape of the ECTI is significantly more complex in squamous cell carcinomas than normal mucosa, and dysplastic lesions are somewhere between the two. Many pathologists would regard such statements as facile, and yet few if any would be prepared to make their diagnoses based on the shape of the ECTI alone (Fig. 1). Instead diagnosis is done by taking into account a variable number of features, most often in a non-systematic way. By contrast, the methods of fractal analysis are systematic and statistically reliable. Other morphometric methods such as the quantification of the number of basal cell pseudopodia [9][16] have been described, but these methods involve electron microscopy and intensive user intervention on what structures to analyse. Further attempts to standardise the shape of the epithelial buds (and consequently the irregularity of the ECTI) in the esophagus yielded a classification of epithelial buds of type I, II or III, depending if they are regular and same sized (type I), regular but differing in size (type II) or irregular (type III) [14]. However this represents another subjective categorical variable.

Scale-bounded measurements (such as maximum length of the ECTI at a determined resolution) are of use only if all the measurements are done with the same resolution [10]; at the moment there is no such standardisation in histopathology. This approach may also lead to wrong assumptions about the assessment of «irregularity» in lesions such as pillomatrixomas, inverted papillomas and basal cell carcinomas (lesions with large and smooth epithelial buds) in which the length of the ECTI is increased but there is no «irregularity». The technique presented here, after the isolation of the epithelial-connective tissue interface (now being investigated using selective staining methods and a digitizer camera), is fast, straightforward and performed at a relatively low magnification. Of course, this is not to say that every single case can be classified on the basis of the fractal dimension of its ECTI. The usual implication of designating a lesion as showing «mild epithelial dysplasia» or «severe epithelial dysplasia» is that the latter is more likely to transform than the former, yet such studies as have been carried out have demonstrated the poor reliability of such a process [17]. Thus, it may be that those lesions that in the present study have been designated as «mild dysplasia» but had a high fractal dimension (the highest in this group was 1.29), or those designated as «moderate/severe dysplasia» but with a high fractal dimension (the highest in this group was 1.37), have in fact a greater risk of transformation than the histological designation would imply. This contention can be appraised only when a large number of cases have been analysed and followed for a number of years, when their ultimate prognosis will be known. The fractal dimensions can then be statistically analysed against that prognosis in order to determine the predictive value of fractal dimension of the ECTI in histopathology.

Acknowledgement:
This study was supported by The Leverhulme Trust grant No. F.94AO.

References

[1] Ashworth CT, Stembridge VA and Luibel FJ: A study of basement membranes of normal epithelium, carcinoma in situ and invasive carcinoma of the uterine cervix utilizing electron microscopy and histochemical methods. Acta Cytol 5: 369–384, 1961.

[2] Barnsley M: Fractals everywhere. San Diego, Academic Press, 1988.

[3] Iannaccone PM. Fractal geometry in mosaic organs: a new interpretation of mosaic pattern. FASEB J 4: 1508–1512, 1990.

[4] Kramer IRH: Computer-aided analyses in diagnostic histopathology. Postgrad Med J 51: 690–694, 1975.

[5] Kramer, IRH, El-Labban N, Sonkodi S: Further studies on lesions of the oral mucosa using computer-aided analyses of histological features. Br J Cancer 29: 223–231, 1974.

[6] Landini G, Rippin JW: Fractal fragmentation in replicative systems. FRAC-TALS (In press), 1993.

[7] Mandelbrot BB: The fractal geometry of nature. San Francisco, Freeman; 1982.

[8] McKinney RV, Singh BB: Basement membrane changes under neoplastic oral mucous membrane. Ultrastructural observations, review of the literature, and a unifying concept. Oral Surg 44: 875–888, 1977.

[9] Okagaki T, Clark BA, Twiggs LB: Measurement of number and cross-sectional area of basal cell pseudopodia: a new morphometric method. J Cell Biol 91: 629–636, 1981.

[10] Paumgartner D, Losa G, Weibel ER. Resolution effect on the stereological estimation of surface and volume and its interpretation in terms of fractal dimensions. J Microsc 121(1): 51–63, 1981

[11] Pindborg JJ: Oral cancer and precancer. Bristol, John Wright & sons Ltd.; 1980.

[12] Rigaut JP: An empirical formulation relating boundary lengths to resolution in specimens showing «non-ideally fractal» dimensions. J Microsc 133 Pt1: 41–54, 1984.

[13] Rigaut JP: Fractals, semi-fractals et biometrie. In: Fractals, dimensions non entieres et applications. G. Cherbit ed. Paris, Masson, 1987.

[14] Rubio CA, Liu F-S, Zhao H-Z: Histological classification of intraepithelial neoplasias and microinvasive squamous carcinoma of the esophagus. Am J Surg Pathol 13: 685–690, 1989.

[15] Selby P, Buick RN, Tannock I: A critical appraisal of the «Human Tumor Stem-cell Assay». N Engl J Med 308, 129–134, 1983.

[16] Twiggs LB, Clark BA, Okagaki T: Basal cell pseudopodia in cervical intraep-ithelial neoplasia; progressive reduction of number with severity: a morpho-metric quantification. Am J Obstet Gynecol 139: 640–644, 1981.

[17] Wied GL, Bartels PH, Bibbo M, Dytch HE: Image analysis in quantitative cytopathology and histopathology. Hum Pathol 20: 549–571, 1989.

Modelling

Modelling HIV/AIDS Dynamics

Philippe Blanchard
Theoretische Physik and BiBoS
Universität Bielefeld
D-33615 Bielefeld

1   Introduction

The main purpose of this paper is to give a short account of two mathematical models dealing with the HIV dynamics. Details and proofs of these results are presented in [1], [2], [3], [8]. As well known AIDS epidemics has particularities which made it quite different from previous epidemics. First there are wide variations in human sexual behaviour (heterogeneity both of the individuals and the contacts they make). Secondly the time-lag between infection by HIV and the appearance of overt AIDS is both long and variable (median roughly about 10 years for homosexual men). Thirdly the degree of infectiousness is also time dependent. To include all these facts discrete stochastic models are really necessary and needed. For this reason we have proposed and developed in recent years in Bielefeld a new type of stochastic modelling based on Random Graphs which allows both precise mathematical predictions and better adaptation to the available behavioural, epidemiological and biological data. This model will be presented in Section II.

Despite intensive study a lot of uncertainty still surrounds the interaction mechanism between immune system and HIV-viruses in infected patients. Many different hypotheses have been discussed and suggested to explain the decay of the immune system and the induced increased susceptibility of HIV infected patients to opportunistic infections. In [4] M. Nowak, R. Anderson and R. May use a simple differential equation model to try to explain the long latent period on the basis of the extreme sequence diversity of HIV. According to the model, AIDS is the result of an increase in the number of genetically distinct virus strains throughout the incubation period. The authors show that when the number n of variants exceeds a critical value n_c one observes a dramatic increase of the total number of viruses and the collapse of the host immune system. In other words according to this model, HIV infection is an evolutionary process that determines the time scale from infection to disease.

A mathematical model describing how infection agents can persist in reservoirs of latently infected cells, the larger of which being the reservoir of follicular dendritic cells (FDC) and showing why progression to AIDS occurs in a sudden transition will be presented in section III. With hindsight it is not surprising that the lymph nodes should be sites at which the presence of HIV is most important;

that, after all, is where most T-cells usually reside. The conclusion must be that the apparently latent period between infection with HIV and the overt symptomes of AIDS is not clinical latency but rather a period in which the replication of the virus proceeds apace in the lymph nodes or some other tissues of the immune system (the spleen, adenoid glands, ...). It is clear that each model of the interaction of the HIV with parts of the immune system is a very simple caricature of the true tremenduous complexity but it can help to reduce the ignorance about how HIV causes the diseases and for this reason mathematical modelling can play a serious role in the advancement of the knowledge of the pathogenesis of HIV infection.

Let us mention briefly a third very interesting aspect of HIV dynamics namely the study of the evolutionary history of HIV and AIDS. How old is HIV? Where does the HIV virus come from? Using the new method of statistical geometry in sequence space developed by M. Eigen, R. Winkler-Oswatisch and A. Dress [5], M. Eigen and K. Nieselt-Struwe obtain answers to questions of age and kinship of HIV by comparative sequence analysis of the viral genomes or parts thereof, concluding that the present form of AIDS is probably not older than 50–100 years [6].

2   The Spread of HIV on Random Graphs

The main use of mathematics in epidemiology is to understand the relative importance of factors influencing the spread of a disease. A topical and central question is to obtain insight into the relation between mechanics operating on the individual level and phenomena resulting on the level of population. The last decade has seen enormous interest in the mathematical modelling of the spread of AIDS. The wast majority of papers have been concerned with deterministic models.

The basic probability space we consider in our model is a space of random graphs encoding mathematically a set of behavioural and demographical properties of a given community. Let $V = \{x_1, \cdots, x_N\}$ be the set of vertices (individuals) which split into several relevant groups V_i (expressing e.g. the risk behaviour and the heterogenity of the population), $V = \bigcup_{i \in I} V_i$. Since edges between vertices describe a sexual contact, the basic given data generating the underlying graphs are the degree distributions inside the V_i as well as the coupling between pairs (V_i, V_j) of subgroups of the vertices. All graphs G which satisfy these given distributions of contact rates are the elements of the random graph space $\mathcal{G}$ which become a probability space by giving each element $G \in \mathcal{G}$ the same probability $P(G) = \frac{1}{|\mathcal{G}|}$, where $|\cdots|$ denotes the total number of elements.

To describe the epidemics dynamics we introduce on each graph with vertex set $V = \{x_i\}_{i \in I}$ a sequence $\{\varphi\}_{n \in \mathbb{N}}, \varphi_n : V \to V$ of pairing (matching) functions. The φ_n tells us at time n the individuals which are potentially able to propagate an «infection» by having a sexual contact i.e. $\varphi_n(x)$ represents if $\varphi_n(x) \neq x$ the sexual partner of x at time n. Now we turn to the description of the spread of the virus. First of all for all $n \in \mathbb{N}$ we define a mapping

$$X_n : V \to \{0, 1, \cdots, k\} \quad k \in \mathbb{N}$$

$X_n(x_i) = \ell$ being the function which assigns to individual x_i its state of health ℓ at time n. We consider for instance the following progression: $X_n(x) = 0 \Leftrightarrow x$ is healthy at time n, $X_n(x) = 1 \leftrightarrow x$ is HIV-infected at time n and not infectious, $X_n(x) = 2 \leftrightarrow x$ is HIV-infected at time n and infectious, ...

With this in mind the dynamics of the model can be described for each x using the set of transition probabilities

$$P \cdot [X_{n+1} = i | X_n = j]$$

for $i, j = 0, \cdots, k$. These conditional probabilities express the progression of the disease.

The simplest model of this class is generated by independent matchings and can be interpreted as describing a homosexual community V of constant size and divided in p age groups V_i of size N. At each iteration step $n = 1, 2, \cdots$ we allow complete independent matchings between the individuals. Furthermore each iteration is connected with a shift of the elements of V_i into V_{i+1} (growing old of the population) and with the choice of a sexual partner. The model has only two states: $0 =$ healthy, $1 =$ infected. The probability of infection per partnership (i.e. per contact) is assumed to be $\gamma, 0 \leq \gamma \leq 1$. At each time $n = 0, 1, 2, \cdots$ the process will be in one of the $(N+1)^p$ possible states corresponding to the number of infected individuals in each group, i.e. by a random vector $(I_1^n, \cdots, I_p) = I^n$ and the matrix of all transition probabilities $P[I^{n+1} = \cdots | I^n = \cdots]$ can be explicitly given. Moreover the asymptotic behaviour of this Markov chain can be analysed [9]. For this model we can introduce a discrete dynamical system for the iterated expectations of the number of infected individuals, which leads for large N to a good approximation. The type of the globally attractive stationary solutions depends on the value of a critical parameter R which is nearly equal to $\frac{\gamma(p-1)}{2}$. This threshold parameter $R \approx \frac{\gamma p-1}{2}$ has a epidemiological as well as a graph theoretical interpretation. R is on the one side nothing else than the basic reproduction ratio giving the expected number of secondary cases generated by a typical individual during its infectious period.

For $R > 1$ there exists therefore a unique non trivial globally attractive endemic state. On the other side the initial degree of our graphs was p. Let us know introduce an effective degree by

$$d_{eff} = \gamma \frac{p-1}{2} + 1.$$

On the graphs the condition $d_{eff} > 2$ can be interpreted as the connectivity threshold condition for the existence of a «giant» component corresponding to a non trivial endemic state.

For independent matching models in the limit where the population size $N \rightarrow +\infty$ it can be shown that the effective graph restricted to any finite subset of vertices is almost surely a tree and that the epidemic dynamics can be described by a branching process with time replaced by graph distance. We studied also in the framework of independent matching models an age structured population with age dependent choice of the partners and non constant infectivity. In the limit $N \rightarrow +\infty$ the critical parameters are obtained by studying a $(p-1)$-type branching process (p being the number of age groups). The overcritical case corresponds to the situation where $\lambda_{max} > 1$, λ_{max} being the largest eigenvalue of the transition matrix for the corresponding $(p - 1)$-type branching process. $\lambda_{max} = 1$ defines the critical manifold in the parameters space (transmission probability γ, length of partnership, age dependent coupling matrix, time dependent infectivity). Our result shows that time dependent infectivity can indeed correspond to the worst case in spread of HIV. We refer to [3], [13] for a more detailed discussion.

The principal advantage of this modelling approach is, of course, to have incorporated in the mathematical framework an object representing the network of sexual partnerships in its important structural properties. This property distinguishes the random network approach to epidemiological modelling and standard differential equation models and explain their differing predictions for the course of the epidemics at least for certain subpopulations. The model simulates single individuals, making it particularly well adapted to study certain prevention scenarios (condom use, contact tracing, change of behaviour), to include different characteristic properties for single individuals (like predisposition, cofactors) as well as time dependent infectivity and different virus mutants.

In [2] we studied a reference scenario for which the graph consists of 100.000 vertices, divided into 5 groups, corresponding respectively to homosexual males, bisexual males, heterosexual males, heterosexual females without and with contact to bisexual males. 4% of the males are homo/bisexual. Partner numbers integrated over a period of one year are Poisson distributed with an average of 5 for the homosexuals and 1.22 for heterosexuals, corresponding to 83% living in monogamous relationships with an average of 8 life time partners.

We use for our discrete process a time step of length one day and in each time step only one edge is activated per vertex. The epidemic dynamics inside the heterosexual population follows a different pattern inside the homosexual population, and this difference persists over time. The homosexual epidemics shows the familiar pattern of an exponential increase turning into saturation. The number of heterosexual cases, apart from a very short transient period, raises linearly with time. Due to the low number of partner the graph inside the effective heterosexual population consists almost exclusively of linear or almost linear structures like chains or small trees with few branchings, which implies that the epidemics in the heterosexual part of the population is undercritical. It is very important to realize that HIV-infection cannot persist inside the heterosexual population as a self-sustaining epidemics for sets of parameters typical for North European societies without constant influx from those subgroups where it is overcritical or endemic. But on the other hand even if the heterosexual epidemic is undercritical we should be aware of the possibility of large number of cases from purely heterosexual transmission, since slight modifications in the assumptions about transmission probabilities will increase the absolute number of heterosexual cases to or beyond the level of homosexual cases in the long term. In comparison with differential equation models the spread of HIV from the highly promiscuous part of the homosexual population into the heterosexual population with low partner numbers is considerably slowed down by the discrete properties of the interface between bisexual man and the heterosexual population. For more details the reader is referred to [1], [2].

3 Follicular Dentritic Cells as Major HIV Reservoir

HIV infection causes progressive depletion of CD4/HIV receptor positive T helper lymphocytes, ultimatively leading to AIDS. The major HIV reservoir of T-helper cells infection in lymphoid tissues however has remained poorly defined. In [7] H. Spiegel et al. have identified the FDC cells as a major reservoir for HIV-1 in lymphoid tissues facilitating infection of $CD4^+$-T-helper cells. Since T-helper cells travel through the FDC meshwork during the migration within lymphoreticular tissues, it appears likely that HIV-replicating T cells may infect FDC, which then infect now T cells, thus causing a gradual dissemination of the virus to cell FDC and thereby a steadily increasing infection of T-helper/memory cells. This results in $CD4^+$-T cell depletion and ultimatively in immunodeficiency. Two other groups have also shown that the long and variable latency period after HIV infection is explained by replication of virus in the lymph nodes [11], [12].

The FDC cells can be directly infected by HIV and can stay infected for a long time, corresponding to a immunological silent way to transmit the virus. We start by formulating a mathematical model in the framework of interacting particle systems. Consider an finite lattice $\Omega \subset \mathbf{Z}^d, d = 2, 3$, whose sites $\omega \in \Omega$ are occupied by cells (representing the FDC). At each discrete time $t \in \mathbf{N}$ i particles (representing the T4 lymphocytes) are emitted by the source with probability p_i and each particle is supposed to undergo a random walk in Ω up to the (random) time it reaches the boundary $\partial \Omega$ of Ω and leaves Ω. Let T be the mean sojourn time of a particle in Ω. Each site of Ω is therefore occupied by 1 or 2 particles; so if two particles simultaneously make the site occupied a cell-particle contact takes place. We consider only two possible states $\{0, 1\}$ for the particles and the cells. The states 0 and 1 have the following meanings: 0 = healthy, 1 = infected. With A_0 we denote the set of cells infected at time 0 and with A_t the set of infected cells at time t. Infection can only take place during a contact cell-particle. Let α be the infection probability by a contact infected cell-particle and β the infection probability by a contact infected particle-cell. Our attention focuses on the following problems:

1) Describe the asymptotic behaviour of $|A_t|$ for $t \to +\infty$ starting from a finite set $|A_0|$ of infected cells.

2) Study the behaviour of P_t, P_t being the conditional probability that a particle entering healthy in Ω at time t will leave Ω infected.

The main results says that there is a critical time $t_c \equiv t_c(\alpha, \beta, T)$ so that each particle entering healthy in Ω will leave Ω with probability one

– healthy for $t - T < t_c$

– infected for $t > t_c$.

In other words P_t obeys almost a $0 - 1$ law and for $t = t_c$ «percolation occurs» and the system undergoes a phase transition. The occurence of a «critical phenomena» is central to the appeal of percolation. The transition time Δt for $\alpha \geq \beta$ is much smaller that t_c i.e. $\Delta t \ll t_c$. For $\alpha > \beta$ if we suppose that $t_c \approx$ years then $\Delta t \approx$ months. At percolation time t_c the critical density $\rho_c = \frac{|A_{t_c}|}{|\Omega|}$ of infected cells is nearly zero and for $t > t_c$ ρ_t can remain low. These results can give an explanation for the very long incubation time by HIV infection. Indeed intuitively the results above say that the FDC meshwork behaves like a switch: In the subcritical phase when $t < t_c$ the T4 lymphocytes leave the lymphe nodes Ω with very probability non infected while in the supercritical phase when $t > t_c$ a very high proportion of the T4 lymphocytes becomes infected in the meshwork of FDC cells.

Acknowledgements:
This work has profited from collaborations with G. Bolz, T. Krüger and B. Voigt and discussions with S. Albeverio, K. Dietz, A. Dress and H. Heesterbeck. I wish to thank them for their invaluable contributions.

References

[1] Ph. Blanchard, G.F. Bolz, T. Krüger, Modelling AIDS-Epidemics or any venereal diseases on random graphs, in «Stochastic Processes in Epidemic Theory», Eds. J.P. Gabriel, C. Lefevre, P. Picard, Lecture Notes in Biomathematics 86, Springer 1990.

[2] G. Bolz, Ph. Blanchard, T. Krüger, Stochastic Modelling on Random Graphs of the Spread of Sexually transmitted diseases, in «Progresses in AIDS-Research in the Federal Republic of Germany», Proceedings of the 2nd Statusseminar, BMFT Research Program on AIDS, Ed. M. Schauzu, MMV-Verlag München 1990.

[3] Ph. Blanchard, G. Bolz, T. Krüger, Modelling contact structure and the dynamics of HIV/AIDS on random graphs, to appear in Mathematical Biosciences.

[4] M. Nowak, R. Anderson, R. May, The evolutionary dynamics of HIV-1 quasispecies and the development of immunodeficiency diseases, AIDS (1990) 4 1095–1103.

[5] M. Eigen, R. Winkler-Oswatisch, A. Dress, Statistical geometry in sequence space: A method of comparative sequence analysis, Proc. Natl. Acad. Sci. USA 85 5913–5917 (1988).

[6] M. Eigen, K. Nieselt-Struwe, How old is the immunodeficiency virus? AIDS 1990 4 (Suppl. 1) 85–93.

[7] H. Spiegel, H. Herbst, G. Niedobitek, H.D. Foss, H. Stein, Follicular Dendritic Cells Are a Major Reservoir for Human Immunodeficiency Virus Type 1 in Lymphoid Tissues Facilitating Infection of $CD4^+T$−helper cells, American Journal of Pathology 140 15–22 (1992).

[8] Ph. Blanchard, T. Krüger, B. Voigt, A stochastic model for the HIV-infection of follicular dendritic cells, BiBoS-Preprint 1993.

[9] S. Albeverio, Th. Stahlmann, Discrete Stochastic Processes on Random Graphs and the Spread of HIV, BiBoS-Preprint 525/92.

[10] G. Bolz, Analyzing the Structure of Contact Graphs used in Epidemiological Models of AIDS, in «Dynamics of Complex and Irregular Systems», Eds. Ph. Blanchard, L. Streit, M. Sirugue-Collin, D. Testard, World Scientific (1993) 264–277.

[11] G. Pantaleo, C. Graziosi, J. F. Demarest, L. Butini, M. Moutroni, C. H. Fox, J. M. Orenstein, D. P. Kotler, A. S. Fanci, HIV infection is active and progressive in lymphoid tissues during the clinically latent stage of disease, Nature 1993, 362, 355–358.

[12] J. Embretson, M. Zupanzic, J. L. Ribas, A. Burke, P. Racz, K. Tenner-Racz, A. T. Haase, Massive covert infection of helper T lymphocites and macrophages by HIV during the incubation period of AIDS, Nature 1993, 362, 359–362.

[13] Ph. Blanchard, T. Krüger, Independent matching models: The worst case in a model for HIV-infection with age structured population, pair formation and non-constant infectivity, BiBoS Preprint 1993.

Morphological Diagnosis Turns from Gestalt to Geometry

Vittorio Pesce Delfino, Teresa Lettini,
Michele Troia, and Eligio Vacca
Consorzio di Ricerca Digamma
Università di Bari, Italy

In 1917, W. D'Arcy Thompson had posed the question of size-shape relationship in terms of «Growth and Form» by linking the size-shape problem to the following two fundamental concepts: on the one hand, that is possible to base morphological classifications on a non dimensional type of mathematics (analytical functions), on the other hand, it can be postulated that this approach would allow the verification of the «principle of discontinuity», with particular relevance if this principle occurs in morphogenesis. It is by a Sistema Naturae rethinking D'Arcy Thompson (the «great» D'Arcy Thompson, in our opinion), and not by more popular D'Arcy Thompson's proposal of coordinate trasformation method which is, actually, unrealable and uneffective because it uses discrete landmarks that require to assign an arbitrary meaning of biological homology to the given landmarks, so loosing information on intervals among landmarks.

On this legacy, the problem of allometric finding and shape measurement became particularly relevant. There was a full awareness that differential growths, although per se purely quantitative variations, because they are responsible for allometric changes, are still the fundamental mechanism for shape differentiation in biological structures. The problem became so relevant that R. Holloway (1981) strongly but efficaciously said «measurements such as lenght, width, height, wheter in chords or arcs, only describe space. . . and further run into the abyss of allometric correction. . . If additional information (shape?) to size is expected, some methods of allometric correction must be used».

Measures of development length, both along straight line segments and along boundaries however irregular, together with the definition of areas and angles, represent the primary data in size evaluation. We must fear redundancy and other elements of confusion which may be introduced by use of derived fractions, usually calculated in a great number, but may be submitted to criticism:

- in the first place any index between the two primary measures does not obviously increase the information already available;

- moreover, the derived fractions, typically indexes or percentage, and which are the results of a division, tend to make uncontrollable and hardly interpretable the behaviour of the relative numerators and denominators;

– finally, the derived fractions cannot be considered as independent on the measures used to calculate them and, therefore, their overall use for statistical comparisons or discriminations is not advisable.

Attempts to obtain deterministic valuations of volume beginning from single determinations of surface are frequently misleading; in fact in these cases a reference to regular solids considered as sufficiently approximating to the biological structure examined, whose real irregular shape cannot be submitted to such a reduction, is inevitable. It is also necessary to pay attention to the fact that, in general terms, the difference between the linear and surface measurements is not merely computational, but can be conditioned by intrinsic characteristics of the investigated components.

As overall example: the sum of the perimeters of the transverse sections of two blood vessels considered after a bifurcation does not allow one to go back to the perimeter from the original trunk section, while this is possible when the surfaces are being considered; such behaviour can clearly produce misleading effects on the statistical evaluations. Also the probabilistic evaluation of volumes, starting from a mean value of the measured object areas in a section, must be submitted to some critical considerations: the statement, very important in stereology, that a volume fraction may be calculated from a mean area fraction, is based on the principle of Delesse who applied it in geology to an analogous problem of components of sedimentary layers. The principle of Delesse assumes that in a defined object the different components have an isotropic distribution, i.e. that they are distributed randomly and with the same density in each part of the examined object. When this is not the situation, as Aherne and Dunnill correctly say «morphometric methods which have been devised for randomly disposed or isotropic elements cannot be validly applied to anisotropic element or tissue components...». From a practical point of view the inadequacy of this approach is revealed in another passage, of the same Authors, about the anisotropy of biological structures: «a cursory knowledge of anatomy will at once reveal that this (isotropy) is a property possessed by very few structures and tissues». It means that each time that a fundamental stereological principle is applied, the hypothesis that in different compounds the component distribution is istropic must be tested in a convincing way. This statement, apart from the difficulties in correct data handling, runs each time into a general reflection. In geology the «morphogenetic factors» of a sediment are certainly less numerous and more simple than in biological structures; for example a water sediment formed in undisturbed conditions gets effects only of the regulating vector which is represented by gravity. For this reason different granules, hypothetically present during the sedimentation, will be ranged according to a gradient which is the function of their mass. Therefore, for the same mean value of the mass one can expect an isotropic distribution of the granules. The distribution of the biological structure components, on the other hand, arises from numerous morphogenetic factors which are of different types (structural, functional, active, passive, stable, provisory) and oppose isotropic distribution. In fact, stereology uses widely different types of reference grid to perform the calculus of surface function; some of these grids are conceived to count the oriented components (and so not the isotropic ones). But this refers only to the optimization of the counting procedure to define the surface fraction and does not modify in any way the limits of the validity of valuation of the relative volume fraction.

The techniques of analytical morphometry, in our opinion, overcome the limitations of discrete measurements, and are effective in describing shapes.

This approach approximates the natural logic with which we usually identify a person known to us. We base recognition on the physiognomy of this person — his or her dimensions, color, and, above all, the shape of constituent elements (nose, forehead, or any other profile, narrow areas or wide regions). The trend of anatomical profile (its shape) undoubtedly contains more information than any other evaluation made in the same region. For example, any dimension measurable between landmarks on the glabella, the nasal spine, and the extremity of the nasal bones will not tell us anything about the shape of the nose. However, the profile between those points represents the characteristic with physiognomic value — this forms the basis for any identification operation. It is a case of the classic morphological problem of the relationship between «size» and «shape». Actually, our visual perception is very effective in understanding and using the information of «shape». Furthermore our mind is very effective in shape recognition but very poor in shape description. Yet, in contrast, methods that quantify morphology have already produced relatively simple and effective procedures for dimensional evaluations by calling for specific solutions for numerical descriptions of the «shape» in order to make them good parameters for multivariate statistics. Since the «shape» of an anatomical region must be reduced to the characteristics of the profiles that compose it and since these profiles can be fully considered as curves, the solutions that can describe them must revert to analytical geometry. Once we have progressed from the anatomical term «profile» to the analytical term «curve», many descriptive quantifications are possible.

It is, however, necessary to include the algorithms in a strictly defined logical architecture. The logical architecture of the S.A.M.(Shape Analytical Morphometry) ® system, (Pesce Delfino et al., 1990) originally developed to address the anthropological problems of evolutionary morphology, and subsequently expanded to numerous domains of medical morphological diagnosis, assumes that each irregular shape contains elements of two distinct logical domains: gross distortions that interest the contour as a whole, and its minor local variations. In fact, any irregular object is a displaying of a complex mixing of such two kinds of characters.

Thus, the information is processed separately by appropriate procedures so as to acquire independent parameters both on the logical level and the numerical one; the best prerequisite for multivariate statistics.

Any review of morphological reports shows the frequent occurence of terms such as «irregular shape», «more or less regular shape», «dysmorphic», «polymorphic», «pleomorphic», «distorted», «asymmetric» etc. These are all terms and statements related to the biological object shape (cells, nuclei, particles, tissues, organs, etc.) and are considered so relevant as to be expressly mentioned in support of the diagnosis.

However, this particular aspect at present lacks non only of adequate and easily available instruments, but also of an organised methodologic reference framework. Moreover, it appears to suffer from limited, approximate and sometimes incorrect solutions.

The S.A.M. (Shape Analytical Morphometry) approach is based on the extractions of shape-descriptive parameters, by using procedures which do not require any previous information about the object to be studied in an image and which supply numerical parameters with deterministic significance, i.e. characterized by a not ambiguous relation between the obtained numerical value, the described morphological characteristic and the adopted algorithm. Such parameters, used for a description able to reduce the share of subjectivity, moreover overcome the impossibility of teaching the logic of the diagnostic thinking and may be easily used in multivariate statistics.

The logic of the S.A.M. architecture assumes that the relevant objects from a histological feature are considered as closed figures to be studied in terms of less or more stressed distortions of the whole figure, combined with more or less marked local irregularities of the outline itself; these characteristics may reach extreme degrees by following one another or by variously combining indentations, extroversions, and offshoots which may have a high degree of recursiveness.

When the morphologist uses the terms as above in the diagnosis, he tries to distinguish somehow between the two components and to evaluate them. The aim is thus to quantify the two features independently. As already mentioned, the methods which are traditionally available are not very effective and merely give one very simple parameter called Form Factor (contour index, shape index or roundness factor) that relates the perimeter length (P) to the area value (S) of the corresponding domain by the relationship:

$$FF = (4\pi S)/p^2 \tag{1}$$

This parameter is quite insufficient because there are many different shapes for which the value of this ratio is the same, and, moreover, because it does not express in any way what proportion of the profile length is supported by great distortions of the whole figure and what is affected by the fine contour irregularities. The form factor frequently turns out to be misleading because, when the morphometric machines produce it under this name, the user is led to insert it in his statistical matches and he is convinced that it is just that information relative to «shape» which he, when carrying out the observation and the non morphometric evaluation of the investigated object, extracts by a logical procedure which is practically its opposite; because the observer tries to distinguish between the overall object shape and the local contour perturbations, while the form factor evaluates, with all the numerical limits of a two measure fraction, the result of their combination. As already written just above, the architecture of the software

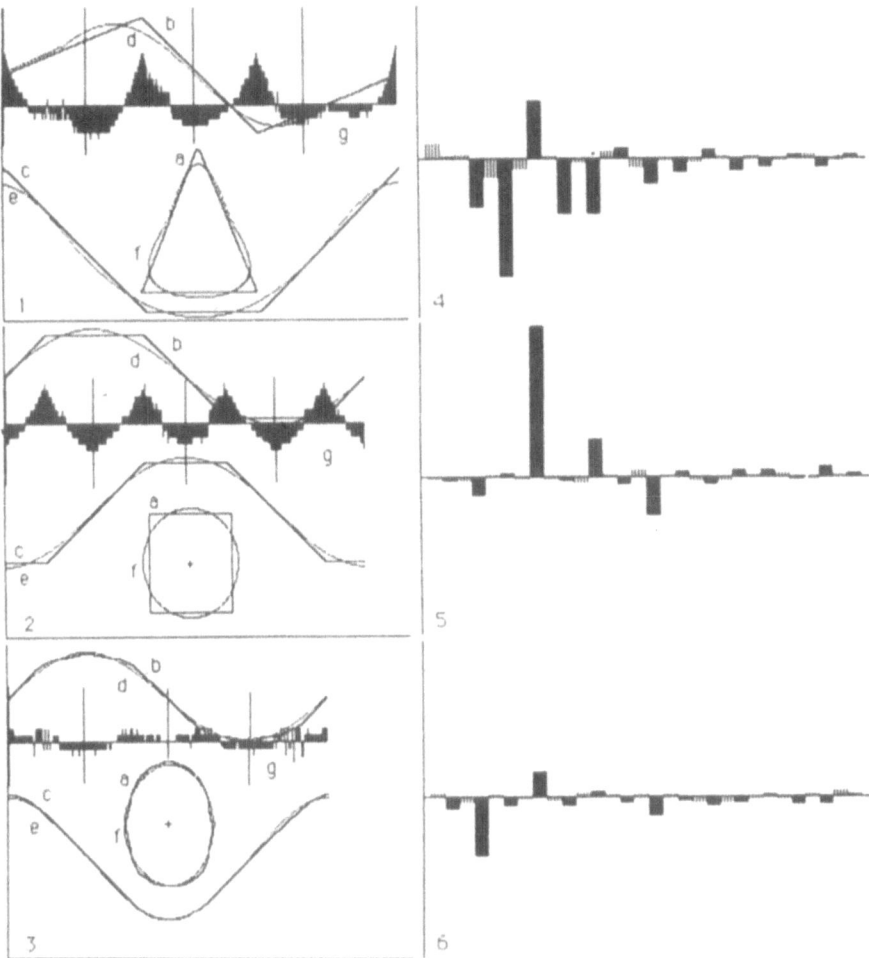

Fig. 1

system S.A.M. supplies parameters with which it is possible to describe independently the morphological features carrying the information about figure symmetry and about local perturbations.

The step in which the separation of these two class scales of information occurs corresponds to the synthesis of the so-called «fundamental shape». Afterwards two separate working channels are defined to parametrize the corresponding characteristics. Before the fundamental shape synthesis the dimensional normalization (dimensional evaluations are performed separateley since the shape parameters are typically independent of the dimensions) and standardization operations are performed. The standardization concerns the detection of the point to be assumed as

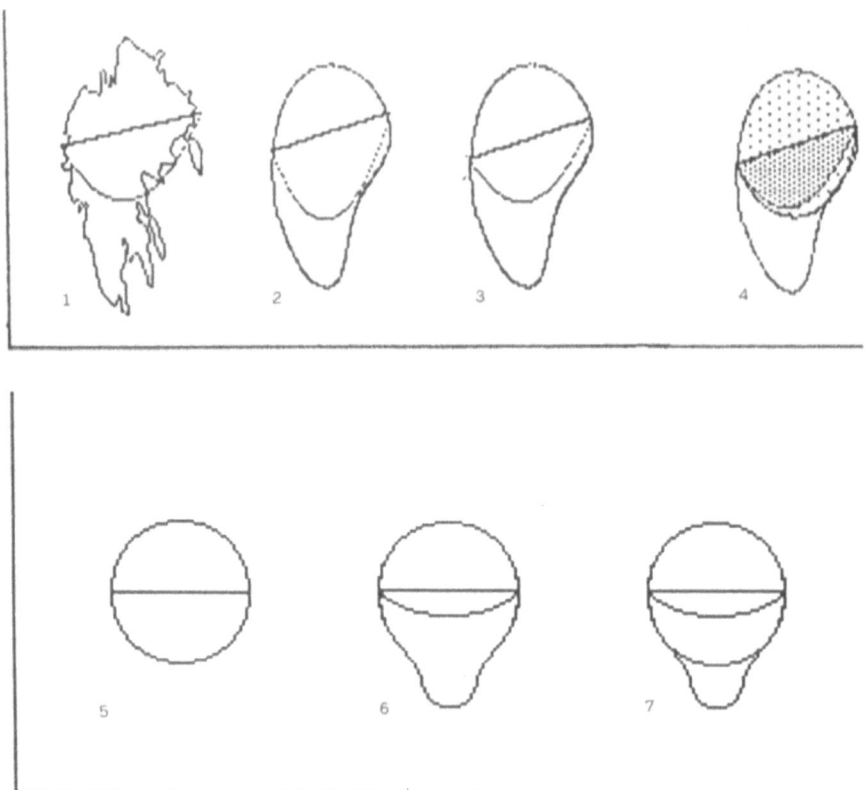

Fig. 2

the starting point for the analytical processing of the object contour. In fact, when investigating a nucleus or a cell contour there is no intrinsic rule for finding a unique point which could be taken as standard; however the numerical analysis, using equations, requires the validity of the equation itself to be considered as definite.

Actually, since the «shape» of a district must be referred to the characteristics of the performance of the profiles which compose it, and since these profiles can be fully considered as curves, the solutions able to describe them must be referred to the mathematical domain of analytical geometry (Pesce Delfino and Ricco, 1983).

The starting point is determined in an arbitrary way which is nevertheless strictly bounded by a procedure that will be described later and that, moreover, produces a group of parameters describing some characteristics.

This procedure requires the detection of the figure's centre of gravity since it is necessary to apply the algorithm for the following rotation steps in relation to that point. By the term «fundamental shape» we mean a new curve approximating the original profile by introducing a smoothing effect.

The need for such an operation derives from the need to obtain, on the one hand, a simplified model of the investigated shape which is easier to submit to classificative schemes and, on the other hand, from the need to have at our disposal a reference curve for further computation.

This procedure supplies some synthetic parameters of contour irregularities in terms of reciprocal punctual error between the function curve and the original profile.

The procedure operates as follows (tab. 1, figs.1-2-3): the abscissa and ordinate profile point values are considered separately (tab.1, figs. 1-2-3: b,c), as dependent variables from which to calculate the coefficients of an upper degree polynomial (according to the least square method), type:

$$f(x) = b_0 + \sum_{k=1}^{D} b_k x^k \tag{2}$$

where D is equation degree, x is the independent variable (each time the abscissa and the ordinate value), bo is a constant. For each profile the method is thus applied twice and the independent variable is, in both cases, the series of positive integers with unitary incremental stepping and which is included between 1 and N, ie number of points into which the profile has been subdivided (tab.1, figs. 1-2-3: d,e). The two polynomials are then merged so to obtain a new smoothed closed curve — fundamental curve — (tab. 1, figs. 1-2-3: f) that can be matched with the original curve (tab. 1, figs. 1-2-3: a).

This procedure has been adopted because, whatever the original profile complication entity may be, two values series without any ambiguity relative to the point coordinates and consequently the analytical procedure applicability are always obtained.

Concerning the gross distortions of whole figure, we adopted a procedure we named S.A.E. (Shape Asymmetry Evaluator); it consists of the interpolation of the figure with a parabola, type:

$$f(x) = a + bx + cx^2 \tag{3}$$

and in the construction of a set consisting of a parabola arc defined each time (which may vary from a straight line segment to a more or less convex arc) and of a straight line segment joining its ends (chord). A barycentric rotation for a 180° total range with constant angular step of a sufficiently small value (10° represents a typical value) is being carried out. This way figure exploration is carried out and the place of the arc/chord complex for the maximum difference of development between the arc and the chord (maximum distortion condition) and the minimum difference is found.

For every rotation step, development of the parabola, length, chord intercept and angular coefficient, surface given by the arc and the chord (allometric area)

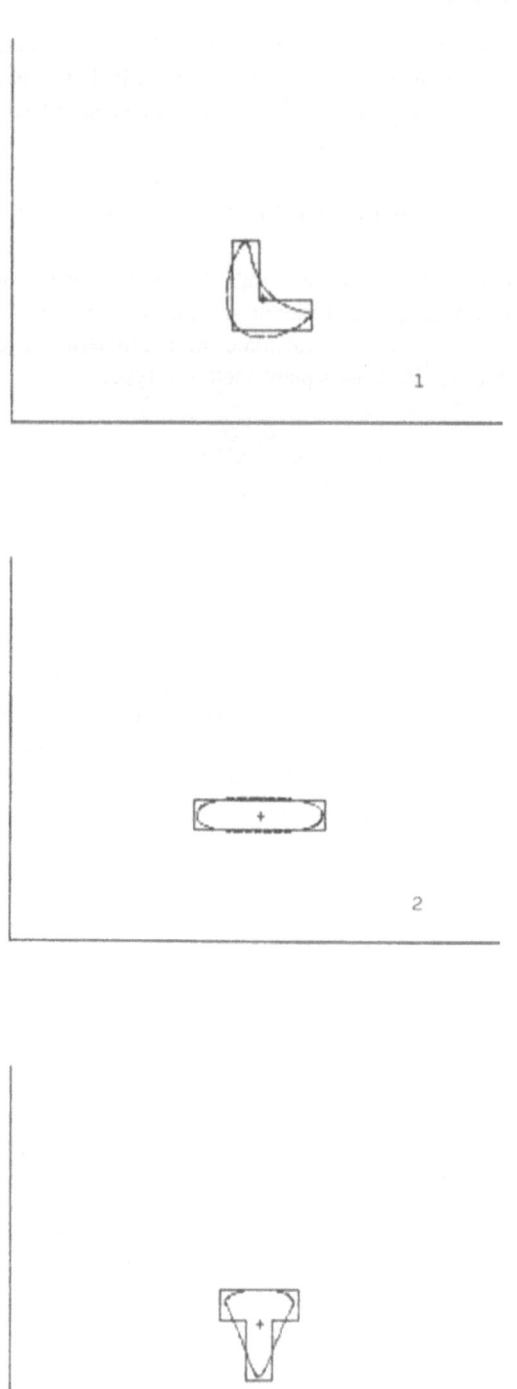

Fig. 3

and surface of the domain between the chord and the profile of the figure on the side opposite to the one to which the parabola convexity is turned (isometric area) are computed. The parabola convexity expresses, for each rotation step, the relative prevalence of that part of the profile towards which the convexity itself is orientated.

For such a prevalence a vectorial representation is given. If the chord and the arc coincide, the allometric area will be equal to zero (in fact the distortions are lacking) and the domains on the two sides of the chord will be equal.

It, typically, happens for any rotation step of the circle. Tab. 2 shows an irregular figure (fig. 1) fitted by the parabola, its fundamental curve (fig. 3) fitted the same way. Fig. 5 refers to a circle and fig. 6 to a circle bearing a distortion.

The isometric area represents the fraction that could be duplicated in a specular way on the opposite side of the chord and which will give a figure that may be irregular but without distortions (figs. 4, 7).

At the end of the whole rotation the mean values, standard deviations, variability, minimum, maximum and the interval for the relationship arc/chord and for the surface values are given.

This procedure, applied to the fundamental shape, is also used to define the starting point of the profile on the original shape; this point is located in the intersection of the profile itself with the chord in the condition of minimum value of the length difference between the arc and the chord; of the two intersections the one whose distance from the barycentre is smaller is chosen.

Concerning local contour perturbations, we used Fourier harmonic analysis.

The Fourier harmonic analysis has been put forward (Lestrel, 1974; Rhole and Archie, 1984; Diaz et al., 1989) as an effective procedure in form analysis. The values used as the dependent variable in the trigonometric polynomial type:

$$f(x) = a + \sum_{k=1}^{N/2-1} b_k \sin(kx) + c_k \cos(kx) \qquad (4)$$

are represented relative to a series of angular values which are ranged at equal distance in the interval 2π, by distance of profile single points from an original (centre of gravity or mid point of maximum diameter) and which may be decreased by the minimum found distance. The result is represented by the Fourier sine-cosine coefficient couple series and by values which may be obtained from these coefficients, of amplitude and phase of each single harmonic contributor.

The procedure gives the exact matching of the investigated figure. It also enables the synthesis of the figure itself or of its relevant patterns simply using subsets of the harmonics.

The reported version, may, however, be applied only to the figures without recursive trends of the profile parts in relation to the other one and without convexities towards the barycentre. In different cases the procedure cannot be applied

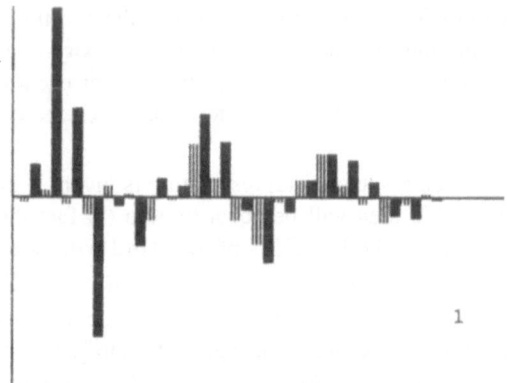

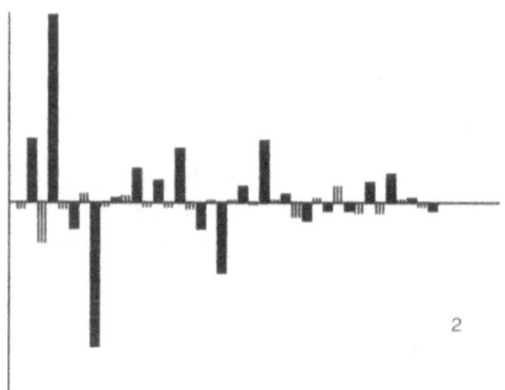

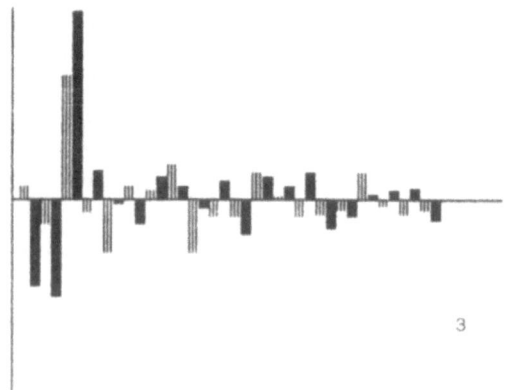

Fig. 4

because it will obviously occur that the same vector radius will intersect the profile more than once and, thus, different values of the dependent variable will result for the same angular value of the independent variable. Recursiveness situations are very frequent in biological objects. The S.A.M. system not only provides for the application to figures which do not present such ambiguous situations, but also for a solution of such conditions. As seen above, the polynomials calculated separately for the abscissa and ordinate values, to define the fundamental shape, eliminate any ambiguous situation; the function approximating the curve obtained from the polynomials is being used as reference line for the description, by means of harmonic analysis, of trends which the original values curve expresses in relation to the function curve. To do that it is necessary to rectify the polynomial function curve which becomes the reference line (zero line) and to perform the necessary transformations of the corresponding original curve.

In this way a series of values with an irregular but periodic trend is obtained that presents a particularly suitable format (difference graph) for submission to harmonic analysis (tab. 1, figs. 1-2-3: g).

Because the analysis is made on the contour distances from the fundamental shape and not on the distances of the radius vector from the barycentre, the sinusoidal components which describe the contour irregularities are typically stressed, while the amplitude value of the fundamental (1st degree) harmonic is very small. Most of the information is contained in the low degree harmonics.

The result of the harmonic analysis is typically represented by the Fourier spectrum: a bar graph where, for all the harmonic contributors disposed in a rising order from left to right, the sine-cosine coefficient amplitude with positive values up and negative values down are reported.

It is typical for the Fourier harmonics to be numerically independent and therefore the phase and amplitude values calculable by coefficients represent very effective variables from the statistical point of view.

It is easy to draw (tab. 1) three regular shaped figures — a triangle (fig. 1), a rectangle (fig. 2), an ellipse (fig. 3) — characterized by the same area, but with very dissimilar shapes.

The harmonic which represents the maximum amplitude is different for percentage values and above all for the order for each of the analyzed shapes. Therefore, there is no possible confusion between the graphic of Fourier coefficients which describes the triangle and which is characterized by the maximum amplitude value in correspondence with the third order harmonic (fig. 4), and the spectrum which describes the rectangle with the maximum amplitude values for the fourth harmonic (fig.5) or for the ellipse with the maximum amplitude on the second harmonic (fig. 6).

The information on shape differences due to the different ratios between sides or diameters is expressed by the amplitude and phase values of the other harmonics, so that only in the case of the equilateral triangle, the analysis of

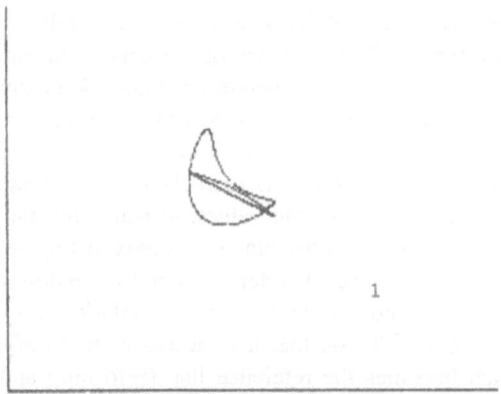

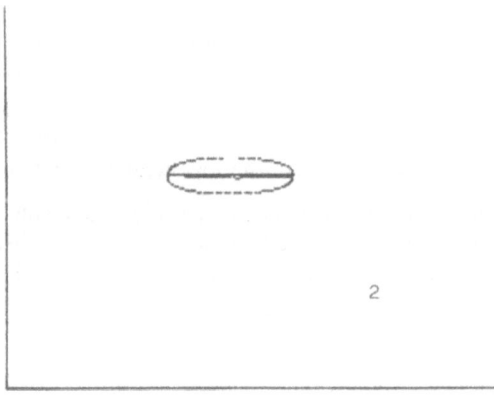

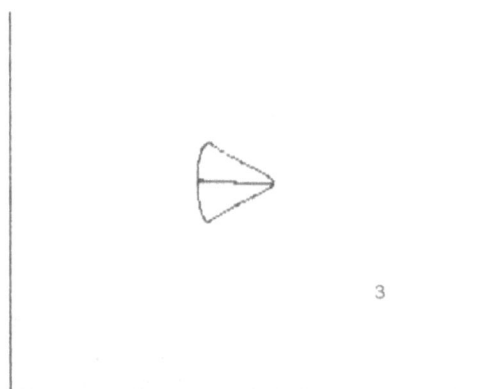

Fig. 5

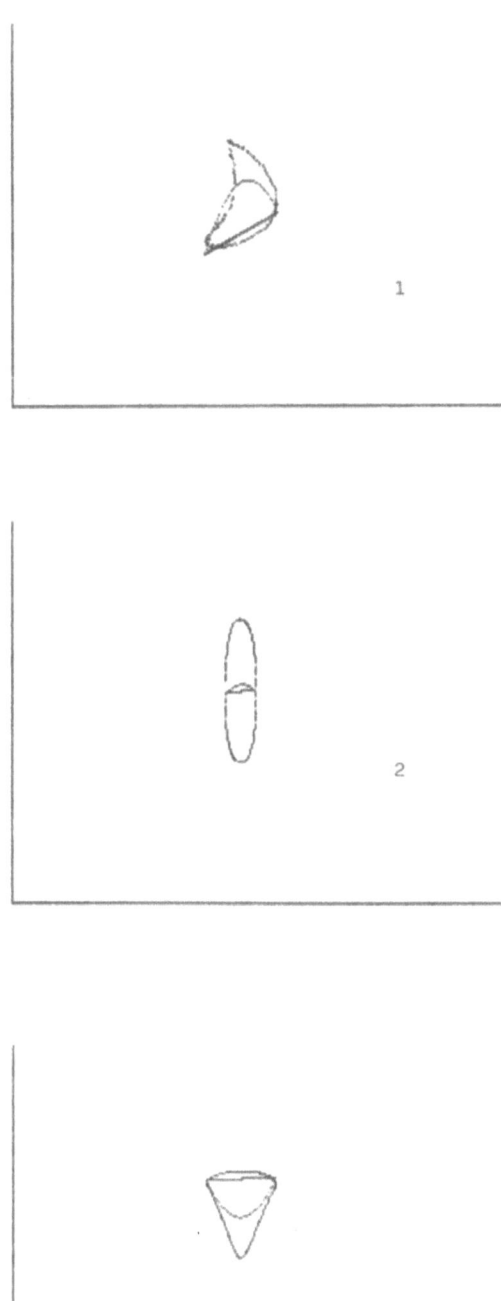

Fig. 6

a)

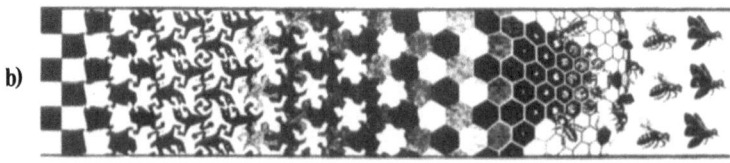

b)

Fig. 7

Fourier recognizes the actual presence of a 3th order harmonic and only in the case of the square, the analysis of Fourier recognizes the actual presence of a 4th order harmonic.

Concerning the symmetry conditions, in the triangle, where this peculiarity is extreme, there is a considerably high allometric average rate in respect to the rectangle and to the ellipse which have similar fundamental shapes and which in any case have the same allometric rate for the same rotation step.

It's also easy to draw figures with different shapes («L», «I», «T», shaped objects), but with the same perimeter, area and so called form factor; three example are reported (tabs. 3-4-5-6).

S.A.M. system gives the fingerprints of the three figures by combination of parameters given from different fundamental shape, symmetry analysis and Fourier analysis.

  – «L» shaped figure (tab. 3, fig.1) shows a Fourier spectrum driven by the second order harmonic (tab. 4, fig.1) and very pronounced asymmetry for both the minimum value (tab. 5, fig 1) and the maximum one (tab. 6, fig. 1).

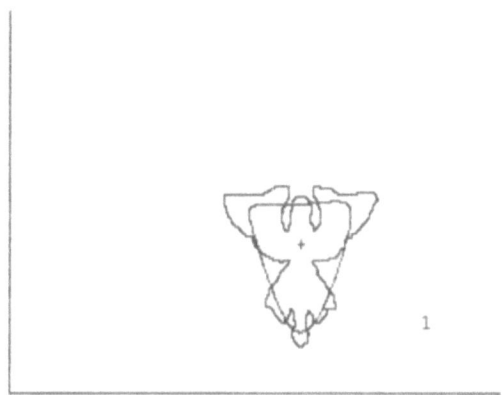

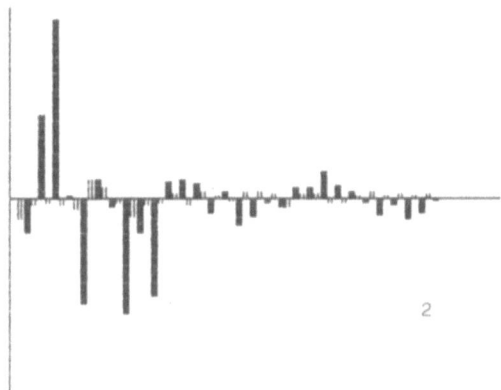

Fig. 8

- «I» shaped figure (tab. 3, fig. 2) shows a Fourier spectrum driven by the second order harmonic (tab. 4, fig. 2) but shows a very low asymmetry condition in orthogonal examination (tab. 5, fig. 2; tab. 6, fig. 2).

- «T» shaped figure (tab. 3, fig. 3) shows a Fourier spectrum driven by the third order harmonic (tab. 4, fig. 3) and symmetry exploration reveals absence of asymmetry (tab. 5, fig. 3) for a positioning and very high asymmetry for a 90 degrees rotation (tab. 6, fig. 3).

The S.A.M. software system does solve M.C. Escher's «Circle limit IV» (tab. 7, fig. 1) and «Metamorphose» (tab. 7, fig.2) tables.

In the «Circle limit IV» table, our perception is able to discriminate Angels from Devils only by a judgement on the profile shape. Values of perimeter (450), area (3050), abscissa projection (87), ordinate projection (81), roundness factor

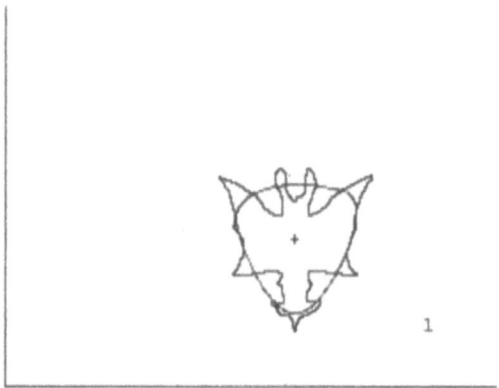

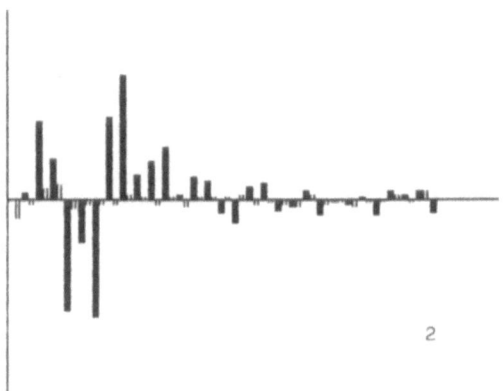

Fig. 9

(.18) are equal for the two figures and, thus, completely unable to distinguish Angels from Devils in spite of well evident shape differences.

By a gestaltic approach, sharp-cut, pointed, hooked and hard features are referred to Devils (takete, malignant), on the contrary rounded, soft, smooth, gently sloped features are referred to Angels (maluma, benign). This should be considered a popular example of our current image culture; also morphologist works with these cultural suggestions and in his judgments on the shape of biological objects widely uses gestaltic terminology in diagnosis as already reported.

S.A.M. system discriminated very well the angel (tab. 8, fig. 1) from devil (tab. 9, fig. 1) by order and amplitude of maximum Fourier harmonic: third order with 6822 amplitude for angel (tab. 8, fig. 2) and eighth order with 4731 amplitude for devil (tab. 9, fig. 2).

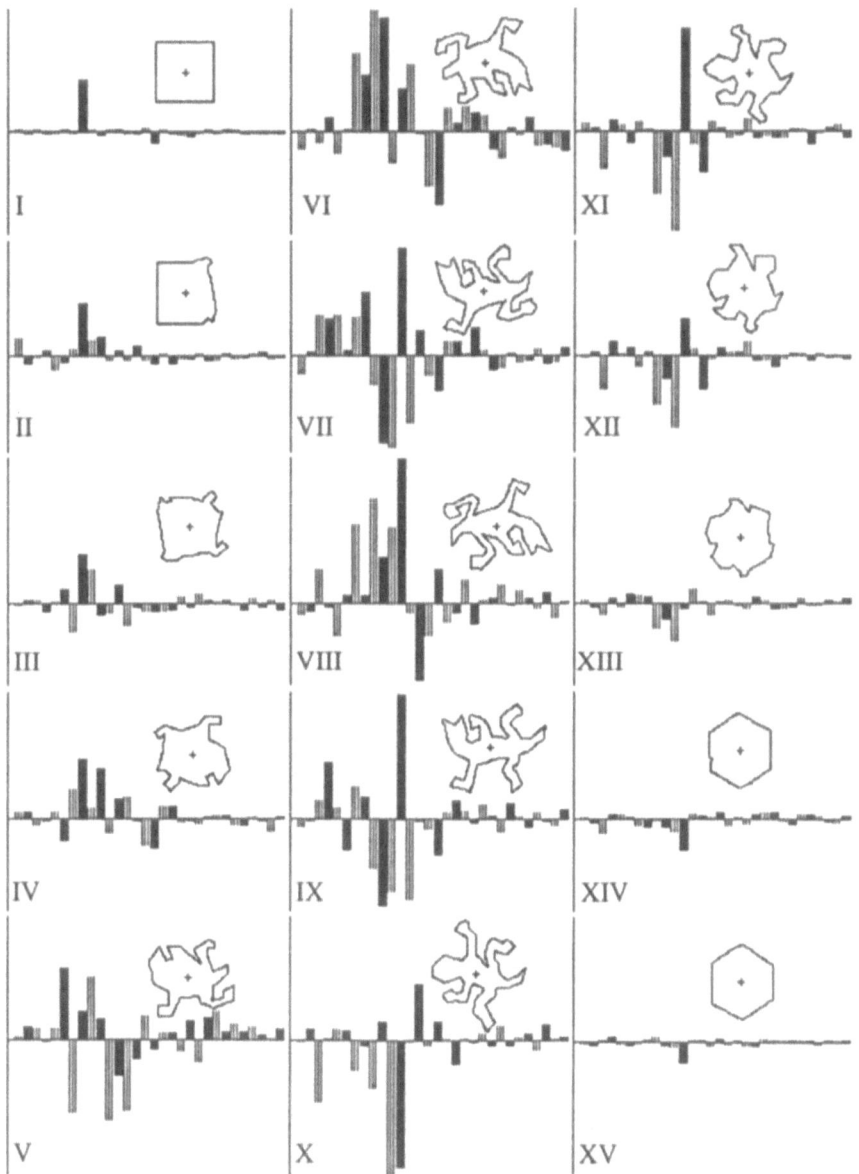

Fig. 10

Moreover an index computed by the match of the original profiles and their fundamental curves gives .43 value for angel and .96 for devil. The problem of «Circle limit IV», an example for a global view of the morphometrical analytical procedures, applied on artificial but intriguing figures, consisted in discriminating Angel from Devil: a «taxonomic» goal. The problem of «Metamorphose» consisted in monitoring step by step shape changes from a regular pattern (square) to another one (hexagon) through irregular forms: a «morphogenetic» goal. In this strip the morphological continuous transformation ends where a wasp flies out from the hexagon of a comb (where a morphological discontinuity is overcome by a logic continuity).

The S.A.M. system monitors the transformation giving non ambiguous fingerprint for each figure (tab. 8, figs. I–XV).

For the square (fig.1), Fourier analysis gave a pure fourth order harmonic spectrum, whereas for the hexagon (fig.XV) Fourier analysis gave a pure sixth order harmonic spectrum, with smaller amplitude.

First four intermediate irregular forms (figs.II–V) were driven by the fourth order harmonic: modulated square patterns. Last eight intermediate irregular forms (figs. VII–XIV) were driven by the sixth order harmonic: modulated hexagon patterns. The turning point was located at the sixth intermediate irregular form (fig. VI) where maximum amplitude Fourier harmonic was of fifth order.

B. Mandelbrot (1983) suggests a «Fractal Geometry of the Nature» where fractal, among other specifications, is a shape so made that, at smaller scales, the form exhibits self-similarity.

Whereas for artificial fractals, for instance Koch curves, the self-similarity occurs in a deterministic and exact way, for several natural objects at smaller scale of observation, we observe a statistical self-similarity; in other terms not the same pattern is repeated, but the degree of the complexity (Voss, 1988).

But, subtle differences whitin the same degree of complexity could be considered the marks of the individuality and if a specific recognition is required, at that level the crucial information lies.

It occurs, commonly, in medical morphology diagnosis where an individual case has to be attributed to a class (the patterns related to a disease).

The geometry of nature may be or may be not fractal but, anyway, for many purposes we need of a method to measure the geometry of nature, in most cases trying to discover specific characters in order to describe, but also to classify and discriminate, the objects.

Tab. 11 shows two closed Kock curves less (fig.1) and more (fig.2) iterated; they have the same fractal dimension.

They were worked by S.A.M. system giving, form the left to right, symmetry evaluation, fundamentral shape and Fourier spectrum; fig. 3 shows the same figure as in fig. 1 after distortion.

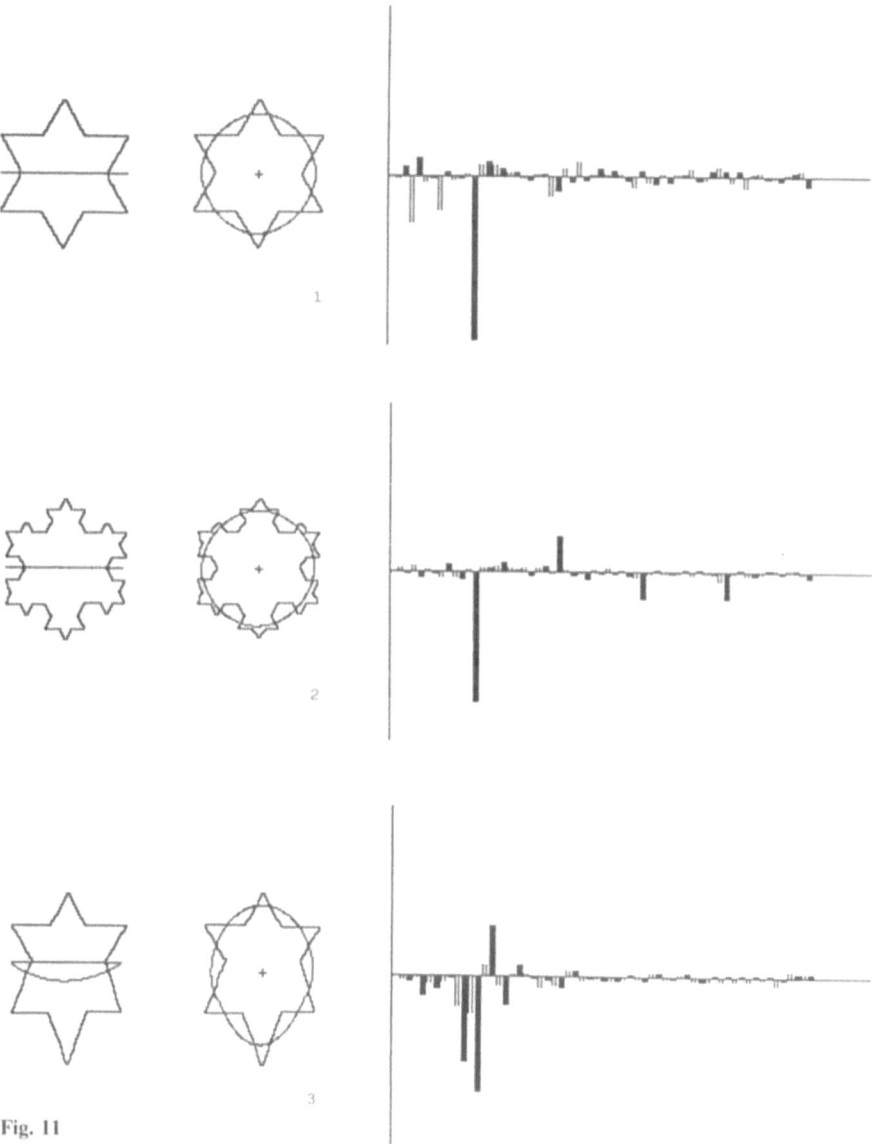

Fig. 11

Figs. 1 and 2 are symmetrical whereas in fig. 3 the parabola fitting finds asymmetry, Fourier spectrum is driven by sixth order harmonic for all three figures, but for fig. 1 the sixth order harmonic is together with second order and fourth order; for fig. 2 the sixth order harmonic is together with harmonic contributors of order twelve, eighteen and twenty-four; for the fig. 3 the driving sixth order harmonic is together with a composite subset that expresses the modulation due to the loss of symmetry.

Fractal dimension measures fractal objects alone; we try to give a tool able to work all objects, including fractal ones, even distorted (but is it still a fractal?).

Acknowledgement:
The Authors wish to thank Dr. E. Ferrara, F. Potente, P. Ragone and Mr. D. Vinci for their valuable contributions to this work.
® S.A.M. (Shape Analytical Morphometry) is registered trade mark of Metamorphosis s.r.l. – Bari, Italy.

References

[1] Aherne W.A. and Dunnill M.S. Morphometry. Edward Arnold, 1982.

[2] Diaz G., Zuccarelli A., Pelligra I., Ghiani A., Elliptic Fourier analysis of cell and nuclear shapes. Comp. Biom. Reas 22: 405–414, 1989.

[3] Holloway, R.L., Exploring the dorsal surface of Hominoid brain endocasts by stereoplotter and discriminant analysis. Phil. Trans. R. Soc. Lond. 292, 155–166, 1981.

[4] Lestrel P.E., Some problems in the assessment of morphological size and shape differences. Yearbook of physical anthropology, 18, 140–162, 1974.

[5] Mandelbrot B.B., The Fractal Geometry of Nature. 2nd edition. W.H. Freeman and Company, New York, 1983.

[6] Pesce Delfino V., Ricco R., Remarks on analytical morphometry in biology: procedure and software illustration. Acta Stereol. 2, 458–464, 1983.

[7] Pesce Delfino V., Potente F., Vacca E., Lettini T., Ragone P., Ricco R., Shape Evaluation in Medical Image Analysis. European Microscopy and Analysis, 21–24, September, 1990.

[8] Rhole F.J. & Archie J.W., A comparison of Fourier method for the description of wing shape in Mosquitoes (Diptera: Culicidae). Syst. Zool. 33, 302–317, 1984.

[9] Voss R.F., fractals in Nature: from characterization to simulation. In the Science of Fractal Images. H. Peitgen and D. Saupe, Eds. Springer-Verlag, New Yor, 1988.

Fluorescence Recovery after Photobleaching Studied by Total Internal Reflection Microscopy: An Experimental System for Studies on Living Cells in Culture

Torsten Mattfeldt[1], Theo F. Nonnenmacher[2],
Armin Lambacher[1], Walter G. Glöckle[2] and Otto Haferkamp[1]

1) Institute of Pathology, Oberer Eselsberg M23, University of Ulm, D-89069 Ulm, Germany
2) Department of Mathematical Physics, Oberer Eselsberg O25, University of Ulm, D-89069 Ulm, Germany

Abstract. After bleaching of cultured cells labelled with a fluorescent dye with laser light, one observes fluorescence recovery due to lateral motion of dye molecules from adjacent unbleached domains (Fluorescence Recovery After Photobleaching, *FRAP*). In the usual *FRAP* systems, the laser beam is directed vertically onto the cells, hence one measures the average fluorescence kinetics of *all* membranes throughout the full depth of the cell. This experimental design was modified by an inclination of the laser beam at an oblique angle, which was selected in such a manner that total reflection of the laser light near the cell surface resulted (Total Internal Reflection Fluorescence Microscopy, *TIRF*). Despite of *TIRF*, the irradiated region is bleached because a small fraction of the beam energy is transferred into the cell, which exponentially decays to values near zero already within a penetration depth in the order of magnitude of 50–150 nm. This combination of *FRAP* and *TIRF* for the study of living cells is new and permits a largely selective measurement of the fluorescence kinetics in the outer cell membrane, due to the selective superficial bleaching process. As the theory of the classical design is not fully appropriate in the present context, a generalized theoretical model of *FRAP* was developed which considers angular-dependent effects. Experiments on lateral mobility of the low molecular weight lipid analogon Nile Red (9-Diethylamino-5H-benzo[α]phenoxazin-5-on) on confluent cultures of HeLa cells were performed at bleaching periods of 200–1000 msec with an Argon laser of wavelength 488 nm. From the resulting data, we could quantify the speed of diffusion in terms of time for half recovery of the fluorescence intensity, determine whether the investigated molecule was completely mobile or partially fixed to the membrane, and explore whether its lateral mobility is due to diffusion only or also to active transport mechanisms.

1   Introduction

According to the well-known «fluid mosaic model» of membrane structure, the fundamental component of biomembranes is a double layer of phospholipids, in which lipids and proteins are either freely mobile or firmly anchored [10]. In addition to the transmembranous mobility *perpendicular* to the membrane surface, it was shown that intramembranous lipids and proteins can move *parallel* to it, travelling within the membrane plane. This phenomenon is denoted as *lateral mobility*, in the special case of a purely passive movement following a concentration gradient we have *lateral diffusion*. The speed of lateral diffusion is closely related to *membrane fluidity*. This important concept describes the viscous properties of the membranes, indicating whether the membrane is in a more fluid or in a more rigid state [6,9]. Several studies have shown that lateral diffusion and membrane fluidity play a key role in the modulation of important biological processes at the cellular level. Receptor-mediated membrane interactions, the immunogenity of the membrane, cellular temperature adaptation, membrane permeability, and activity of intramembranous enzymes may be mentioned in this context [4]. Studies on lateral diffusion are therefore of paramount interest for the investigation of the effects of drugs and toxins at the cellular level. We have developed an experimental system which allows to explore lateral diffusion *largely selectively in the plasma membrane* (i.e. the outer cell membrane) in living cells in culture. The system consists only of simple, commercially available building blocks (Argon laser, inverse microscope, PC), which makes its assembly easily reproducible in other laboratories. In the present contribution we describe the practical realization of the method. The theoretical principles are given in the companion paper [5].

2   Methods

2.1 Biophysical foundations

When living cells are bleached with a laser beam after incubation with a fluorescent dye, one observes subsequently a recovery of fluorescence in the bleached domain of the culture plate, which results from transport of fluorescent molecules from neighbouring regions into the irradiated area. This phenomenon is denoted as *Fluorescence Recovery After Photobleaching (FRAP)* [3]. The fluorescent molecules are irreversibly lost due to chemical reactions during the bleaching process. Therefore a gradient of fluorescent dye concentration to the adjacent domains is induced, which can be totally or partially reversed by diffusion of intact fluorescent molecules from unbleached regions. A mathematical analysis of the observed fluorescence kinetics in such experiments provides useful information on the lateral mobility of the investigated molecule. In a classical *FRAP*-experiment, a vertical laser beam traverses the full depth of the cell. The resulting fluorescence kinetics represent the global recovery of all membranes irradiated by the laser light, i.e. the plasma membrane, the nuclear membrane, mitochondrial membranes, and so on. We have tried to construct an apparatus for the selective investigation of the plasma membrane, which is the first target of drugs and toxins when these reach the cell. This purpose was achieved by the combination of *Total Internal Reflection Fluorescence Microscopy (TIRF)* with the *FRAP* principle, which was applied to living cells for the first time (Fig. 1a,b). In a *TIRF*-design, the laser beam does not traverse the whole cell, but there is total reflection at the glass/cell interface [1,2,11]. At the area of reflection, the laser beam transfers a small amount of energy to the cell, which is propagated as an electromagnetic wave whose intensity decays exponentially. Because of this property, the latter is denoted as the *evanescent wave*. The energy of the evanescent wave is nevertheless sufficient to bleach the fluorescent molecules irreversibly. As the penetration depth of the evanescent wave is in the order of magnitude of 0.1 to 0.3 of the wavelength of the laser light ($\approx$ 50–150 nm; see eq. (9) in the companion paper [5]), a largely selective

Fluorescence Recovery After Photobleaching (FRAP)

Experimental system

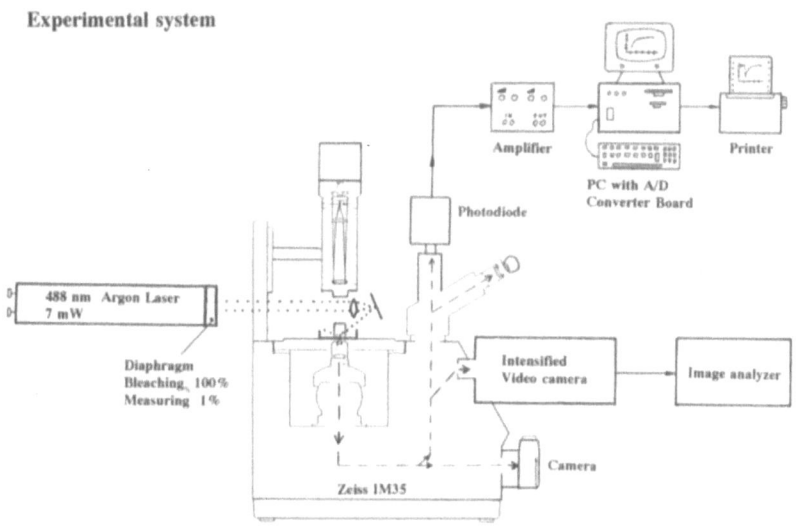

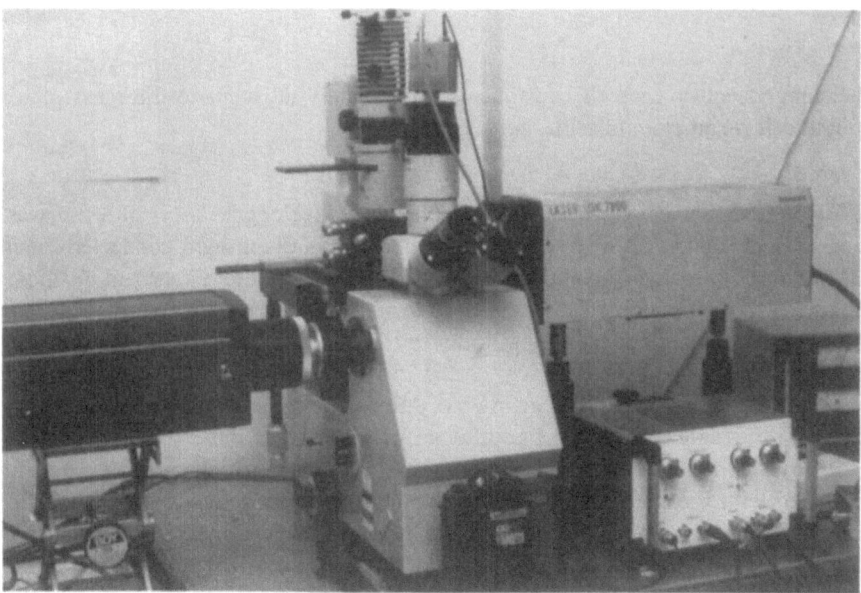

Fig. 1 **a)** Schematic overview of the complete experimental system. **b)** Schematic drawing of the construction by which *TIRF* is achieved at the interface of the coverslip and the cells.

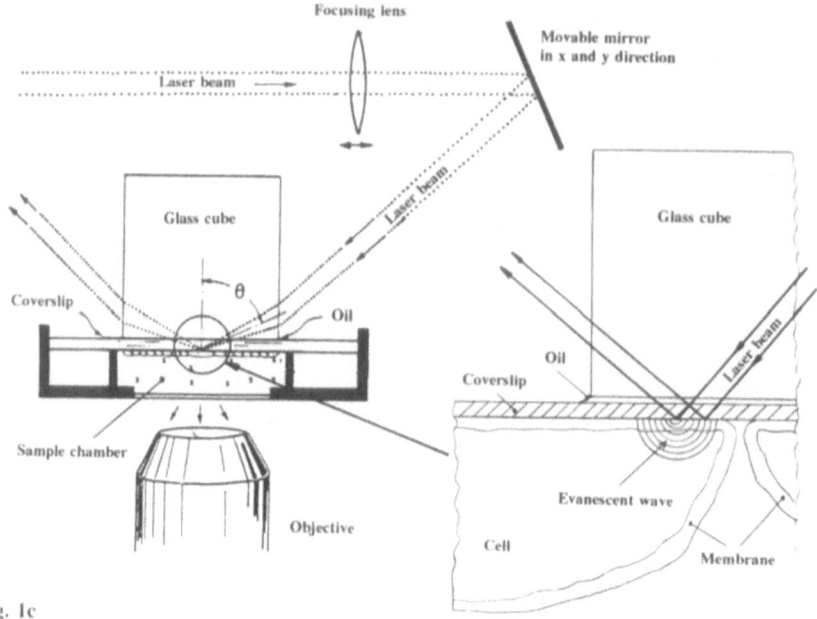

Fig. 1c

bleaching effect on the cell constituents immediately adjacent to the glass plate, i.e. the cell membrane, is achieved.

2.2 Experimental design

A 7 mW Siemens Argon laser LGK 7800 with wavelength $\lambda_0 = 488$ nm was used as light source. This device was equipped with a diaphragm connected to an IBM-compatible personal computer AT-286. According to the diaphragm position, the intensity of the emitted laser beam is switched between 100% and 1%. The full intensity is used for bleaching, the weak intensity is used for measuring the cellular fluorescence before bleaching and in the recovery period; the bleaching effect of the attenuated beam is negligible. After passing a focusing lens, the beam reaches a movable mirror and thereafter traverses a glass cube (Fig. 1a,b). This cube is positioned on the coverslip of the sample chamber which contains the cells incubated with the fluorescent dye [7]. The cube is separated from the coverslip only by a thin layer of immersion oil and can be moved horizontally. The laser beam is adjusted in such a manner that total reflection takes place at the interface of the coverslip and the cells. These are bathing in culture medium, thereby attached to the coverslip from below, hanging downwards like bats (Fig. 1a,b). An inverse microscope Zeiss IM35 is focused with a 40× objective lens on the plane where the cells adhere. The microscope transfers the fluorescence of the cells to a photodiode, which converts the photons into electric signals. The resulting voltage is amplified and fed into an A/D-converter board which is located within

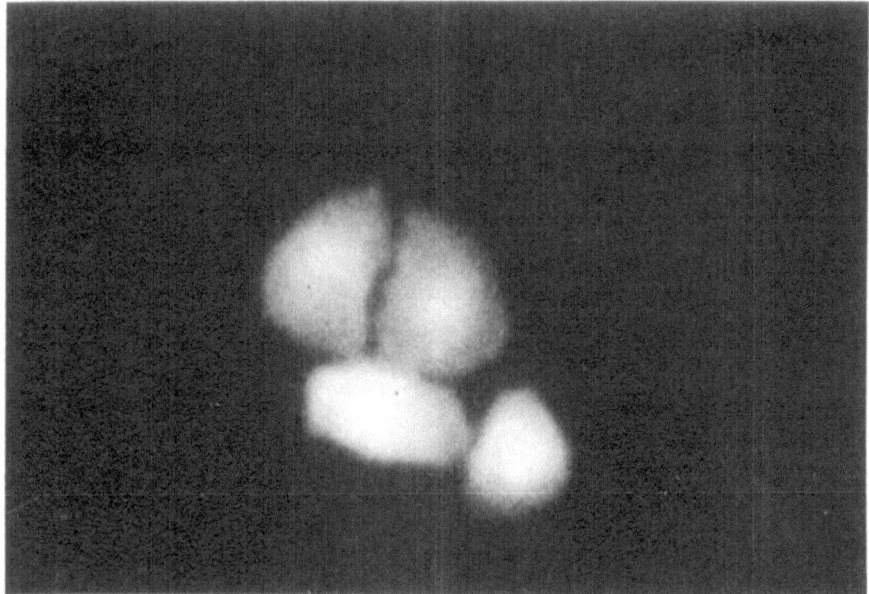

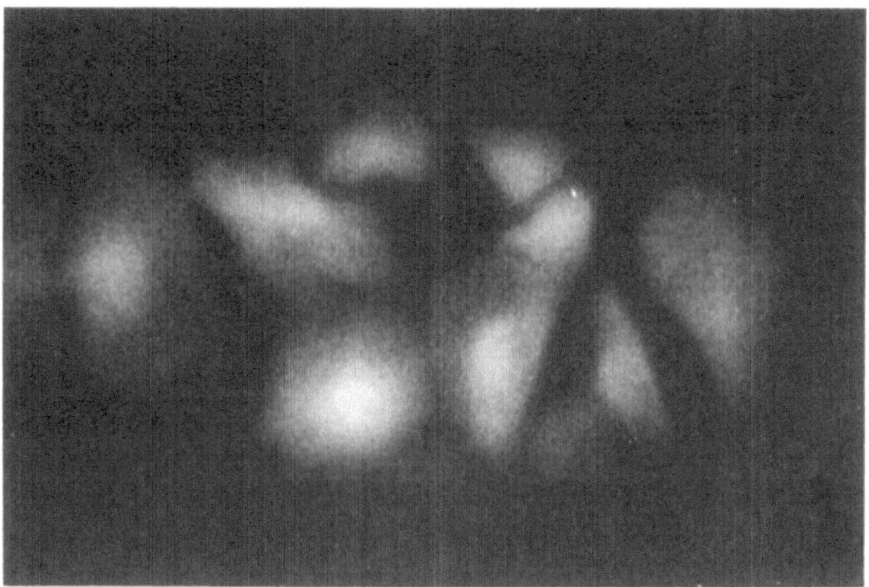

Fig. 2 **a)** A group of 4 HeLa cells, after incubation with NileRed, shows specific fluorescence labelling. The cells appear as 4 bright spots whereas the adjacent medium is dark. **b)** In this sample, more than 90% of the area is occupied with HeLa cells, hence nearly the whole illuminated area shows fluorescence. The elliptical profile of the obliquely incident laser beam is clearly visible.

the aforementioned computer. Using self-written programs in Turbo-Pascal, the PC thus controls the laser intensity by positioning the diaphragm, measures the fluorescence of the cells according to the voltage from the photodiode, and stores the data on the harddisk. An intensified video camera is attached to the microscope. This camera is connected to the image analysis system Kontron IBAS 2000, where the stored microscopical image of the fluorescent cells remains available for subsequent quantitative stereological studies.

2.3 Evaluation of measurements

When irreversible bleaching of the irradiated molecules, vertical direction of the laser beam, pure two-dimensional diffusion of the dye, and Gaussian energy distribution within the beam profile may be surmised, the fluorescence intensity $F(t)$ at time t after bleaching is given by the well-known equation [3]:

$$F(t) = \bar{F} \; \nu \; \tilde{K}^{-\nu} \; \Gamma(\nu) \; \chi^2(2\tilde{K}|2\nu) \tag{1}$$

where $\nu = (1 + 2t/\tau_D)^{-1}$, $\bar{F}$ represents the initial intensity of the fluorescence before bleaching, and $\tilde{K}$ and τ_D are parameters which have to be estimated for each experiment by nonlinear regression of $F(t)$ on t. In a combined *FRAP/TIRF* design, however, the requirement of vertical direction of the laser beam is not met, hence a generalized theory allowing for arbitrary oblique incident angles of the beam is needed. The corresponding theory, presented as eq. (18) in the companion paper [5], shows that the following equation holds for general values of $\gamma = \cos \theta$:

$$F(t) = \bar{F} \sum_{n=0}^{\infty} \frac{(-\tilde{K})^n}{n!} \{(1 + n + 2n\gamma^2 t/\tau_d)(1 + n + 2nt/\tau_d)\}^{-1/2} \tag{2}$$

The generalized approach reduces to eq. (1) for vertical incidence as a special case, when the angle θ between the laser beam and the vertical axis is $0°$. To determine θ in our experimental apparatus, the fluorescent profile of the laser beam was stored by the image analyzer. As the true laser beam has a circular profile of radius r, the eccentricity of the visible elliptical beam profile with semiaxes b, r is related to θ by $\cos \theta = (r/b)$, according to elementary trigonometry. The actual values are $r = 29\mu m$, $b = 69\mu m$, hence $\theta \approx 65°$. The area of the beam profile at the level of the cells amounts to $\approx 6286\mu m^2$. Hence the irradiated area is considerably larger than the projected area of an average cultured cell; from this reason a bleaching experiment represents an average value for a sample of some cells hit by the laser beam (Fig. 2a,b). The speed of diffusion can be indicated by the *time for half recovery* $T_{1/2}$. This is the time after which one half of the bleached fluorescence has recovered according to the fitted curve. The *percentage of fluorescence recovery* was determined from the ratio of the asymptotic end value of the fitted curve to the mean intensity value before the bleaching impulse. In addition, the *initial voltage decline* immediately after bleaching was recorded in each experiment.

Voltage (mV)

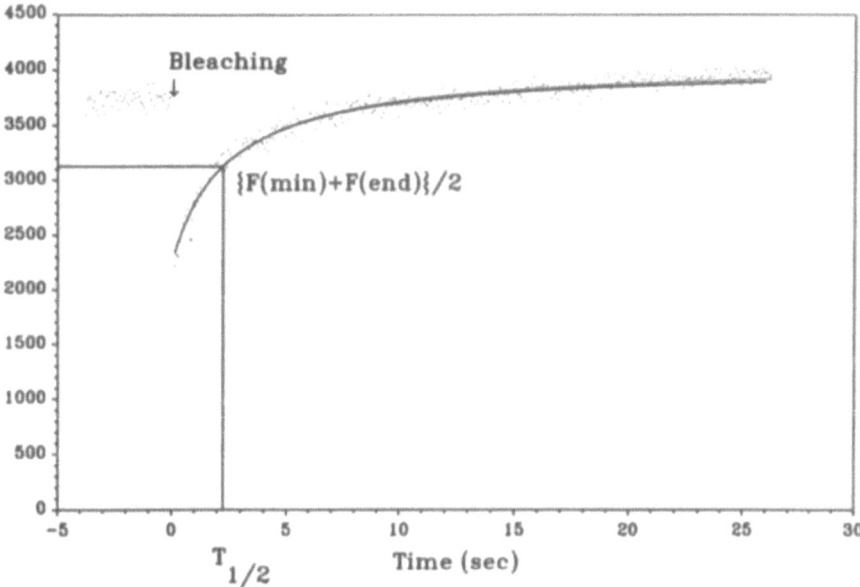

Fig. 3 The diagram shows the fluorescence intensity (here measured as voltage of the photodiode) as a function of time t. Negative values of t represent the time before bleaching, $t = 0$ is the moment of bleaching. The curve was fitted by nonlinear least square regression of voltage on t, according to eq. (2) with $\theta = 65°$.

2.4 Cell cultures and protocols

Confluent cultures of HeLa cells in RPMI 1640 medium on coverslips were used for the studies. Nile Red, a lipid analogue of low molecular weight (9-Diethyl-amino-5*H*-benzo[α]phenoxazin-5-on), which localizes largely specifically to the hydrophobic inner domains of biomembranes [8], was used as fluorescent dye. Incubation with the dye was begun at 37° C on the coverslips 1 hour before the bleaching experiments at a concentration of 6.25 μl Nile Red per ml medium. The chambers were inserted into the experimental system at room temperature (Fig. 1a,b). The fluorescence was measured under steady state conditions for 3.75 sec; after bleaching, the fluorescence recovery was measured for 26.25 sec. Thus a single bleaching experiment lasts 30 seconds. The voltage is recorded every 75 msec, which provides a series of 400 measurements per experiment. It is possible to bleach a constant domain of the culture repetitively for 3–4 times without detectable changes of the fluorescence kinetics, thereafter a new domain of the cell culture should be investigated. To obtain reliable data, at least 6, but more often 7–10 positions were tested, each with 3 consecutive bleaching experiments. Therefore 18–30 *FRAP*-curves are available in a typical study. Mean values of initial voltage decline, % recovery, and time for half recovery were determined between replications within positions. Finally, means and standard deviations of these average values per position were determined.

2.5 Experiments

First the effect of the areal density of the cells, i.e. the ratio of the area covered by cells to the total area of the laser beam profile, on the fluorescence kinetics was explored. At different time points of culture growth the areal density of the cells shows considerable variations, in addition there is spatial variability within a culture at a given moment. Therefore *FRAP*-experiments were performed at 9 different positions on untreated HeLa cells. After a series of 3 replicated experiments at each position, the image of the elliptical profile of the laser beam with the cells inside was stored each time on disk. The areal density was evaluated interactively by tracing the cell profiles and the laser beam profile. This was done after the whole series of bleaching studies on a culture was completed, hence there was no need to keep the cells longer in the sample chambers than usual. Next, we intended to explore the effect of increasing bleaching periods on the *FRAP* kinetics. For this purpose, 3 positions on untreated HeLa cell cultures with bleaching times of 200, 400, 600, 800, and 1000 msec were evaluated with 3 replications each. Bleaching times below 200 msec or above 1000 msec duration did not provide additional information at the selected, fixed voltage amplification, as the initial voltage decline was too small, or the initial voltage decline was so large that the resulting voltage difference exceeded the range of values that can be measured by the A/D converter board.

3   Results

The key parameter $T_{1/2}$ could be determined with good accuracy within a cell culture; typical coefficients of variation were below 5%. The estimate of $T_{1/2}$ in the order of magnitude of a few seconds corresponded to the quick lateral mobility, which could be expected in a low molecular weight lipid analogon (Fig. 3). The kinetics of fluorescence recovery after photobleaching could be fitted by a pure 2D diffusion process in very good approximation according to eq. (2) (Fig. 3). Furthermore, all experiments showed 100% recovery of fluorescence after bleaching. Increasing bleaching periods from 200 to 1000 msec caused an increasing initial voltage decline, i.e. the initial voltage jump at $t = 0$ was enhanced (correlation coefficient $r = +0.71$, $p < 0.005$) and a concomitant increase of $T_{1/2}$ (Fig. 4a,b). The positive correlation between bleaching time and $T_{1/2}$ is also statistically significant ($r = +0.84$, $p < 0.0001$). Furthermore, a positive correlation could be detected between the areal density of the HeLa cells and the initial voltage decline ($r = +0.84$, $p < 0.05$) (Fig. 5).

Initial voltage decline (mV)

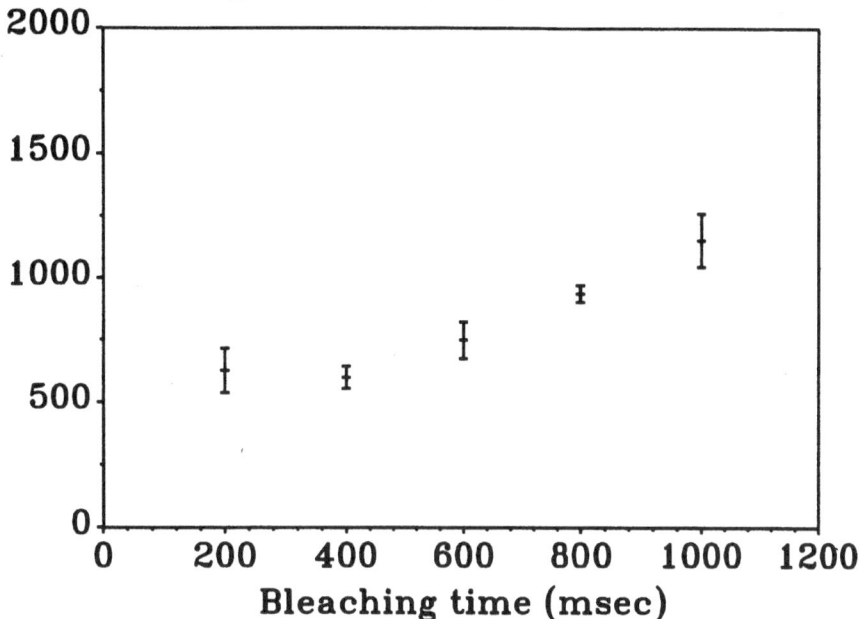

Time for half recovery (msec)

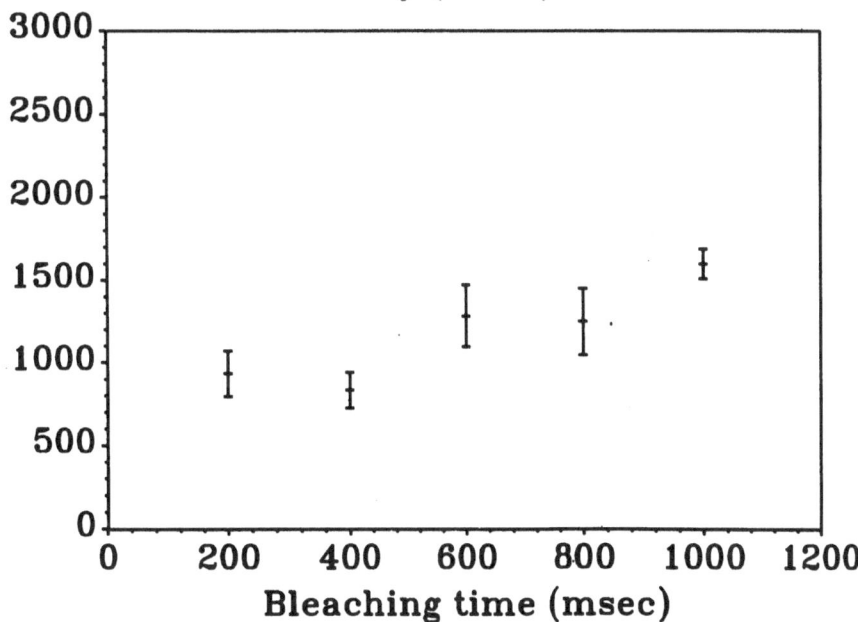

Fig. 4 **a)** The initial voltage decline (voltage decay immediately after bleaching) increases with bleaching time. **b)** Mean time for half recovery also increases with bleaching time. Both plots represent mean values with standard errors.

Initial voltage decline (mV)

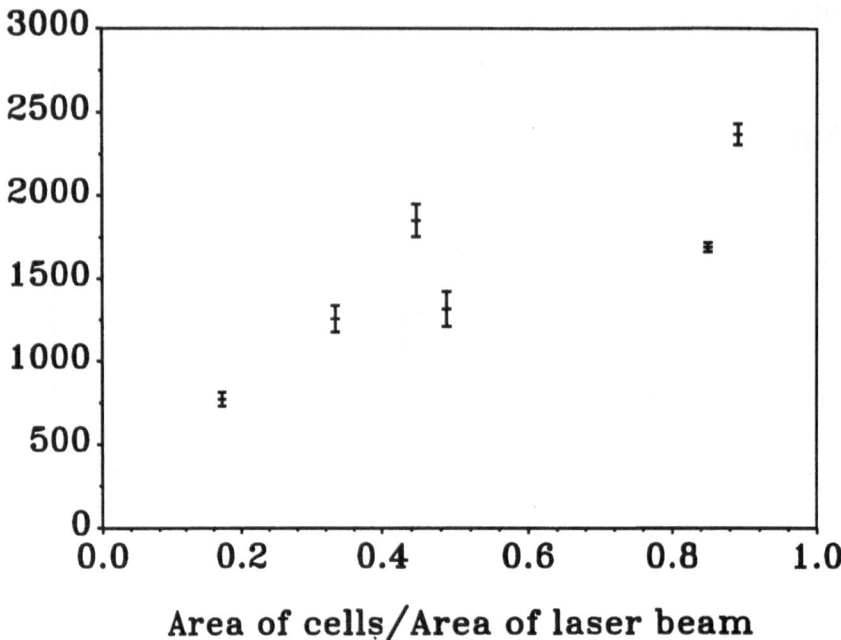

Area of cells/Area of laser beam

Fig. 5   The initial voltage decline increases with the areal density of cells within the laser beam profile. The plot represents mean values with standard errors.

4   Discussion

The finding of 100% fluorescence recovery indicates complete lateral mobility of the lipid molecules. An «immobile fraction», which would lead to an incomplete fluorescence recovery, could not be detected. Considerable immobile fractions have been described for membrane proteins (e.g. [12]). Excellent agreement was obtained when curves were fitted to the experimental data according to eq. (2). Therefore, the whole observed lateral lipid mobility may be explained in terms of diffusion. In other words, the data do not indicate active transport mechanisms for the lipid in a relevant order of magnitude. An interesting finding is the dependence of $T_{1/2}$ on the duration of bleaching. The effect is apparently caused by the fact that more molecules are bleached when the laser beam is longer exposed. When lipid fluidity and the coefficient of lateral lipid diffusion remain constant, a longer time will subsequently be necessary until the induced gradient of concentration is fully reversed, and thus $T_{1/2}$ will tend to increase, too. Presumably, the increased initial voltage jump with increasing areal density of HeLa cells is due to the same reason: when more cell surface area containing dye molecules is exposed to the laser beam, more molecules will be bleached, which enhances the initial voltage decay. A potential drawback of the method, as it stands now, is the fact that single cells are not studied selectively. Instead, global *FRAP* curves are obtained for clusters of cells. On the other hand, the theory as given above is *strictly* correct only for studies where only a small spot on the surface of a single cell is bleached, although it shows such a good agreement with the experimental data. To make further progress here, the following modifications are possible. In order to examine the behaviour of single cells, the laser beam could be focused more narrowly, or a central segment of the emitted beam could be evaluated selectively with a photomultiplier after insertion of a diaphragm at the detection side of the beam path. Alternatively, the experimental apparatus might remain as it stands, and the theory could be further generalized by allowing for multi-cell effects with multiple recovery compartments. From our point view, such refinements are necessary to estimate the parameters $\tilde{K}$ and τ_D in absolute terms; in fact, this is why we have restricted ourselves to a presentation of $T_{1/2}$, initial voltage decline and percentage of recovery, i.e. parameters which merely require a good curve fitting for their estimation. In summary, we have described an experimental device linking the principles of *TIRF* and *FRAP*, which allows accurate and reproducible measurements of lateral mobility of membrane components in living cells. From the resulting data, the experimenter can quantify the speed of diffusion, determine whether the investigated molecule is completely mobile or partially fixed to the membrane, and explore whether its lateral mobility is due to diffusion only or also to active transport mechanisms. It should be possible to extend the method to other interesting molecules, e.g. long-chain lipids, tunnel proteins and receptors labelled with fluorescent dyes.

References

[1] Axelrod, D. (1981) Cell-substrate contacts illuminated by total internal reflection fluorescence. J. Cell Biol. 89, 141–145.

[2] Axelrod, D. (1989) Total internal reflection fluorescence microscopy. Methods Cell Biol. 30, 245–270.

[3] Axelrod, D., Koppel, D.E., Schlessinger, I., Elson, E., Webb, W.W. (1976) Mobility measurement by analysis of fluorescence photobleaching recovery kinetics. Biophys. J. 16, 1055–1069.

[4] Boullier, J.A., Melnykovych, G., Barisas, B.G. (1982) A photobleaching recovery study of glucocorticoid effects on lateral mobilities of a lipid analog in S3G HeLa cell membranes. Biochim. Biophys. Acta 692, 278–286.

[5] Glöckle, W.G., Mattfeldt, T., Nonnenmacher, T.F. (1993) Anomalous diffusion and angle-dependency in the theory of fluorescence recovery after photobleaching. Proceedings of the First Symposium on Fractals in Biology and Medicine, ed. Nonnenmacher, T.F., Losa, G. and Weibel, E.R. Basel: Birkhäuser (this volume, in press).

[6] Lenaz, G., Castelli, G.P. (1985) Membrane fluidity: molecular basis and physiological significance. *In:* Structure and Properties of Cell Membranes, Vol. I, ed. Benga, G. Boca Raton: CRC Press, pp. 93–136.

[7] Pentz, S., Hörler, H. (1992) A variable cell culture chamber for «open» and «closed» cultivation, perfusion and high microscopic resolution of living cells. J. Microsc. 167, 97–103.

[8] Sackett, D.L., Wolff, J. (1987) Nile Red as a polarity-sensitive fluorescent probe of hydrophobic protein surfaces. Analyt. Biochem. 167, 228–234.

[9] Shinitzky, M. (1984) Membrane fluidity and cellular functions. *In:* Physiology of Membrane Fluidity, Vol. I, ed. Shinitzky, M. Boca Raton, CRC Press, pp. 1–51.

[10] Singer, S., Nicolson, S. (1972) The fluid mosaic model of the structure of cell membranes. Science 175, 720–731.

[11] Thompson, N.L., Burghardt, T.P., Axelrod, D. (1981) Measuring surface dynamics of biomolecules by total internal reflection fluorescence with photobleaching recovery or correlation spectroscopy. Biophys. J. 33, 435–454.

[12] Vaz, W.L.C., Goodsaid-Zalduondo, F., Jacobson, K. (1984) Lateral diffusion of lipids and proteins in bilayer membranes. FEBS Letters 174, 199–207.

Anomalous Diffusion and Angle-Dependency in the Theory of Fluorescence Recovery after Photobleaching

Walter G. Glöckle[1], Torsten Mattfeldt[2] and Theo F. Nonnenmacher[1]

1) Department of Mathematical Physics, University of Ulm, Albert-Einstein-Allee 11, D-89069 Ulm, Germany

2) Institute of Pathology, University of Ulm, Albert-Einstein-Allee 11, D-89069 Ulm, Germany

Abstract. Fluorescence recovery after photobleaching (FRAP) is an experimental method for the investigation of transport processes of fluorescent molecules. Under total reflection conditions a selective measurement of processes in thin surface layers is achieved. We present theoretical calculations demonstrating the influence of the angle under which the detection beam is directed to the probe on the fluorescence signal. Apart from the angle dependency, we consider anomalous diffusion processes with the mean square displacement $\langle r^2 \rangle \propto t^\alpha$. Modeling the process by a fractional order diffusion equation, we find that the relaxation of the fluorescence to the equilibrium value is asymptotically $\propto t^{-\alpha}$ for large t.

1 Introduction

The study of fluorescence recovery after photobleaching (FRAP) provides a method for measuring transport and reaction processes of fluorescent molecules in biological systems like living cells [1][5]. In the experiment, a short, intense laser pulse irreversibly bleaches the fluorophore in a small region. Transport coefficients can be determined by measuring the rate of fluorescence recovery resulting from the transport of fluorescent molecules from non-illuminated regions of the system into the bleached spot. Under total reflection conditions, only the fluorescence of molecules in a thin superficial layer is excited by the exponentially decaying evanescent wave [7]. Thereby a largely selective monitoring of protein or lipid molecules labelled with fluorescent dyes in cell membranes is achieved. By analyzing the fluorescence recovery after photobleaching, the lateral diffusion of the molecules in the membrane can be studied.

Theoretical calculations were performed for several intensity profiles of the bleaching and probing beam and for different reaction and diffusion processes [1][7]. Here, we present results of the dependency of the fluorescence recovery on the angle of illumination for a Gaussian beam profile. Furthermore we discuss the influence of cooperative effects if the fluorescent molecules do not behave like Brownian particles but show correlated motion. In this case, the transport process with long-range correlations is modeled by a fractional order diffusion equation.

2   General Formula for FRAP

The formula for the fluorescence recovery after photobleaching has to take three processes into account: (i) the bleaching by a short laser pulse, (ii) the transport process by which unbleached fluorescent molecules migrate into the bleached area, and (iii) the detection of the fluorescence signal.

(i) According to ref. [1] we assume that photobleaching of the fluorescent molecules is a simple first-order reaction process with a rate constant proportional to the intensity of the bleaching beam I^b. Then the concentration of the unbleached fluorophore $C(\vec{r}, t)$ is governed by

$$\frac{d}{dt} C(\vec{r}, t) = -\alpha I^b(\vec{r}) C(\vec{r}, t) \ . \tag{1}$$

With the equilibrium concentration $\bar{C}$, the concentration after a short bleaching time T is given by

$$C(\vec{r}, 0) = \bar{C} \exp\left(-\alpha T I^b(\vec{r})\right) \tag{2}$$

which defines the initial state, at which the fluorescence recovery starts.

(ii) By bleaching, fluorescent molecules are destroyed. Because of the concentration gradient, molecules from unbleached areas migrate into the bleached spot. The most simple case of such a transport process is diffusion

$$\frac{\partial}{\partial t} C(\vec{r}, t) = D \Delta C(\vec{r}, t) \tag{3}$$

with a constant diffusion coefficient D. If $G(\vec{r}, t)$ is the Green's function of the transport process, the concentration of unbleached particles is given by the convolution integral

$$C(\vec{r}, t) = \int d\vec{r}' \, C(\vec{r}', 0) G(\vec{r} - \vec{r}', t) \ . \tag{4}$$

(iii) The property which is detected is the total fluorescence in the detection area. It is proportional to the intensity of the probing beam and the fluorophore concentration and, thus, given by

$$F(t) = Q \int d\vec{r} \, I^p(\vec{r}) C(\vec{r}, t) \tag{5}$$

where integration is performed over the whole detection area. In eq. (5), Q is a constant containing the quantum efficiencies of the light absorption and emission. If $F(t)$ is considered to be the electrical signal of the detector, attenuation and amplification of the optical and electrical systems are also included in Q.

Collecting the three components (2), (4), and (5), we arrive at

$$F(t) = Q\bar{C} \int d\vec{r}' \int d\vec{r} \, \exp\left(-\alpha T I^b(\vec{r}')\right) G(\vec{r} - \vec{r}', t) I^p(\vec{r}) \tag{6}$$

which is valid for arbitrary transport processes.

3 Fluorescence recovery and relaxation function

In total internal reflection measurements the laser beam is directed from a medium 2 (glass) with refraction index n_2 to a medium 1 (cell) with the index $n_1 < n_2$. If the angle θ between the direction of the beam and the optical axis (z direction) is greater than the angle of total reflection

$$\theta_c = \arcsin\left(\frac{n_1}{n_2}\right) \tag{7}$$

the beam is totally reflected. In medium 1 only an exponentially decaying evanescent wave with the intensity

$$gl8I(z) = I(0)\exp(-z/d) \tag{8}$$

penetrates. The penetration depth

$$d = \frac{\lambda_0}{4\pi}\left(n_2^2\sin^2(\theta) - n_1^2\right)^{-1/2} \tag{9}$$

depends on the angle θ and the vacuum wavelength λ_0. For typical values $n_1 = 1.33$ (H_2O) and $n_2 = 1.5$ (glass) the angle of total reflection is $\theta_c \approx 62.5°$. For values of θ more than about 5 degrees above θ_c, the value of the penetration depth d is nearly constant between $0.1\lambda_0$ and $0.2\lambda_0$. Fluorescence is therefore only excited in a thin surface layer. If the fluorescent molecules are solved in lipid membranes, molecules in the outer cell membrane are largely specifically detected. Furthermore the mobility of the molecules in the lipid membrane is restricted to lateral processes, i.e. $G(\vec{r}, t) = G(x, y, t)$, and thus the method provides a selective tool to study lateral transport in cell membranes. In the thin layer the dependency of the beam intensity (8) on z can be neglected.

The intensity of a laser beam is well described by a Gaussian beam profile J. In the $(\bar{x}, \bar{y})$-plane perpendicular to the beam direction, the intensity is given by

$$I(\bar{x}, \bar{y}) = I_0\exp\left(-\frac{2(\bar{x}^2 + \bar{y}^2)}{w^2}\right) = I_0 J(\bar{x}, \bar{y}) . \tag{10}$$

In the (x, y)-system of the membrane we have

$$I(x, y) = \gamma I_0\exp\left(-\frac{2(\gamma^2 x^2 + y^2)}{w^2}\right) = \gamma I_0 J(\gamma x, y) \tag{11}$$

with the geometric factor $\gamma = \cos(\theta)$. Hence, if lateral processes in the outer cell membrane are measured by FRAP under total internal reflection conditions, the fluorescence recovery function (6) takes the form

$$F(t) = Q\bar{C}I_0^p\gamma\int dx\,dy\int dx'\,dy'\,e^{-\bar{K}J(\gamma x', y')}G(x - x', y - y', t)J(\gamma x, y) . \tag{12}$$

In eq. (12), the bleaching parameter $\tilde{K} = \alpha T \gamma I_0^b$ is introduced which is a measure for the strength of bleaching.

An alternative representation considers the relaxation function $G(t) = \bar{F} - F(t)$ describing the approach of $F(t)$ to the equilibrium value $\bar{F} = F(t \to \infty)$. Using the Fourier transform

$$\hat{G}(k_x, k_y, t) = \int dx\,dy\, e^{i(k_x x + k_y y)} G(x, y, t) \tag{13}$$

of the Green's function $G(x, y)$, then

$$G(t) = -Q\bar{C}I_0^p \gamma \frac{1}{(2\pi)^2} \int dk_x dk_y \hat{G}((k_x, k_y) \int dxdy J(\gamma x, y) e^{-ik_x x - ik_y y}$$
$$\times \int dx'dy' \left(1 - e^{-\tilde{K}J(\gamma x', y')}\right) e^{ik_x x' + ik_y y'} \tag{14}$$

is obtained from eq. (12). For a Gaussian beam profile J, we find

$$G(t) = -QI_0^p \bar{C} \frac{w^4}{16\gamma} \sum_{n=1}^{\infty} \frac{(-\tilde{K})^n}{n!} \frac{1}{n}$$
$$\times \int dk_x\,dk_y\, \hat{G}(k_x, k_y, t) \exp\left(-\frac{(n+1)w^2}{8n}\left(\frac{k_x^2}{\gamma^2} + k_y^2\right)\right) \tag{15}$$

by series expansion of $\exp(-\tilde{K}J(\gamma x', y'))$ and integration.

4   Ordinary diffusion

Now we can use models of the lateral transport process in the membrane. The normal diffusion equation (3) in the (x, y)-plane has the Green's function

$$\hat{G}(k_x, k_y, t) = \exp\left(-(k_x^2 + k_y^2)Dt\right) \tag{16}$$

in the Fourier space and

$$G(x, y, t) = \frac{\pi}{Dt} \exp\left(-\frac{x^2 + y^2}{4Dt}\right) \tag{17}$$

in the ordinary (physical) space, resp. From eq. (12) the fluorescence recovery

$$F(t) = \bar{F} \sum_{n=0}^{\infty} \frac{(-\tilde{K})^n}{n!} ((1 + n + 2n\gamma^2 t/\tau_d)(1 + n + 2nt/\tau_d))^{-1/2} \tag{18}$$

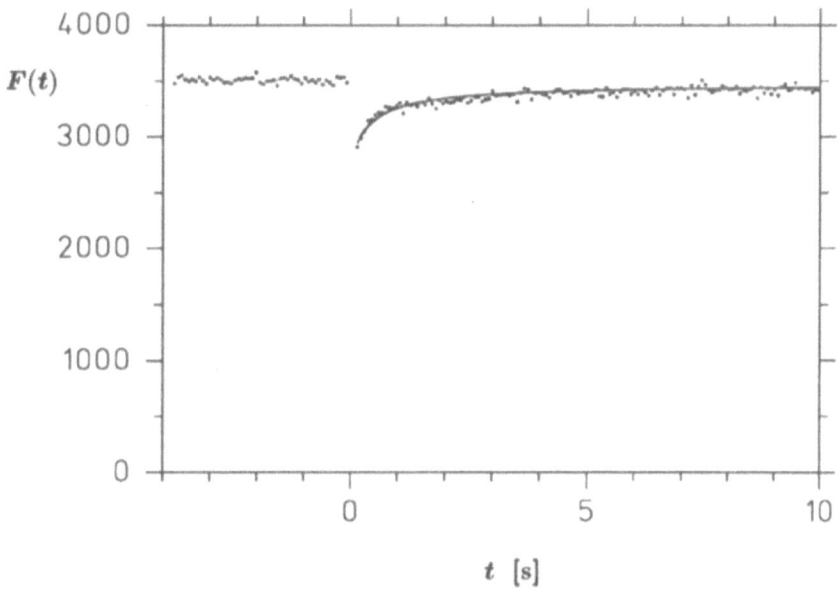

Fig. 1 Data points from a typical FRAP experiment [5] fitted by (18) with $\tilde{K} = 0.47$, $\tau_d = 0.41$s, $\bar{F} = 3464$, $\gamma = 1$ ($\theta = 0°$) as well as by (18) with $\tilde{K} = 0.49$, $\tau_d = 0.21$s, $\bar{F} = 3470$, $\gamma = 0.5$ ($\theta = 60°$); the curves are not distinguishable.

is found in terms of a convergent series expansion. In eq. (18)

$$\bar{F} = \frac{1}{2}\pi w^2 I_0^p Q \bar{C} \tag{19}$$

is the equilibrium fluorescence value, $\tau_d = w^2/(4D)$ is a characteristic time constant, and $\tilde{K}$ describes the strength of bleaching. In $\gamma = \cos(\theta)$ the dependency on the angle of incidence of the beam is included. For a perpendicular incident beam ($\gamma = 1$), the sum can be expressed in closed form, e.g. by the χ^2-probability function

$$F(t) = \bar{F}\nu\tilde{K}^{-\nu}\Gamma(\nu)\chi^2(2\tilde{K}|2\nu) \tag{20}$$

with $\nu = w^2/(w^2 + 8Dt)$ [1]. The relaxation function is given by

$$G(t) = -\bar{F}\sum_{n=1}^{\infty}\frac{(-\tilde{K})^n}{n!}\left((1+n+2n\gamma^2 t/\tau_d)(1+n+2nt/\tau_d)\right)^{-1/2} \tag{21}$$

where for $\tilde{K} < 1$ the first term dominates.

A typical experimental result [5] is shown in Fig. 1. For negative times we have constant fluorescence. By the bleaching at time zero the fluorescence becomes smaller and afterwards it tends to the original value because of the diffusion of

$G(t)/G_0$

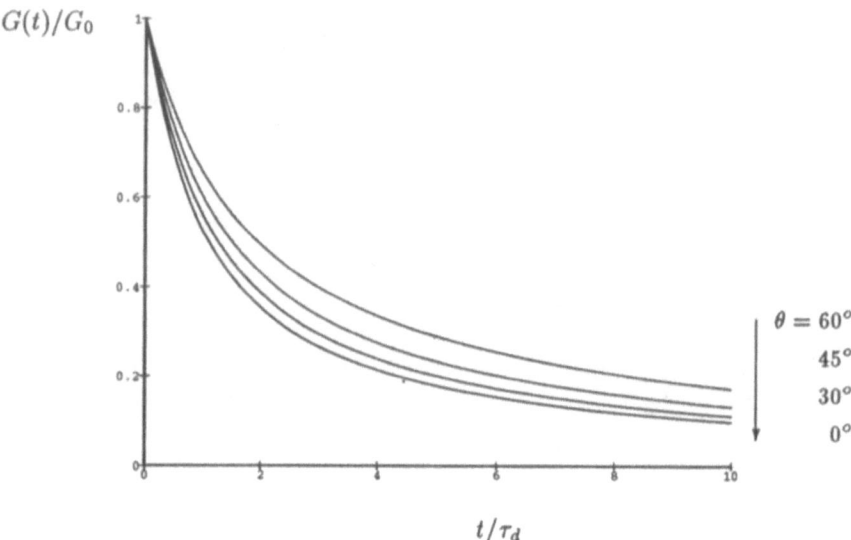

t/τ_d

Fig. 2   Relaxation function (22) for various values of the angle θ under which the laser beam is directed to the probe ($\gamma = \cos \theta$).

unbleached molecules into the detection area. Experimental data sets of FRAP measurements under total internal reflection conditions are well represented by eq. (18) with the given angle $\theta \approx 65°$. In Fig. 1 the data points were fitted with eq. (18) corresponding to $\theta = 0°$ and $\theta = 60°$. The two curves are scarcely to distinguish. They fit the data points within the accuracy of the measurement, but the time constants and thus the diffusion coefficients differ by a factor 2. Fig. 2 shows the dependency on the angle θ of the normalized relaxation function. One recognizes that with increasing θ the relaxation decreases.

5   Anomalous diffusion

Diffusion processes in systems with long-range correlations or in 'disordered media' deviate from the standard diffusion (3) [3][4]. For anomalous diffusion the mean square displacement is given by

$$\langle r^2 \rangle = -\frac{\partial^2}{\partial \vec{k}^2} G(\vec{k}, t)|_{\vec{k}=0} \propto t^\alpha \tag{22}$$

($\alpha \neq 1$) instead of the result $\langle r^2 \rangle = 2Dt$ of eq. (3). An equation leading to such a behavior is the fractional diffusion equation [6]

$$C(\vec{r}, t) - C(\vec{r}, 0) = D \, _0D_t^{-q} \Delta C(\vec{r}, t) \tag{23}$$

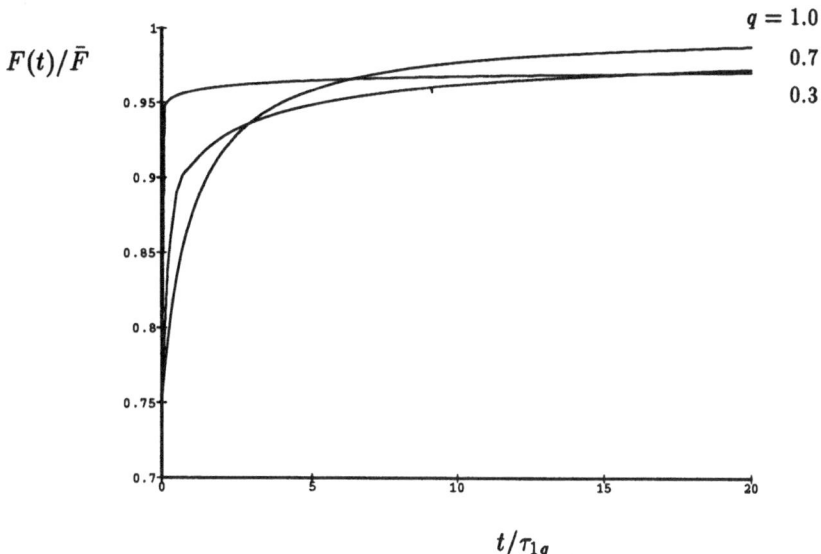

t/τ_{1q}

Fig. 3   Flourescence recovery for anomalous diffusion ($q = 0.3$ and $q = 0.7$) compared with normal diffusion ($q = 1.0$) for $\tilde{K} = 0.5$ (only the $n = 1$ term is considered).

where the operator ${}_0D_t^{-q}$ is a fractional Liouville-Riemann integral operator of the order q ($0 < q < 1$) defined by

$$
{}_aD_t^{-q} = \frac{1}{\Gamma(q)} \int_a^t (t - t')^{q-1} f(t') dt' . \tag{24}
$$

The Green's function, Fourier transformed in the space coordinates and Laplace transformed in the time, takes the form

$$
\hat{G}(k_x, k_y, p) = \frac{p^{-1}}{1 + Dp^{-q}(k_x^2 + k_y^2)} \tag{25}
$$

leading to the exponent $\alpha = q$ in eq. (22).

With eq. (25), the relaxation function

$$
G(t) = -\bar{F} \frac{\beta}{\gamma} \sum_{n=1}^{\infty} \frac{(-\tilde{K})^n}{(n+1)!} \frac{1}{q} H_{2,2}^{2,1} \left(\frac{\tau_{n,q}}{t} \middle| \begin{array}{c} (1, 1/q), (1, 1) \\ (1, 1/q), (1, 1/q) \end{array} \right) \tag{26}
$$

is obtained in terms of the Fox function $H_{2,2}^{2,1}$ by applying Laplace-Mellin transform techniques [2]. The time constants $\tau_{n,q}$ are given by

$$
\tau_{n,q} = \left(\frac{w^2(n+1)}{8Dn\beta} \right)^{1/q} \tag{27}
$$

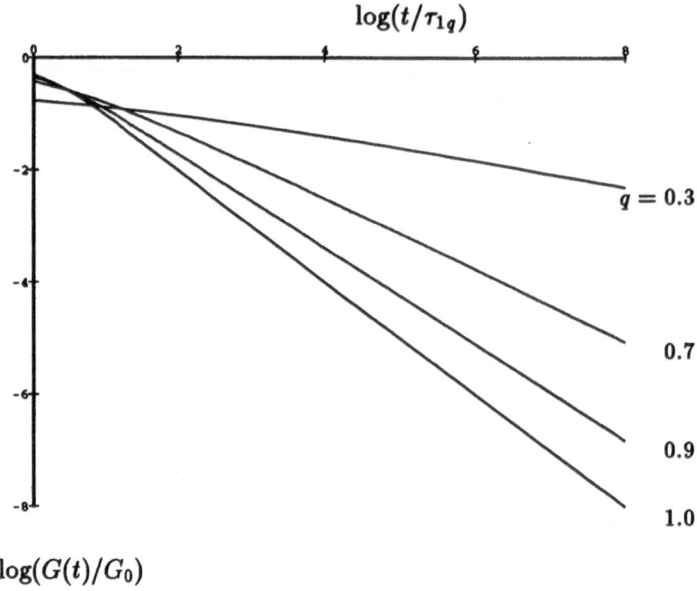

$$\log(t/\tau_{1q})$$

$$\log(G(t)/G_0)$$

Fig. 4 Asymptotic behavior of the fractional relaxation function (27) for various values of q in the case of $\tilde{K} < 1$ (i.e. only the $n = 1$ term is considered).

and β is a numerical constant ($\beta \in [1, 1/\gamma^2]$) which is $\beta = 1$ for $\gamma = 1$. For the H-functions series and asymptotic expansions can be derived [2]. Especially in the case (26) we find

$$G(t) \sim \frac{\bar{F}\beta}{\gamma} \sum_{n=1}^{\infty} \frac{(-\tilde{K})^n}{(n+1)!} \left(\frac{\gamma_e + q\Psi(1-q) + q\log(\tau_{n,q}/t)}{\Gamma(1-q)} \left(\frac{\tau_{n,q}}{t} \right)^q + O\left(t^{-2q} \right) \right)$$

$$(28)$$

for large t. Here, γ_e is the Euler constant and Ψ is the logarithmic derivation of the Γ-function. In Fig. 3, the recovery function $F(t)/\bar{F}$ in the cases of anomalous diffusion ($q = 0.3$ and $q = 0.7$) is compared with the result for the ordinary diffusion ($q = 1.0$). It demonstrates that the value of q determines the velocity with which the plateau value $\bar{F}$ is reached. The relaxation function (26) is plotted for various values of the fractal parameter q in Fig. 4. In the log-log plot one recognizes the power law decay $G(t) \sim t^{-q}$ for anomalous diffusion and $G(t) \sim t^{-1}$ for normal diffusion ($q = 1$). Especially for large protein molecules we expect the occurrence of diffusive transport with long-range correlations leading to a relaxation function of FRAP of the type (26).

References

[1] Axelrod, D., Koppel, D.E., Shlessinger, J., Elson, E., Webb, W.W. (1976) Mobility measurement by analysis of fluorescence photobleaching recovery kinetics. Biophys. J. 16, 1055–1069.

[2] Glöckle W.G., Nonnenmacher, T.F. (1993) Fox function representation on non-Debye relaxation processes. J. Stat. Phys. 71, 741–757.

[3] Havlin, S., Ben-Avraham, D. (1987) Diffusion in disordered media. Adv. Phys. 36, 695–798.

[4] M. Isichenko, B. (1992) Percolation, statistical topography, and transport in random media. Rev. Mod. Phys. 64, 961–1043.

[5] Mattfeldt, T., Nonnenmacher, T.F., Lambacher, A., Glöckle, W.G., Haferkamp, 0. (1993) Fluorescence recovery after photobleaching studied by total internal reflection microscopy, in this volume.

[6] Schneider, W.R., Wyss, W. (1989) Fractional diffusion and wave equations. J. Math. Phys. 30, 134–144.

[7] Thompson, N.L., Burghardt, N.P., Axelrod, D. (1981) Measuring surface dynamics of biomolecules by total internal reflection fluorescence with photobleaching recovery or correlation spectroscopy. Biophys. J. 33, 435–454.

List of Speakers

Peter Achermann
Institute of Pharmacology
University of Zürich
Winterthurstr. 190
CH-8057 ZÜRICH
Switzerland

Werner Arber
Abt. Mikrobiologie
Biozentrum
Universität Basel
Klingelbergstraße 70
CH-4056 BASEL
Switzerland

Gerd Baumann
Abteilung für
Mathematische Physik
Universität Ulm
Albert Einstein Allee 11
D-89069 ULM
Germany

C.L. Benhamou
CHR Orléans-La Source
B.P. 6709
F-4507 ORLEANS-Cedex 2
France

Philippe Blanchard
Theoretical Physics and BiBoS
Universität Bielefeld
Universitätsstraße 25
Postfach 8640
D-33615 BIELEFELD
Germany

Curtis B. Caldwell
Departments of Medical
Biophysics and Radiology
University of Toronto and
Department of Radiological Sciences
Sunnybrook Health Science Centre
2075 Bayview Avenue
North York, Ontario
Canada M4N 3M5

Giuseppe Damiani
Università di Pavia
Dipartimento di
Genetica e Microbiologia
Via Abbiategrasso 207
I-27100 PAVIA
Italy

Carl J. G. Evertsz
Center for Complex
Systems & Visualisation
University Bremen
Bibliothekstraße 1
Postfach 330440
D-28334 BREMEN
Germany

Zeno Földes-Papp
Section for Polymers
Universität Ulm
Albert-Einstein-Allee 11
D-89069 ULM
Germany

Walter Glöckle
Department of
Mathematical Physics
University of Ulm
Albert-Einstein-Allee 11
D-89069 ULM
Germany

Ary L. Goldberger
Department of Cardiology
Harvard Medical School
Beth Israel Hospital
330 Brookline Ave
BOSTON, MA 02215
USA

Ricardo Gutfraind
Group Matière
Condensée et Materiaux
Bâtiment 11B
Université Rennes I
35042 RENNES Cedex
France

Hiroko Kitaoka
Department of Internal Medicine
Kitaoka Hospital
1031 Meijimachi
Kurayoshi Tottori 632
Japan

Klaus-D. Kniffki
Institute of Physiology
University of Würzburg
Röntgenring 9
D-97070 WÜRZBURG
Germany

Haymo Kurz
Image Analysis and
Stereology Laboratory
Anatomisches Institut II
Albertstraße 17
D-79001 FREIBURG
Germany

Gabriele Landini
Oral Pathology Unit
The Dental School
University of Birmingham
St. Chad's Queensway
BIRMINHGHAM B4 6NN
England UK

Gabriele A. Losa
Laboratorio Patologia Cellulare
Istituto Cantonale di Patologia
Via in Selva 4
6604 LOCARNO
Switzerland

Martin Lüneburg
Universität Bielefeld
Fakultät für Mathematik
Universitätsstraße 25
D-33615 BIELEFELD 1
Germany

Benoit B. Mandelbrot
IBM Research Division
Thomas J. Watson Research Center
P.O.Box 218
YORKTOWN HEIGHTS
N.Y. 10598 USA

Raphaël Marcelpoil
Equipe de Reconnaissance des
Formes et Microscopie Quantitative
Université Joseph Fourier
CERMO
BP 53X.38041
GRENOBLE Cedex
France

Shu Matsuura
Tokai University
School of High Technology
for Human Welfare
Numazu
Nishino 317
SHIZUOKA 410-03
Japan

Torsten Mattfeldt
Institute of Pathology
University Ulm
Oberer Eselberg M23
D-89069 ULM
Germany

Theo F. Nonnenmacher
Abteilung für
Mathematische Physik
Albert-Einstein-Allee 11
Universität Ulm
D-89069 ULM
Germany

Vittorio Pesce Delfino
Consortio di Ricera DIGAMMA
C.So.A. de Gasperi 449/a
I-70125 BARI
Italy

Fançois Rothen
Institut de Physique Expérimentale
Université de Lausanne
BSP Dorigny
CH-1015 LAUSANNE
Switzerland

Bernard Sapoval
CNRS Ecole Polytechnique
Laboratoire de Physique de
la Matière Condensée
F-91128 PALAISEAU Cedex
France

Manfred Sernetz
Institut für Biochemie
und Endokrinologie
Justus-Liebig-Universität
Frankfurterstraße 100
D-35392 GIESSEN
Germany

Tom G. Smith, Jr.
Laboratory of Neurophysiology
National Institute of Health
BETHESDA, MD 20892
USA

Petre Tautu
German Cancer Research Center
Department of Mathematical Models
Neuenheimer Feld 280
D-69009 HEIDELBERG
Germany

Ewald R. Weibel
Anatomisches Institut
Universität Bern
Bühlstraße 26
CH-3000 BERN 9
Switzerland

Bruce J. West
Physics Department
University of North Texas
DENTON, TX 76203
USA

List of Participants

Dr. Peter Achermann
Institute of Pharmacology
University of Zürich
Winterthurstr. 190
CH-8057 ZÜRICH
Switzerland

Prof. Gianluigi Agnoli
Dipartimento di Matematica
Università di Bologna
40138 BOLOGNA
Italy

Prof. Werner Arber
Abt. Mikrobiologie
Biozentrum
Universität Basel
Klingelbergstrasse 70
CH-4056 BASEL
Swizerland

Dr. Gerd Baumann
Abteilung für
Mathematische Physik
Universität Ulm
Albert Einstein Allee 11
D-89069 ULM
Germany

Dr. C.L. Benhamou
CHR Orléans-La Source
B.P. 6709
45067 ORLEANS-Cedex 2
France

Dr. Bernasconi Giuliano
Inst. de Physique Exp.
Bât. Sciences Physiques
Université Lausanne
CH-1015 LAUSANNE-DORIGNY
Switzerland

Prof. Philippe Blanchard
Theoretical Physics and BiBoS
Universität Bielefeld
Universitätsstrasse 25
Postfach 8640
D-33615 BIELEFELD 1
Germany

Dr. Bonnaz Didier
Institut de Physique Expérimentale
UNIL, BČt. Sci. Physiques
CH-1015 LAUSANNE-DORIGNY
Switzerland

Dr. Urs Braschler
Rehetoblstrasse 5
9000 ST GALLEN
Switzerland

Dr. Hans-Jürgen Bubenzer
Diabetes Forschung
Biochem. Abteilung
Auf'm Hennekamp 65
D-40225 DÜSSELDORF
Germany

Dr. Curtis B. Caldwell
Departments of Medical
Biophysics and Radiology
University of Toronto and
Department of Radiological Sciences
Sunnybrook Health Science Centre
2075 Bayview Avenue
North York, Ontario
Canada M4N 3M5

Dr. Giovanna Codipietro
Dip. Sci. Ambiente Forestale
Università della Tuscia
Via S. Camillo de Lellis
I-01100 VITERBO
Italy

Dr. Giovanni Creton
Clinica Villa Flaminia
Via Bodio 58
I-191-ROMA
Italy

Prof. Giuseppe Damiani
Università di Pavia
Dip. Genetica e Microbiologia
Via Abbiategrasso 207
27100 PAVIA
Italy

Dr. Angelo Destradis
Dip. di Chimica
Università della Basilicata
Via Nazario Sauro 85
8510 POTENZA
Italy

Dr. Marc Dupuis
Institut de Biochimie
Université de Lausanne
Ch des Boveresses 155
CH-1066 EPALINGES
Switzerland

Cristoforo Dürig
Contrada Maggiore 30
CH-6613 LOSONE
Switzerland

Dr. Carl Eugster
Marigen S. A. 40
Hackbergstrasse
CH-4125 RIEHEN
Switzerland

Dr. Carl J.G. Evertsz
Center for Complex
Systems & Visualisation
University of Bremen
Bibliothekstrasse 1
Postfach 330440
D-28334 BREMEN
Germany

Dr. George Feron
GSF
Ingolstadter Landstrasse 1
D-91465 NEUHERBERG
Germany

Dr. Josef Feyertag
Dept. for Medical Physiology
University of Vienna
Schwarzspanierstrasse 17
A-1090 VIENNA
Austria

Dr. Zeno Földes-Papp
Section for Polymers
Universität Ulm
Albert Einstein Allee 11
D-89069 ULM
Germany

Dr. Elio Giroletti
Servizio di Radioprotezione
Palazzo dell' Università
C.so Strada Nuova, 65
I-27100 PAVIA
Italy

Dr. Walter Glöckle
Department of
Mathematical Physics
University of Ulm
Albert Einstein Allee 11
D-89069 ULM
Germany

Prof. Ary L. Goldberger
Dept. of Cardiology
Harvard Medical School
Beth Israel Hospital
330 Brookline Av.
BOSTON MA 02215
USA

Dr. Riccardo Graber
Laboratorio di Patologia Cellulare
Istituto Cantonale di Patologia
CH-6600 LOCARNO
Switzerland

Dr. Guerreiro Joaquim
Institut de Physique Expérimentale
UNIL, Bât. Sci. Physiques
CH-1015 LAUSANNE-DORIGNY
Switzerland

Dr. Massimo Gulisano
DIBIT-Istituto HS Raffaele
Via Olgettinago
I-20123 MILANO
Italy

Dr. Ricardo Gutfraind
Group Matière
Condensée et Materiaux
Bâtiment 11B
Université Rennes I
35042 RENNES Cedex
France

Dr. Rachid Harba
CHR Orléans-La Source
B.P. 6709
45067 ORLEANS-Cedex 2
France

Dr. Achakri Hassan
Laboratoire de Génie Médical
Dept. Physique EPFL
Champs-Courbes 1
CH-1024 ECUBLENS
Switzerland

Dr. Armin Herold
Schering Company
Central Biological Research
Müllerstrasse 171
D-13353 BERLIN 65
Germany

Dr. Hubert Jacot-Guillarmod
Hoffmann La Roche
Grezucherstrasse 124
CH-4000 BASEL
Switzerland

Dr. Hiroko Kitaoka
Dept. of Internal medecine
Kitaoka Hospital
1031 Meijimachi, Kurayoshi
TOTTORI 682
Japan

Prof. A.K. Kleinschmidt
University of Ulm
Albert Einstein Allee 11
D-89069 ULM
Germany

Prof. Dr. Klaus-D. Kniffki
Institute of Physiology
University of Würzburg
Röntgenring 9
D-97070 WÜRZBURG
Germany

Dr. Christian Kocourek
Dept. for Medical Physiology
University of Vienna
Schwarzspanierstrasse 17
A-1090 VIENNA
Austria

Dr. Martin Kunz
Institut de Physique Expérimentale
Bât. Physique UNIL
CH-1015 DORIGNY-LAUSANNE
Switzerland

Dr. Haymo Kurz
Image Analysis and
Stereology Laboratory
Anatomisches Institut II
Albertstrasse 17
D-79001 FREIBURG
Germany

Dr. Gabriele Landini
Oral Pathology Unit
The Dental School
Birmingham University
St. Chad's Queensway
BIRMINHGHAM B46NN
UK

Dr. Leoni Lorenzo
Laboratorio di Patologia Cellulare
Istituto Cantonale di Patologia
CH-6600 LOCARNO
Switzerland

Dr. E. Lespessailles
CHR Orléans-La Source
B.P. 6709
45067 ORLEANS-Cedex 2
France

Dr. Petra Leuchtenberg
Fak. Biologie
Zoologie Neurobiologie
Postfach 102148
Universitätstr. 150
D-44799 BOCHUM-1
Germany

Dr. Vehel Levy
Institut National de Recherche
en Informatique et Automatique
INRIA B. P. 105
78153 LE CHESNAY Cedex
France

Prof. Gabriele A. Losa
Laboratorio Patologia Cellulare
Istituto Cantonale di Patologia
Via in Selva 4
6604 LOCARNO
Switzerland

Dr. Martin Lüneburg
Universität Bielefeld
Fakultät für Mathematik
Universitätsstrasse
D-33615 BIELEFELD
Germany

Prof. Benoit B. Mandelbrot
IBM Research Division
Thomas J. Watson Research Center
P.O.Box 218
YORKTOWN HEIGHTS, N.Y. 10598
U.S.A

Dr. Raphaël Marcelpoil
Eq. de Reconnaissance des Formes
et Microscopie Quantitative
CERMO-Université Joseph Fourier
F-BP53X.38041 GRENOBLE Cedex
France

Prof. Claudio Marone
Primario Medicina Interna
Ospedale San Giovanni
CH-6500 BELLINZONA
Switzerland

Dr. Corrado Mascia
Clinica di Pediatria
Università la Sapienza
00100 ROMA
Italy

Dr. Shu Matsuura
Tokai University
School of High Technology
for Human Welfare
Nishino 317, Numazu
SHIZUOKA 410-03
Japan

Prof. Dr. Torsten Mattfeldt
Institute of Pathology
University Ulm
Oberer Eselsberg M23
W-89069 ULM
Germany

Prof. Danilo Merlini
CERFIRM
Via Rusca 1
CH-6600 LOCARNO
Switzerland

Dr. Silvia Metzeltin
RTSI
Besso
CH-6900 LUGANO
Switzerland

Dr. Christian Müller
Institut de Biochimie
Université de Lausanne
Ch des Boveresses 155
CH-1066 EPALINGES
Switzerland

Prof. Theo F. Nonnenmacher
Abteilung für Mathematische Physik
Albert Einstein Allee 11
Universität Ulm
D-89069 ULM
Germany

Dr. Frédéric Paycha
Service d'Explorations Fonctionelles
Hôpital Luis Mourier 178
Rue Renouillers
F-92700 COLOMBES CEDEX
France

Dr. G.P. Pescarmona
Dipartimento di Genetica
Bilogia e Chimica Medica
Via Santerna 5 bis
10126 TORINO
Italy

Prof. Vittorio Pesce Delfino
Consortio di Ricera DIGAMMA
C.So.A. de Gasperi 449/a
I-70125 BARI
Italy

Dr. Frédéric Pythoud
Laboratoire de Génie Médical
Dept. Physique EPFL
Champs-Courbes 1
CH-1024 ECUBLENS
Switzerland

Prof. Rosalia Ricco
Istituto di Zoologia
e Antropologra
Università di Bari
I-70124 BARI
Italy

Dr. Daniel Robert
Mikrobiologisches Institut
Frohbergstrasse 3
CH-9000 ST GALLEN
Switzerland

Prof. François Rothen
Institut de Physique Expérimentale
Université de Lausanne
BSP Dorigny
CH-1015 LAUSANNE
Switzerland

Prof. Bernard Sapoval
CNRS Ecole Polytechnique
Laboratoire de Physique de
la Matière Condensée
F-91128 PALAISEAU Cedex
France

Dr. Anne Schulz
Projekt Inhaltion
GSF
Ingolstadter Landstrasse 1
D-91465 NEUHERBERG
Germany

Prof. Manfred Sernetz
Justus-Liebig-Universität
Inst. Biochemie u. Endokrinologie
Frankfurterstrasse 100
D-35392 GIESSEN
Germany

Dr. Tom G. Smith, Jr.
Laboratory of Neurophysiology
National Institute of Health
BETHESDA, MD 20892
U.S.A

Prof. Petre Tautu
German Cancer Research Center
Dept. Mathematical Models
Research Program Bioinformatics
Neuenheimer Feld 280
D-69120 HEIDELBERG
Germany

Dr. Ernst Wehrli
Lab. for Electron Microscopy
ETH Institut Zellbiologie
Schmelzbergstrasse7
CH-8029 ZURICH
Switzerland

Prof. Ewald R. Weibel
Anatomisches Institut
Universität Bern
Bühlstrasse 26
CH-3000 BERN 9
Switzerland

Prof. Bruce J. West
Dept. Physics
University of North Texas
DENTON TX 76203
USA

Dr. Tiziano Zanin
Servizio Istologia
Ospedali Galliera
Via Mura delle Cappuccine 14
I-16128-GENOVA
Italy

Index